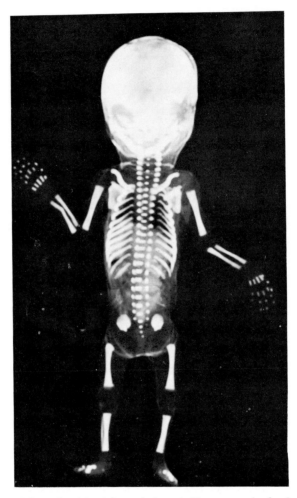

Frontispiece A 16-week fetus. Special staining techniques hold great promise for the investigation of abnormal skeletal development.
(Courtesy of Professor B. Cremin, Cape Town)

Inherited Disorders of the Skeleton

Peter Beighton
M.D., Ph.D., F.R.C.P.(Ed.), D.C.H.
Professor of Human Genetics, University of Cape Town
Consultant in Medical Genetics, Groote Schuur Hospital,
Cape Town, South Africa

FOREWORD BY

Alan E. H. Emery
M.D., Ph.D., D.Sc., F.R.C.P., M.F.C.M., F.R.S.(E)
Professor of Human Genetics, University of Edinburgh

CHURCHILL LIVINGSTONE
EDINBURGH LONDON AND NEW YORK 1978

CHURCHILL LIVINGSTONE
Medical Division of Longman Group Limited

Distributed in the United States of America by
Longman Inc., 19 West 44th Street, New York,
N.Y. 10036, and by associated companies,
branches and representatives throughout
the world.

First published 1978

ISBN 0 443 01724 7

British Library Cataloguing in Publication Data
Beighton, Peter
    Inherited disorders of the skeleton.
    1. Bones – Diseases – Genetic aspects
    I. Title
    616.7′1′042   RC930   77–30319

Printed in Great Britain by Adlard & Son Ltd, Dorking, Surrey, England

TO THE WOMEN IN MY LIFE;
Greta, Victoria and Mary Violet

Foreword

'Evolution . . . is a change from an indefinite, incoherent homogeneity, to a definite coherent heterogeneity'.
(Herbert Spencer's *First Principles*, 1862).

There is little doubt that over the last few years the importance of genetics in clinical medicine has become increasingly recognised. This has partly stemmed from developments in genetics itself, but at the same time there has been a growing interest among physicians in the problem of resolving genetic heterogeneity whenever it is seen to exist. The resolution of such heterogeneity is not simply an enjoyable academic exercise, which it is, but it has important practical implications particularly with regard to prevention. Clinically similar disorders may be inherited differently and have different prognoses, the recognition of which is a *sine qua non* of genetic counselling. Further, seemingly similar but genetically different disorders may be due to different biochemical defects and if expressed in cultured amniotic fluid cells the recognition of this biochemical heterogeneity is essential for reliable antenatal diagnosis. But it is important to realise that there is a singular pitfall in this approach to disease – that heterogeneity may be sought where none exists. This is therefore not a province for the inexpert and nowhere is this more so than in the case of the skeletal dysplasias which is a notoriously treacherous field. Professor Beighton with his extensive personal knowledge and his combined background of clinical medicine and genetics is one of the very few who have the experience and training to tackle this difficult problem. The result is a scholarly yet highly readable account. I am sure it will prove of considerable value to all those who are professionally involved in the management and counselling of families with these disorders.

Edinburgh 1978   Alan E. H. Emery

Preface

The dramatic manifestations of the inherited disorders of the skeleton have always attracted the attention of connoisseurs of rare syndromes. More than 200 genetic skeletal dysplasias have been delineated and there is little doubt that many others await recognition. Equally, it can confidently be anticipated that a number of conditions which are at present regarded as distinct entities will ultimately turn out to be heterogeneous. Although individually rare, these dysplasias are collectively quite common. In the period 1971–1976 more than 1000 patients with disorders of this type have been investigated by the author in genetic clinics in Southern Africa. Their case details, which have been recorded in the skeletal dysplasia registry of the Department of Human Genetics, Medical School, University of Cape Town, have formed the basis for this book. The illustrations are derived from the same source.

These disorders have been discussed from the point of view of the practice of medical genetics. Pertinent clinical and radiological features are described and illustrated in order to provide a balanced perspective. However, their detailed consideration is outside the scope of this book, and no attempt has been made to review them in depth. Similarly, management and pathogenesis are mentioned only when strictly relevant. Further information on topics of this nature is available in several excellent monographs, such as *The Atlas of Constitutional Disorders of Skeletal Development* (Spranger, Langer and Wiedemann, 1974), and Fairbank's *Atlas of General Affections of the Skeleton* (Wynne-Davies and Fairbank, 1976). Other relevant works include books by McKusick (1972), Carter and Fairbank (1974), Maroteaux (1974), Gorlin, Pindborg and Cohen (1976), and Smith (1976). These are listed in the appendix.

The establishment of bone dysplasia registries, such as those of Maroteaux in France and Spranger in Germany, has facilitated the accumulation of material and the delineation of many of these conditions. McKusick and his group, working with the 'Little People of America' have made notable contributions. The catalogue of genetic disorders, *Mendelian Inheritance in Man* (McKusick, 1975) has made an enormous impact. It is an indispensable aid to any researcher engaged in the investigation of these conditions. Much of the accumulated data has been presented at the annual Conferences on the Clinical Delineation of Birth Defects, sponsored by the 'National Foundation – March of Dimes'.

Categorisation of the skeletal dysplasias poses many problems. The early development of the nomenclature owes much to the vast clinical experience of Sir Thomas Fairbank. Ruben's anatomical approach to the classification of these disorders represented a further advance. The terminology was eventually standardised in 1970 in the 'Paris Nomenclature for Constitutional Disorders of Bone'. The arrangement of the Nomenclature has been followed in this book, although adaptations have been made in the light of modern knowledge and with regard to the underlying purpose. The nature of these osseous disorders precludes precise separation and overlap is inevitable. For this reason, the contents of any particular section represent an attempt at a reasonable compromise.

The terminology of the inherited disorders of the skeleton is complex and some have been given a different label in each publication in which they have been reported. It is not without significance that the philosopher Immanuel Kant commented 'physicians often think that they do a lot for a patient when they give his disease a name'. For the sake of simplicity and clarity, the designations used in this book are those which have gained general acceptance. In accordance with modern usage, eponyms are avoided whenever possible, but, where inclusion is warranted by custom, the possessive form of the eponym has been discarded. Technical terms which are applicable to the intrinsic diseases of bones are defined in the text and listed in the glossary.

It is assumed that the reader has a working knowledge of medical genetics and no attempt has been made to explain established concepts such as the various Mendelian modes of inheritance. Similarly, it is taken for granted that the consequences of any particular form of genetic transmission, with regard to counselling and antenatal diagnosis, are understood. In this respect, the 'burden of the disease', as it influences the quality of life of an individual with any particular condition, is of paramount importance.

Apart from a few metabolic disorders where the defect is detectable in cultured amniotic fluid cells, the majority of the inherited skeletal dysplasias cannot be diagnosed antenatally in early pregnancy. However, new sophisticated radiographic methods may have a place in these investigations. Fetoscopy would certainly permit the early intrauterine recognition of many structural defects, and it is to be anticipated that this technique will have an important role in the future. Little is known of the chromosomal localisation and linkage relationships of these conditions, but antenatal diagnosis by fetal blood sampling and the identification of 'marker' genes may eventually be possible.

The references have been chosen to include key reviews and papers with a genetic slant. Every effort has been made to ensure that these are up-to-date. However, it is axiomatic that the period of time necessary for the preparation of a book always exceeds the author's initial forecast. The fact that a number of the children portrayed in the illustrations have now reached adulthood is ample testimony to the truth of this adage!

Cape Town, 1977   Peter Beighton

Acknowledgements

I am grateful to numerous friends and colleagues for assistance in many ways:

to Dr C. O. Carter of London and Professor V. A. McKusick of Baltimore for fostering my early interest in clinical genetics and to Professor Alan E. H. Emery of Edinburgh for his continued guidance and encouragement.

to Professor J. Spranger of Germany for facilitating the development of the Cape Town Skeletal Dysplasia Registry by consistently reaching the correct diagnosis in the difficult cases which I have referred to him.

to Mr F. Horan, orthopaedic surgeon, my friend and collaborator over many years, for his amicable criticisms and his valuable comments.

to orthopaedic surgeons for allowing access to their patients: Professor C. E. L. Allen and Mr B. Jones, Princess Alice Orthopaedic Hospital and Groote Schuur Hospital, Cape Town: Professor L. Solomon and Mr J. Handelsman, University of the Witwatersrand, and Mr C. A. Bathfied, Baragwanath Hospital, Johannesburg.

to Professor B. Cremin for access to radiographic material

to Dr S. Goldberg for his good natured advice from his vast knowledge of bone disorders and to Professor H. Hamersma for patient referrals.

to clinical photographers for the illustrations – Messrs R. A. de Méneaud, C. Russ, R. C. Clow, E. Norman and Mrs S. Henderson.

to genetic nursing sisters for their assistance with patients – Miss Lecia Durr, Mrs Elizabeth Napier, Mrs Rosemary Duggan, and Mrs Lorraine Groeneveldt of Cape Town; Mrs Pam Otto, Miss Ann Williams and Mrs Judith Mathee of Johannesburg.

to the South African Medical Research Council and the University of Cape Town Staff Research Fund for generous financial support for the skeletal dysplasia clinics.

to Dr M. Nelson and Dr W. S. Winship for their helpful suggestions on the contents of the text.

Most of all to Greta, who prepared the manuscript with enthusiasm and dedication which far exceeded her professional and marital obligations.

Contents

1. Skeletal dysplasias without significant spinal involvement

For practical purposes, the osteochondrodysplasias of unknown pathogenesis which cause dwarfism can be divided into two groups; those in which spinal changes are minimal or absent, and those in which the spine is significantly involved. The following conditions fall into the former category.

1. Achondroplasia
2. Hypochondroplasia
3. Achondrogenesis
4. Thanatophoric dwarfism
5. Asphyxiating thoracic dysplasia (Jeune)
6. Chondroectodermal dysplasia (Ellis–van Creveld)
7. Multiple epiphyseal dysplasia
8. Chondrodysplasia punctata — autosomal dominant type
9. Chondrodysplasia punctata — autosomal recessive type
10. Metaphyseal chondrodysplasia
11. Mesomelic dwarfism
12. Rhizomelic dwarfism
13. Campomelic dwarfism.

Delineation of the osteochondrodystrophies was initially undertaken on a clinical basis. Nelson (1970) described the development of the concept of heterogeneity of these conditions and Dorst, Scott and Hall (1972), Kozlowski (1976) and Kaufman (1976) reviewed their radiographic features. The specific ultrastructural changes which have now been identified in many of these disorders have been discussed by Rimoin, Silberberg and Hollister (1976). It is likely that histopathological investigations of this type will reveal further heterogeneity. In the future, it is conceivable that nomenclature and classification will be based upon abnormalities which are recognisable at this level.

Radiographically, the epiphyses appear to be small and fragmented in many of the osteochondrodysplasias. However, hip arthrograms in various conditions in this category have revealed that the radiolucent cartilage of the femoral head is surprisingly normal in size and configuration. Lachman, Rimoin and Hollister (1974), who undertook these studies, concluded that defective epiphyseal ossification must be a significant pathogenic factor. These findings are of importance from many points of view, ranging from the investigation of the basic defect to the formulation of regimes of management.

Orthopaedic problems predominate in the osteochondrodysplasias. The management of these complications has been reviewed by Kopits (1976). Prosthetic replacement of the hip and knee joints has been successfully accomplished in a few patients and procedures of this type hold great promise for the future.

The term 'dwarfism' implies an abnormal degree of short stature. However, there is marked racial variation in normal height and it is impossible to give a precise definition of this designation. For this reason an individual's stature must be evaluated on a basis of his ethnic background. The last word on this ambiguous situation undoubtedly remains with the dwarfed fairground exhibitionist who advertised himself as 'the smallest giant in the world'!

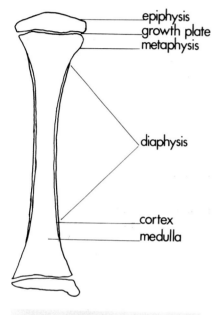

epiphysis
growth plate
metaphysis

diaphysis

cortex
medulla

Fig. 1.1. (left) The components of a tubular bone.

Fig. 1.2. (below left) Genu varum or bowlegs; a feature of many skeletal dysplasias.

Fig. 1.3. (below right) Genu valgum or knock-knees, affected twins.

1. ACHONDROPLASIA

Achondroplasia is by far the commonest and best-known form of short-limbed dwarfism and it has attracted attention since ancient times. For instance, achondroplasts can be recognised amongst the dwarfed goldsmiths portrayed in a frieze on the tomb of Merruka. Similarly, the Egyptian deity Bes was conventionally depicted with the features of achondroplasia. Other classical portrayals include a statue of an achondroplast gladiator dating from the time of the Roman Emperor Domitian (AD 51–96), and Velasquez's portrait of Don Sebastian de Morro, an affected nobleman at the court of Philip V of Spain. It has even been suggested that Attila the Hun, the 'Scourge of God', might have had achondroplasia.

In the past, the term 'achondroplast' was used indiscriminately for any individual with short-limbed dwarfism. However, with increasing diagnostic sophistication, many forms of dwarfism have been delineated as distinct disorders in their own right. Achondroplasia now remains as a specific and well-defined entity.

Warkany (1971) calculated that there are about 5000 achondroplasts in the USA and 65,000 on Earth. A measure of the relatively high frequency of the condition can be gained from the fact that investigators at the Moore Clinic, Johns Hopkins Hospital, Baltimore, USA, were able to collect data concerning 393 cases (Todorov and Bolling, 1974).

Clinical and radiographic features
The characteristic facies, habitus and stance of the achondroplast are unmistakeable. The limbs are disproportionately short and the knees are often bowed, while the

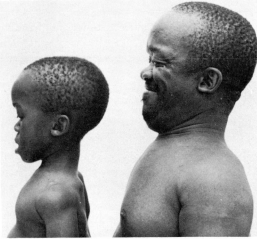

Fig. 1.4. (left) Achondroplasia in a father and his son.

Fig. 1.5. (right) Achondroplasia; the forehead is prominent and the nasal bridge is depressed.

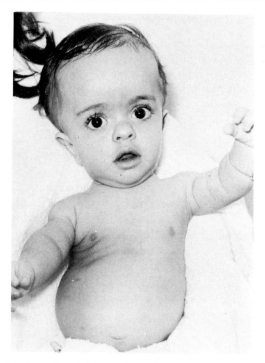

Fig. 1.6. Achondroplasia; shortening of the arms is maximal in the proximal regions (rhizomelia).

lumbar spine is lordotic. The forehead is bossed and the nasal bridge is depressed. Inability to approximate the third and fourth fingers produces a 'trident' configuration of the hand.

General health is good, and the life span is not reduced. Orthopaedic problems include premature osteoarthritis, particularly of the knee joint (Bailey, 1970). Backache may be troublesome and some degree of spinal cord compression may supervene. Restriction of the calibre of the nasal airways predisposes to upper respiratory tract infection. Mentality is usually normal, but intellectual ability is compromised in a minority of achondroplasts, probably due to mild internal hydrocephaly. Infrequently, severe hydrocephalus warrants a shunt procedure.

The skull is relatively large, with frontal prominence and a small foramen magnum. The interpeduncular distances in the lower lumbar spine become progressively narrowed, vertebral pedicles are short, and the disc spaces are wide. The iliac bones have a 'tombstone' configuration, the sacroiliac notch is narrow and the acetabular portions of the iliac bones are horizontal. Limb shortening is predominantly rhizomelic (proximal). The tubular bones have widened shafts and flared metaphyses. The radiographic features of achondroplasia have been comprehensively reviewed by Langer, Baumann and Gorlin (1967).

Genetics
Achondroplasia provides an excellent example of autosomal dominant inheritance, as the gene is invariably penetrant, with consistent clinical expression. The pre-

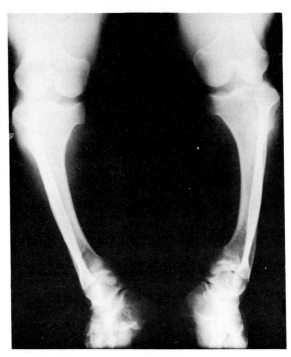

Fig. 1.7. Achondroplasia; anteroposterior radiograph of the knees and shins. The metaphyses are flared and the diaphyses are distorted. The fibula is disproportionately long.

valence per million in the newborn has been estimated as 23 in Denmark (Mørch, 1941), 15 in the USA (Potter and Coverstone, 1948) and 28 in Northern Ireland (Stevenson, 1957). It is likely that dwarfing conditions other than achondroplasia were included in these series, and these figures are probably inflated. Until recently, it was assumed that a significant proportion of achondroplasts were stillborn or died in the neonatal period, and this fact was taken into account when mutation rates were calculated. However, it is now known that achondroplasia is not usually lethal. Harris and Patton (1971) pointed out that the majority of stillborn infants with short-limbed dwarfism do not have achondroplasia (*vide infra*). Using pooled data from four centres, Gardner (1977) identified seven sporadic achondroplasts in 242,257 births and estimated that the mutation rate was of the order of $1 : 4 \times 10^{-5}$. As the diagnostic criteria for achondroplasia are now well established there is little doubt that this figure reflects the true situation.

More than 80 per cent of achondroplasts have normal parents (Scott, 1976). These 'sporadic' patients are assumed to be the result of new mutation of the particular gene, which has taken place before conception. Murdoch *et al.* (1970) have shown that there is a significant increase in the average age of the unaffected fathers of sporadic achondroplasts. Achondroplasia therefore ranks as one of the disorders in which paternal age effect in the genesis of a new mutation has been demonstrated.

Anomalous kindreds have been reported in which achondroplasia was present in two cousins (Opitz, 1969) and in two sisters (Bowen, 1974). The other members of these families were normal. It was proposed that gonadal mosaicism was the most likely explanation for this latter situation. McKusick, Kelly and Dorst (1973) postulated that the achondroplasia and hypochondroplasia genes might be allelic (i.e. at the same locus on a particular chromosome).

In terms of genetic counselling, unaffected parents of an achondroplastic infant can be reassured that the risk of recurrence in further offspring is very low. Conversely, there is an even chance that any child born to an achondroplast of either sex, who is married to a normal individual, will have achondroplasia. Theoretically, 25 per cent of infants born to parents who are both achondroplasts will be homozygous for the abnormal gene, 25 per cent will be homozygous normal individuals and 50 per cent will be heterozygous achondroplasts. Hall et al. (1969) reported two probable homozygotes, while Murdoch et al. (1970) described further cases. These 'doubly affected' individuals had severe respiratory distress and died soon after birth. Their external features resembled classical achondroplasia, although the radiographic changes were of greater severity (Langer et al., 1969).

As with most forms of dwarfism, the pregnant achondroplastic female will usually require Caesarian section for delivery of her baby. At the present time, prenatal diagnosis of achondroplasia at an early stage of pregnancy is not possible. In this respect, radiographic studies have been unhelpful (Globus and Hall, 1974).

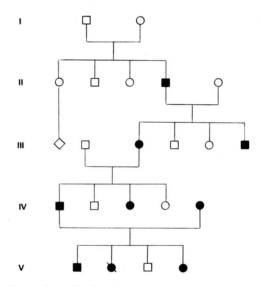

Fig. 1.8. Achondroplasia; a pedigree showing classical autosomal dominant inheritance. The affected male in generation II was the youngest of his sibship. As his parents were normal, it is likely that he represented a new mutation of the gene. In generation IV the affected son married a female with achondroplasia. Two of their children had achondroplasia, one had lethal homozygous achondroplasia and one was normal.

Key to pedigree: □ normal male; ○ normal female; ■ affected male; ● affected female; / deceased.

2. HYPOCHONDROPLASIA

In recent years, hypochondroplasia has been recognised as an entity which is distinct from achondroplasia. Although there have not been a large number of reports, hypochondroplasia is relatively common amongst the osteochondrodystrophies. Specht and Daentl (1975) described six cases and reviewed the manifestations in 35 previously reported patients.

Clinical and radiographic features
Hypochondroplasts resemble achondroplasts, but in general, the stigmata are much milder. In particular, the head and face are unaffected, and the fingers do not have a trident configuration. Birth weight and length are normal and adult height may exceed 150 cm. Skeletal disporportion, mild lumbar lordosis and limitation of full extension of the elbow joints become evident in early childhood. Bowing of the legs may be present in infancy, but this abnormality, which is usually of mild degree, often resolves spontaneously. The only important complication is mental deficiency, which occurs in about 20 per cent of hypochondroplasts. Orthopaedic problems in hypochondroplasia have been reviewed by Scott (1976).

The severity of the condition is very variable. At one end of the spectrum it may be difficult to distinguish patients from true achondroplasts, while at the other extreme they may approach normality. Indeed, one hypochondroplast known to the author plays for an international rugby team! Another enjoys a successful career as a circus strongman.

The radiographic features are similar to those of achondroplasia but of a milder degree. In particular, the lumbar vertebrae may shown some narrowing of the interpeduncular distances and the femoral necks are short. Distal lengthening of the fibula is a useful diagnostic sign (Frydman, Hertz and Goodman, 1974).

Genetics
There have been several well-documented family studies which indicate that hypochondroplasia is transmitted as an autosomal dominant (Beals, 1969). However, there is considerable phenotypic variation and hypochondroplasia could well be heterogeneous. In the author's own experience, only one out of nine un-

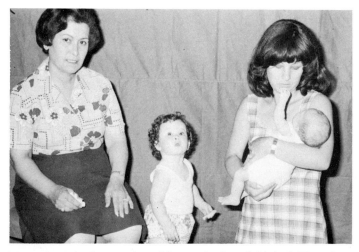

Fig. 1.9. Hypochondroplasia; these members of three generations of a kindred all have the condition. The clinical stigmata are mild, but the diagnosis has been confirmed radiographically.

equivocally affected probands had a parent with the condition. The fact that some affected individuals are virtually normal is of practical importance in genetic counselling. On a basis of the clinical stigmata of a dwarfed child derived from an achondroplastic father and a hypochondroplastic mother, McKusick, Kelly and Dorst (1973) have suggested that the gene for these two conditions might be allelic.

3. ACHONDROGENESIS

Achondrogenesis is one of the conditions which enter into the differential diagnosis of the stillborn short-limbed dwarf. The stigmata have been well documented and recognition is not difficult. The paucity of reported cases probably reflects diagnostic uncertainty rather than genuine rarity.

With the accumulation of experience, it has become evident that achondrogenesis is heterogeneous. Various subclassifications have been proposed, the division into types IA and IB being generally accepted at the present time. The eponym 'Parenti-Fraccaro' has been applied to the former variety and that of 'Saldino' to the latter. The main difference between these disorders is the greater severity of under-ossification of the axial skeleton in the Saldino type (Yang *et al.*, 1974).

The Grebe syndrome, which is sometimes designated 'achondrogenesis type II', is quite distinct from the lethal forms of achondrogenesis. This condition is described in Chapter 16. Confusion between the Grebe syndrome and neonatal achondrogenesis has arisen from the terminology rather than from any similarity of clinical manifestations. It is probable that the term 'Grebe syndrome' will come into use and that achondrogenesis type IA and IB will be redesignated type I and II.

Clinical and radiographic features
Infants with achondrogenesis type IA and IB are usually stillborn, although a few have survived for a limited period of time. The limbs are very short, and the hydropic head and bulging abdomen are disproportionately large. Although the

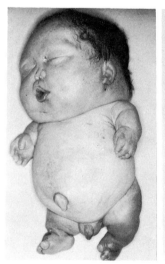 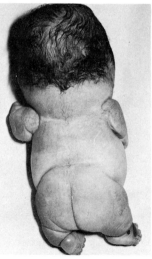

Fig. 1.10. (left) Achondro-genesis in a stillborn infant. The limbs are very short.

Fig. 1.11. (right) Achondro-genesis; the hydropic head is disproportionately large.

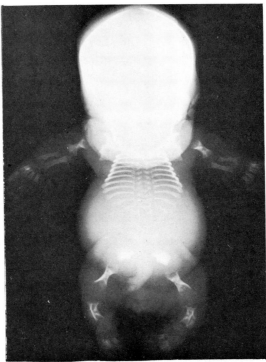

Fig. 1.12. Achondrogenesis; anteroposterior radiograph. The tubular bones are short, with irregular metaphyses. The lumbar vertebrae are undermineralised. (From Cremin, B. J. & Beighton, P. (1974) *British Journal of Radiology*, **47**, 77.)

diagnosis can usually be made clinically, confirmation rests upon recognition of the characteristic radiographic changes.

The bones of the skull are relatively normal, in marked contrast to the swollen surrounding soft tissues. The spine and pelvis are grossly underossified and the tubular bones are short, with expansion and spiky irregularity of their metaphyses.

Genetics

Although precise categorisation of cases mentioned in earlier reports is not always possible, there is little doubt that types IA and IB are inherited as autosomal recessives. Scott (1972) mentioned nine kindreds with multiple affected sibs and normal parents, and a similar situation was described in greater detail by Houston, Awen and Kent (1972). In a report of six infants with achondrogenesis, Wiedemann *et al.* (1974) commented that more than 35 cases would be recognised in the literature. Subsequently, Ornoy *et al.* (1976) reported the histological and electron microscopical changes in three spontaneously aborted affected fetuses from a consanguineous kindred.

Harris, Patton and Barson (1972) used the title 'pseudoachondrogenesis with fractures' in a description of two affected sibs and postulated the existence of a third form of the disorder. McKusick (1975) disagreed, contending that the stigmata in these neonates were in fact compatible with a diagnosis of achondrogenesis type IA.

Reporting yet another lethal form of short-limbed dwarfism, Verma, Bhargava and Agarwal (1975) described a consanguineous Sikh kindred in India, in which six of nine sibs had been stillborn. In addition to severe thoracic constriction, these infants had micromelia, postaxial polydactyly and genital anomalies. As the radio-graphic changes were similar to those of achondrogenesis, the authors suggested that 'achondrogenesis type III' might be an appropriate designation.

Laxova *et al.* (1973) observed intracellular lipid inclusions in fibroblasts from deceased sibs born in a consanguineous marriage. This finding opens up prospects for early antenatal diagnosis, although accomplishment of this procedure has not yet been reported. However, achondrogenesis has been recognised *in utero* at a late stage of pregnancy, by means of standard radiographic techniques (Maroteaux, Stanescu and Stanescu, 1976).

A couple who have produced a child with achondrogenesis are at a one in four risk of recurrence at any subsequent pregnancy. However, the lethality of the condition should be stressed when counselling parents who are contemplating procreation. Equally, it should be emphasised that there is a three out of four chance that any further offspring will be normal.

4. THANATOPHORIC DWARFISM

Thanatophoric dwarfism is probably the most common disorder amongst the poten-tially lethal conditions which present with disproportionate shortening of the limbs in the newborn. The designation, derived from the Greek 'thanatophoras', or 'death bearing', was used in the original case description by Maroteaux, Lamy and Robert (1967).

Clinical and radiographic features

A firm diagnosis cannot be made without radiological studies. Indeed, the short limbs, prominent forehead and depressed nasal bridge can easily be mistaken for the stigmata of achondroplasia, while the narrow thorax is reminiscent of asphyxiating thoracic dysplasia (*vide infra*). Affected individuals are usually stillborn or die of respiratory insufficiency in the neonatal period. However, survival for 10 weeks, with minimal supportive care, has been reported (Moir and Kozlowski, 1976).

The vertebral bodies have an H-shaped appearance in frontal radiographs due to flattening of their mid-portions and relative prominence of their pedicles. The broad pelvis, with horizontal acetabulae, closely resembles that of achondroplasia. The tubular bones are short and broad. The femora are bowed with a 'telephone receiver' configuration, which is pathognomonic of thantophoric dwarfism.

The similarity of the pelvic and limb abnormalities in thanatophoric dwarfism and achondroplasia has prompted speculation than there might be some fundamental relationship between these conditions (Langer *et al.*, 1969). However, there is a significant histochemical distinction, as cultured fibroblasts from these disorders differ in terms of their mucopolysaccharide content (Danes, 1974).

Genetics

The majority of reports have concerned sporadic cases. Following their review of the literature, Maroteaux, Lamy and Robert (1967) postulated that new dominant mutation was the probable genetic basis of the disorder. Subsequently, affected sibs were described (Chemke, Graff and Lancet, 1971; Harris and Paton, 1971), and

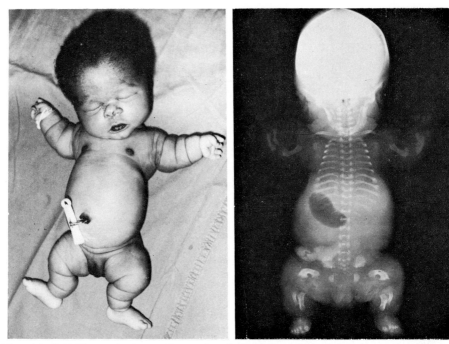

Fig. 1.13. (left) Thanatophoric dwarfism; a short-limbed infant which died six hours after delivery. (From Cremin, B. J. & Beighton, P. (1974) *British Journal of Radiology*, **47**, 77.)

Fig. 1.14. (right) Thanatophoric dwarfism; the 'telephone receiver' configuration of the femur is pathognomonic. (From Cremin, B. J. & Beighton, P. (1974) *British Journal of Radiology*, **47**, 77.)

autosomal recessive inheritance was invoked. This viewpoint gained support when affected triplets were born to consanguineous parents (Sabry, 1974). In a further review of accumulated case reports, Pena and Goodman (1973) suggested that inheritance was polygenic, with a 2 per cent recurrence risk. An alternative explanation, for which there are many precedents, would be heterogeneity.

Bouvet, Maroteaux and Feingold (1974) reviewed the genetic background of thanatophoric dwarfism. No firm conclusions were reached, but these authors suggested that the well-documented but anomalous observation of increased birthrank without increased parental age might be explained by loss of affected fetuses through early spontaneous abortion.

Thanatophoric dwarfism has been diagnosed in late pregnancy by recognition of the characteristic radiographic features in the fetus (Thompson and Parmely, 1971; Bergstrom, Gustavson and Jorulf, 1971), and by ultrasonography (O'Malley *et al.*, 1972; Cremin and Shaff, 1977). The most significant factors in genetic counselling are the small likelihood of recurrence and the lethal nature of the disorder.

The term 'cloverleaf skull' or 'Kleeblattschadel' syndrome has been used synonymously with 'thanatophoric dwarfism' (Young, Pocharzevsky and Lenoidas, 1973). Indeed, using the title 'cloverleaf skull and thanatophoric dwarfism', Partington *et al.* (1971) reported four cases, two of whom were sibs, and postulated autosomal recessive inheritance. Accumulating evidence indicates that a skull with this unusual configuration can exist in isolation and as a component of a variety of disorders (Feingold, Miller and Bull, 1973; Pilz and Swoboda, 1975; Temtamy *et al.*, 1975). In a review of 51 known cases of the Kleeblattschadel anomaly, Hodach *et al.* (1975) pointed out that about 40 per cent have apparent thanatophoric dwarfism. In one report, maternal rubella has been invoked as a pathogenic factor (Widdig, Steinhoff and Guenther, 1974).

5. ASPHYXIATING THORACIC DYSPLASIA (JEUNE)

Asphyxiating thoracic dysplasia was first described in a brother and sister by Jeune, Beraud and Carron (1955). As maldevelopment of the thorax occurs in several other disorders (Kohler and Babbitt, 1970), the eponymous designation 'Jeune thoracic dysplasia' has considerable merit.

Clinical and radiographic features

Narrowness and immobility of the thoracic cage are the major features of the condition. Polydactyly, which is sometimes present, may lead to confusion with the Ellis–van Creveld syndrome. Radiographically, the ribs are short and horizontal and sternal ossification is often incomplete in the neonate. The pelvis is broad and the tubular bones may be shortened.

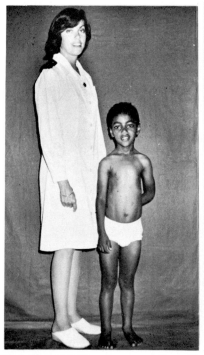 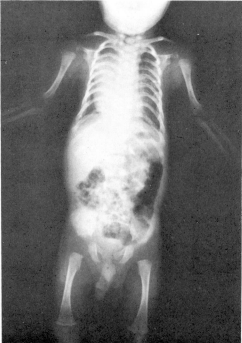

Fig. 1.15. (left) ATD; a nine-year-old boy with moderate thoracic constriction. The chest deformity was much more obvious during infancy. His height is below the third percentile, but apart from recurrent chest infections, he enjoys good health.

Fig. 1.16. (right) ATD; radiograph of an affected infant showing a 'bell-shaped' thorax, due to shortening of the upper ribs. The ilia are broad, with irregular acetabular margins.

Jeune thoracic dysplasia is compatible with life, although respiratory insufficiency and infection are often fatal during infancy. There is considerable variation in the severity of the thoracic constriction, and in some infants it may be of only minor degree. In later childhood the shape of the thoracic cage tends to revert to normality and the threat to life diminishes. Renal failure, with albuminuria and hypertension, supervenes in the second decade (Herdman and Langer, 1968). The only report of an affected adult concerns a 32-year-old American Indian who eventually developed hepatic and renal complications (Friedman, Kaplan and Hall, 1975).

Genetics

Apart from the original report by Jeune, Beraud and Carron (1955), sibs with normal parents have been described by Neimann et al. (1963), Maroteaux and Savart (1964) and Hanissian, Riggs and Thomas (1967). On this basis, it is likely that inheritance is autosomal recessive. Shokeir, Houston and Awen (1971) and Barnes et al. (1971) recognised minor manifestations in parents of affected children and suggested that the gene might occasionally be expressed in the heterozygote. Anomalous patients with thoracic dystrophy and metaphyseal abnormalities have been reported. For instance, Kaufman and Kirkpatrick (1974) studied two brothers, born with narrow thoraces, who developed pancreatic insufficiency, cyclical neutropenia and metaphyseal changes. This syndrome complex is similar to that of the Shwachman form of metaphyseal chondrodysplasia. Ozonoff (1974) described a girl with metaphyseal chondrodysplasia, which resembled the Jansen type of the disorder. This child developed progressive thoracic constriction in early childhood and died of cardiorespiratory embarassment at the age of six.

There is doubt concerning the homogeneity of Jeune thoracic dysplasia. Kaufman and Kirkpatrick (1974) stated that up to 80 per cent of affected individuals die in infancy, thus prompting speculation as to whether survivors have a genetically distinct form of the disorder. This concept of heterogeneity gained further support when Kozlowski and Masel (1976) described two children in whom the diagnosis was reached by chance at the age of four, when radiographs were obtained for unrelated reasons.

Other short rib-polydactyly syndromes

Two potentially lethal disorders, the 'Saldino–Noonan' and 'Majewski' short rib-polydactyly syndromes, have been differentiated from Jeune thoracic dysplasia (Spranger et al., 1974). Naumoff et al. (1977) proposed that these conditions should be respectively designated 'short rib-polydactyly syndrome type I and II'. They also suggested that there might be a distinct type III. Further case reports will be required before this latter subcategory is finally established.

The Saldino–Noonan type, which is comparatively common, is characterised by micromelia, postaxial polydactyly, brachydactyly, thoracic narrowing and abnormalities of the cardiovascular system and genitalia. Significant radiographic changes include short horizontal ribs, small iliac bones, metaphyseal spurs on the tubular bones and deficient ossification of the extremities. The presence of the disorder in two stillborn sibs, reported by Saldino and Noonan (1972), is consistent with autosomal recessive inheritance. In the Majewski type, which is extremely rare,

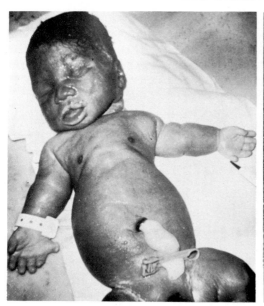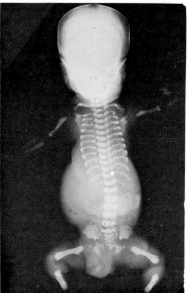

Fig. 1.17. (left) Saldino–Noonan syndrome; limb shortening, brachydactyly, postaxial polydactyly and thoracic constriction are suggestive of the diagnosis in this stillborn infant.

Fig. 1.18. (right) Saldino–Noonan syndrome; the diagnosis is confirmed by the radiographic appearance of short horizontal ribs, small iliac bones and metaphyseal spurs on the tubular bones.

thoracic, digital, limb and visceral abnormalities resemble those of the Saldino–Noonan form of the disorder. However, facial clefting and nose and ear deformity are additional features (Majewski *et al.*, 1971). The condition may be distinguished radiographically by the normal appearance of the pelvis and by marked shortening and an ovoid configuration of the tibia. The genetic basis is unknown.

6. CHONDROECTODERMAL DYSPLASIA (ELLIS–VAN CREVELD)

Chondroectodermal dysplasia is a well-defined entity in which short-limbed dwarfism and polydactyly are associated with structural cardiac anomalies. Since the original report of Ellis and van Creveld (1940), more than 120 cases have been described. The largest series is that of McKusick *et al.* (1964), who identified the disorder in an inbred religious isolate, the Amish of Pennsylvania, USA.

Clinical and radiographic features

Chondroectodermal dysplasia presents as disproportionate dwarfism in the newborn. The components of the syndrome are distal limb shortening (acromelic micromelia), postaxial polydactyly of the hands and dysplasia of the nails and teeth. The thorax may be constricted, and cardiac abnormalities, particularly of the atria, are present in about 50 per cent of patients. A significant proportion of patients die from the consequences of these cardiac or thoracic malformations. Genu valgum is a common complication in later childhood. Adult height is very variable, stature in some adults being virtually normal.

In infancy, the radiographic changes in the pelvis and thorax are very similar to those seen in asphyxiating thoracic dystrophy. Bony fusion in the carpus and hypoplasia of the phalanges serve as distinguishing features.

Genetics
Metrakos and Fraser (1954) recognised the consanguinity of the parents of a child reported by Ellis and van Creveld (1940), reviewed 10 other cases and suggested that inheritance of chondroectodermal dysplasia was autosomal recessive. This mode of transmission was confirmed by McKusick *et al.* (1964), following their studies in the Amish community. Murdoch and Walker (1969) amplified the Amish investigations and ascertained 61 cases in 33 sibships. These individuals were all related to a common ancestral couple who had emigrated to the USA from Europe in 1744.

7. MULTIPLE EPIPHYSEAL DYSPLASIA

In multiple epiphyseal dysplasia (MED) changes are maximal in the epiphyses, involvement of the metaphyses and axial skeleton being minimal or absent. Many cases have been reported and the various types of MED collectively represent one of the most common forms of skeletal dysplasia.

The development of knowledge concerning MED owes a great deal to the work of Sir Thomas Fairbank. In his early descriptions, Fairbank (1935) used the term 'epiphyseal dysplasia' and a decade later he introduced the designation 'dysplasia epiphysealis multiplex'. The clinical and radiological features of 26 patients were depicted in the *Atlas of General Affections of the Skeleton* (Fairbank, 1951).

It is becoming increasingly evident that MED is very heterogeneous and several overlapping classifications have been proposed. This problem has been discussed in detail by Lie *et al.* (1974). The confusion is compounded by the great variation in phenotypic features which may be present in affected members of the same kindred.

Clinical and radiographic features
Diagnosis may be difficult, as the height of some patients is within the normal range. However, shortness of stature, which is of mild degree, usually becomes evident in mid-childhood. In some instances, the hands are broad, with stubby digits and foreshortened nails. The facies and intelligence are normal and no consistent extra-skeletal manifestations have been recorded. Degenerative arthritis with onset in early adult life is a common complication, but general health is otherwise good.

As with many of the genetic skeletal dysplasias, the radiographic features in MED are age related. In the infant, the epiphyses may have a 'stippled' appearance. With advancing age, irregularity and fragmentation of the epiphyses of the long bones becomes increasingly evident. These changes may be particularly marked in the femoral heads. Changes in the digits are very variable but characteristically the tubular bones are shortened while the carpal and tarsal bones may be distorted. Mild spinal abnormalities are sometimes present, the vertebral bodies having irregular surfaces and slight anterior wedging. The skull and axial skeleton are otherwise normal.

2

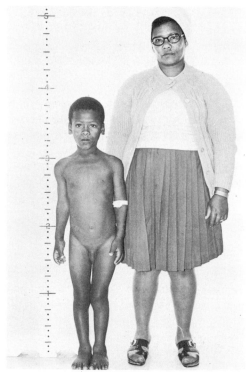

Fig. 1.19. MED; this nine-year-old boy, depicted with his mother, has few problems other than short stature.

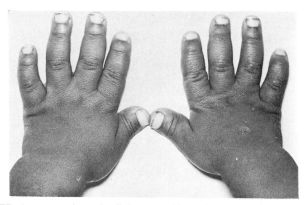

Fig. 1.20. MED; in some patients the digits are stubby and the finger nails are foreshortened.

In adolescence, the radiological changes in the hip joints resemble those of Perthe's disease, and a misdiagnosis may be made if the generalised nature of the articular abnormality is overlooked. Bilateral involvement or radiographic demonstration of widespread epiphyseal changes are useful diagnostic indicators. MED enters into the differential diagnosis in all instances of 'familial' Perthe's disease. Similarly, clinically unrecognised mild forms of MED might be responsible for a proportion of cases of 'idiopathic' osteoarthritis of the hip joint.

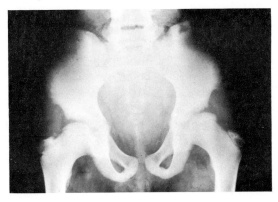

Fig. 1.21. MED; the hip joints show Perthe-like changes, with flattening and irregularity of the femoral capital epiphyses.

Genetics

In terms of clinical manifestations, MED has been divided into severe 'Fairbank' and mild 'Ribbing' types. However, there is considerable overlap of the stigmata, even within the same kindred, and it is likely that this particular subgrouping has no genetic basis. Nevertheless, there is no doubt that MED is very heterogeneous.

Juberg and Holt (1968) reviewed the literature and published a list of reports of families with dominant transmission of MED and of other kindreds where affected sibs with normal parents indicated possible autosomal recessive inheritance. It must be emphasised that in most cases MED is an autosomal dominant trait.

Delineation has been complicated by problems of incomplete ascertainment, particularly when stature has been virtually normal, and by the considerable variability of phenotypic expression within a single kindred. For instance, Diamond (1974) reported a family in which 32 individuals in four generations had a wide variety of changes in the hips and spine. Clinical inconsistency of this nature led Barrie, Carter and Sutcliffe (1958) to postulate that the disorder might be mediated by two separate genes. However, this contention has attracted little support.

Linkage studies undertaken in 12 affected individuals in two generations of a kindred were unfruitful (Hoefnagel *et al.*, 1967), and so far, the MED gene has not been assigned to any particular chromosome.

Epiphyseal dysplasia is a component of a number of rare syndromes. Lowry and Wood (1975) described two brothers with generalised epiphyseal abnormalities, short stature, congenital nystagmus and microcephaly. The authors suggested that inheritance of the condition was either X-linked or autosomal recessive. Pfeiffer *et al.* (1973) reported three brothers, two of whom were monozygous twins, with dysplasia of the femoral head, severe myopia and perceptive deafness. The parents were distantly related, and it is possible that inheritance was autosomal recessive.

Autosomal dominant inheritance of epiphyseal dysplasia which is confined to the hip joint is well recognised (Stephens and Kerby, 1946; Monty, 1962; Wamoscher and Farhi, 1963). This disorder, which is sometimes termed 'familial Perthe's disease', has consistent manifestations within any kindred, and apparently represents a distinct entity. It must be stressed that the genetic component of the common form of Perthe's disease is very small. Indeed, following a survey of the families of 323 affected individuals in South Wales, Harper, Brother and Cochlin (1976) calculated that the risk of recurrence in sibs was under 1 per cent, while the risk to children of affected parents was about 3 per cent.

8. CHONDRODYSPLASIA PUNCTATA — AUTOSOMAL DOMINANT OR CONRADI–HÜNERMAN TYPE

Autosomal dominant and autosomal recessive forms of chondrodysplasia punctata are recognised. The eponym 'Conradi–Hünerman' is conventionally applied to the dominant type, while the designation 'severe rhizomelic' is used for the recessive variety. Other terms which have been used indiscriminately include 'Conradi syndrome', 'stippled epiphyses', 'dysplasia epiphysealis punctata' and 'chondrodystrophia calcificans congenita'. In an extensive review, Spranger, Opitz and Bidder (1971) classified the Conradi–Hünerman form into three subgroups, although they were unable to settle the question of genetic heterogeneity of this type of the condition. Happle, Matthias and Macher (1977) have suggested that the subgroup B might be inherited as an X-linked dominant (*vide infra*).

Stippling of the epiphyses may occur as a harmless isolated anomaly, or in the early stages of conditions such as multiple epiphyseal dysplasia, spondyloepiphyseal dysplasia, hypothyroidism, cerebrohepatorenal syndrome, trisomy 18 and trisomy 21 (Silverman, 1961). For this reason, a distinction must be drawn between the non-specific usage of the term 'stippled epiphyses' as a description of a radiological feature, and its precise application as a name for a distinct disease entity. The situation may be summarised in the following way:

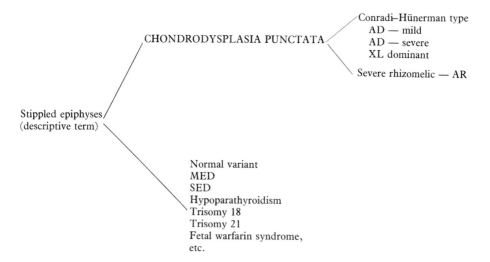

CHONDRODYSPLASIA PUNCTATA
Conradi–Hünerman type
AD — mild
AD — severe
XL dominant
Severe rhizomelic — AR

Stippled epiphyses
(descriptive term)

Normal variant
MED
SED
Hypoparathyroidism
Trisomy 18
Trisomy 21
Fetal warfarin syndrome,
etc.

Clinical and radiographic features

Affected infants have a flat face, with a depressed nasal bridge. Cataracts, alopecia and ichthyosis are inconsistent features. The clinical course is variable but survival to adulthood is usual. Structural abnormalities of the vertebral bodies may lead to spinal malalignment. Asymmetric shortening of the limbs may pose problems in later life, when degenerative osteoarthropathy in weight-bearing joints is a common complication.

In the newborn infant, stippling is radiographically evident in the epiphyses of the long bones and in the larynx, wrists, ankles and spine. Fairbank (1927) likened the radiographic appearances to the effect produced by 'the flicking of paint from a brush on to a clear surface'. The cartilagenous stippling is a transient phenomenon, which disappears by the second year. In the older child, the epiphyses are flattened and widened. Structural anomalies of the vertebral bodies may be an additional feature of the Conradi–Hünerman syndrome.

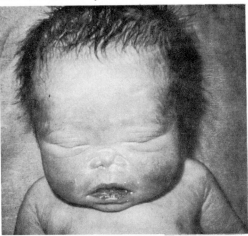

Fig. 1.22. Chondrodysplasia punctata, AD type; a stillborn infant with a flattened face, a depressed nasal bridge and ichthyotic dermal lesions. These non-specific changes are present in all forms of the condition.

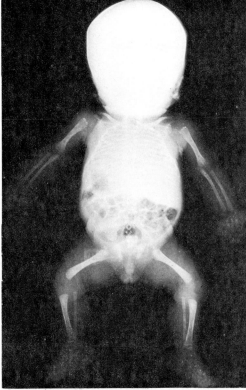

Fig. 1.23. Chondrodysplasia punctata, AD type; stippling is very obvious in the epiphyses and in the lower spine.

Genetics

Generation to generation transmission has been reported by a number of authors, including Vinke and Duffy, (1947), Silverman (1969) and Bergstrom, Gustavon and Jorulf (1972). This autosomal dominant form of chondrodysplasia punctata is comparatively common. As phenotypic expression is variable, it is likely that there is heterogeneity. The mild type which Sheffield et al. (1976) recognised in 23 individuals in Australia might represent a separate entity. Similarly, Happle, Matthias and Macher (1977) have reported an X-linked dominant type of chondrodysplasia punctata, which is lethal in males. The features resemble those of the dominant type, with the addition of widespread atrophic and pigmented dermal lesions. The authors claimed that they could identify nine similar examples in the literature. The accurate delineation of this entity would have important implications for genetic counselling.

Many cases are sporadic, and the absence of clear-cut phenotypic features can make categorisation a difficult matter. Another problem arises from the fact that the stippling might have disappeared by the time that the child is brought to the genetic clinic. In this situation, it can be very difficult to reach a firm diagnosis, and counselling may have to be speculative. The condition has been recognised radiographically in the fetus during the third trimester of pregnancy (Hyndman, Alexander and Mackie, 1976). However, it is unlikely that this technique would permit antenatal diagnosis at an earlier stage.

9. CHONDRODYSPLASIA PUNCTATA — AUTOSOMAL RECESSIVE OR SEVERE RHIZOMELIC FORM

The autosomal recessive form of chondrodysplasia punctata is a clearly defined entity. The prevalence is unknown, but it is certainly rare.

Clinical and radiographic features

Marked symmetrical rhizomelia (proximal limb shortening) is a significant feature, the changes being maximal in the humeri. Lenticular cataracts and dermal abnormalities are often present. The digits are stubby and joint mobility is limited. Death from respiratory complications usually takes place during the first year. Tracheal stenosis may be a precipitating factor (Kaufman et al., 1976).

Radiographically, the epiphyses are stippled, but in distinction to the dominant type of the condition, involvement of the spine is of mild degree. Other distinguishing features are the presence of calcification outside the margins of the epiphyses, widening and fraying of the metaphyses and coronal clefts in the vertebral bodies. The radiographic and pathological features of the rhizomelic form of chondrodysplasia punctata have been reviewed by Gilbert et al. (1976).

Genetics

Affected sibs with normal parents have been reported by several authors, including Fraser and Scriver (1954) and Mason and Kozlowski (1973). Sporadic patients from consanguineous matings were described by Mosekilde (1958) and Melnick (1965). In the latter instance, father–daughter incest had taken place. These observations are all indicative of autosomal recessive inheritance.

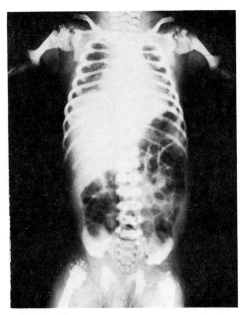

Fig. 1.24. Chondrodysplasia punctata, AR type; the humeri are symmetrical but very short. Stippling extends beyond the margins of the epiphyses, although the spine is relatively spared. The metaphyses are irregular.

Some patients do not fit neatly into the dominant and recessive phenotypic categories, and in the absence of contributory pedigree data, genetic counselling may be difficult.

It has recently become apparent that maternal therapy during early pregnancy with the anticoagulant drug, warfarin, can produce a phenocopy of chondrodysplasia punctata in the fetus. In a review of eight affected infants, Pauli *et al.* (1976) tabulated the clinical and radiographic manifestations, which were virtually indistinguishable from those of the genetic forms of the disorder.

10. METAPHYSEAL CHONDRODYSPLASIA

The metaphyseal chondrodysplasias are a group of conditions in which abnormalities of the metaphyses predominate while the epiphyses, skull and trunk are essentially normal. Immunological incompetence and endocrine dysfunction are important facets of several of these disorders. A number of forms of metaphyseal dysplasia with eponymous or descriptive designations are recognised. The Schmid type is relatively common, the McKusick type is well known but uncommon, and the remainder are rare. The following will be discussed:

  (a) Jansen type
  (b) Schmid type
  (c) McKusick type (cartilage-hair hypoplasia)
  (d) Schwachman (malabsorption and neutropenia).
  (e) Davis (thymolymphopenia)
  (f) Other types of metaphyseal chondrodysplasia.

 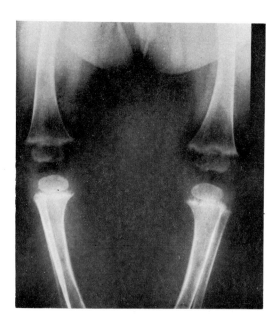

Fig. 1.25. (left) Metaphyseal chondrodysplasia; marked bowing of the legs and shortness of stature are the most obvious clinical features.

Fig. 1.25. (right) Metaphyseal chondrodysplasia; the metaphyseal irregularities are reminiscent of dietary and metabolic rickets.

The bowing of the legs and irregularity of the metaphyses bear a close resemblance to the stigmata of dietary and metabolic rickets. In this context, the differential diagnosis and manifestations of the metaphyseal chondrodysplasias have been reviewed by Spranger (1976). The unusual immunological and endocrine concomitants indicate that there is probably considerable disparity in the pathogenesis of these disorders. Indeed, it could be argued that they have been grouped together on a very arbitrary basis. Once the nature of the underlying defect has been elucidated, it is likely that many of the conditions in this general category will be reclassified.

Metaphyseal chondrodysplasia — Jansen type
The form of metaphyseal chondrodysplasia reported by Jansen (1934) was the first to be clearly delineated and it has received recognition out of all proportion to its prevalence. Less than 10 cases have been described.

Affected individuals are severely dwarfed and disabled, with bowing of the lower limbs and knobbly bone ends. Joint movements are restricted and club feet may be present. Asymptomatic hypercalcaemia is a consistent but unexplained feature. Deafness in adulthood may be the consequence of sclerosis of the base of the skull

(Holthausen, Holt and Stoeckenius, 1975). Mild radiographic changes are present at birth, and the disorder is clinically evident by the age of five. Radiographic abnormalities in the neonate include generalised osteoporosis, uneven expansion of the metaphyses and ribbon-like ribs. By mid-childhood the architecture of the metaphyses is grossly disturbed. However, these changes tend to regress in adult life (Haas, Boer and Griffioen, 1969; Kikuchi *et al.*, 1976).

Apart from an affected mother and daughter (Lenz, 1969) all other cases have been sporadic (Cameron, Young and Sissons, 1954; Weil, 1957; Gram *et al.*, 1959). It is possible that the disorder is a dominant trait and that the isolated individuals represent new mutations.

Metaphyseal chondrodysplasia — Schmid type

This entity, which was first described by Schmid (1949), is by far the most common of all the metaphyseal chondrodysplasias. Genu varum and moderately short stature are the main stigmata. These abnormalities become apparent in early childhood and a misdiagnosis of dietary or hypophosphataemic rickets is not unusual. Radiographically, the metaphyses are expanded and irregular. These changes are maximal at the hip and knee joints, where varus deformities may be evident. The skeleton is otherwise virtually normal.

Several families with generation to generation transmission have been reported and it is evident that the Schmid form of metaphyseal chondrodysplasia is inherited as an autosomal dominant (Maroteaux and Lamy, 1958; Dent and Normand, 1964; Rosenbloom and Smith, 1965; Debray *et al.*, 1975). There is considerable variation in the degree of phenotypic expression in any kindred. Of particular interest is the presence of the condition in more than 40 individuals in four generations of a Mormon family, who were reported by Stephens (1943) under the designation 'achondroplasia'. In keeping with the concept of new dominant mutation, the paternal age effect has been noted in some sporadic cases (Rosenbloom and Smith, 1965).

Metaphyseal chondrodysplasia — McKusick type

This form of metaphyseal chondrodysplasia attains maximum prevalence among the Amish of Pennsylvania, USA (McKusick, 1965). Although originally termed cartilage-hair hypoplasia, it was later redesignated 'metaphyseal chondrodysplasia—McKusick type'. It is uncertain whether the metaphyseal dysplasia reported by Maroteaux *et al.* (1963) represents the same disorder or a separate entity.

Affected individuals have a moderate degree of short stature, with thin sparse hair on the body and head. The finger joints are lax and the digits are stubby, with wide foreshortened nails. Distal prolongation of the fibula may produce inversion deformity of the ankle joint. Cellular immunity is deficient (Lux *et al.*, 1970) and patients are prone to infection, especially during infancy and early childhood. They are at particular risk from chickenpox, which may be severe and lethal. Vaccination against smallpox is contraindicated. An additional unusual feature is a liability to megacolon and intestinal malabsorption. These problems tend to regress in later life.

Radiographically, the metaphyses are irregular, with cystic changes across the entire width of the bone. A central depression in the distal metaphysis of the femur produces a 'scalloped' appearance. The height of the lumbar vertebrae may be increased, but apart from some flaring of the ribs, the axial skeleton is otherwise normal.

In their studies of the Amish, McKusick *et al*. (1966) documented 53 sibships in which at least one individual was affected. These investigators calculated a population prevalence of 1–2 per cent and demonstrated that inheritance was autosomal recessive. Subsequently, Lowry *et al*. (1970) suggested that as there was an excess of females in the published series, phenotypic expression of the gene might be influenced by the sex of the patient.

Metaphyseal chondrodysplasia with pancreatic insufficiency and neutropenia — Schwachman

The Schwachman form of metaphyseal chondrodysplasia is a rare disorder in which short stature is associated with pancreatic insufficiency, malabsorption and neutropenia (Taybi, Mitchell and Friedman, 1969). There is still doubt as to the precise relationship of this condition with other non-skeletal disorders in which pancreatic and marrow dysfunction coexist (Schwachman and Holsclaw, 1972). About 40 cases have now been reported.

Patients are severely dwarfed, with crippling due to hip joint dysplasia. Deficiency of pancreatic enzymes leads to steatorrhoea, which develops during infancy and regresses in later childhood. The neutrophil count may be consistently or intermittently low and anaemia and thrombocytopenia may also occur. Pyogenic infection in various sites is a recurrent problem. Management includes antibiotic therapy for recurrent infections and pancreatic extract for intestinal malabsorption. Femoral osteotomy may be required for the hip deformity. The ultimate prognosis is poor in the inadequately treated patient. Radiographic studies have indicated that the dwarfism is a result of skeletal dysplasia rather than malabsorption. Wide-spread metaphyseal irregularity is a prominent feature. The hip joints are severely affected, and a valgus deformity is initially present. However, slipping and restabilisation of the epiphyses of the femoral heads may ultimately lead to coxa vara.

Genetics
Cases have been described by Bodian, Sheldon and Lightwood (1964) and Burke *et al*. (1967). The condition has been reviewed by Stanley and Sutcliffe (1973). The pedigree data are consistent with autosomal recessive inheritance.

Metaphyseal chondrodysplasia with thymolymphopenia — Davis

Although the original reports were entitled 'Swiss-type agammaglobulinaemia and achondroplasia' (Davis, 1966; Fulginiti *et al*., 1967), it is now evident that this rare condition is a distinct entity which can be classified with the metaphyseal chondrodysplasias.

Lymphopenia and agammaglobulinaemia are associated with ectodermal dysplasia. In particular, the hair and eyebrows are often absent, while ichthyosis and erythroderma are present. The tubular bones are shortened with wide but regular metaphyses. The pelvic acetabulae are horizontal and the vertebral end plates are sclerosed (Alexander and Dunbar, 1968). The axial skeleton and the extremities are otherwise normal. Overwhelming infection may cause death in infancy.

Gatti *et al*. (1969) reported two sibs with normal parents. Other cases have been sporadic. The genetic basis is uncertain but inheritance is probably autosomal recessive.

Other types of metaphyseal chondrodysplasia

The Spahr form of metaphyseal chondrodysplasia is clinically and radiographically indistinguishable from the Schmid type. However, in the single kindred which was reported by Spahr and Spahr-Hartmann (1966), four affected sibs had consanguineous parents, and inheritance was presumed to be autosomal recessive.

Vaandrager (1960) observed a form of metaphyseal dysplasia in four members of two generations of a kindred. Kozlowski and Sikorska (1970) reported similar cases under the designation 'Vaandrager–Pena type of metaphyseal chondrodysplasia'. However, the affected brother and sister studied by Pena (1965) seem to have had yet another distinct disorder. Similarly, the mild metaphyseal dysplasia described by Kozlowski (1964) may represent a separate entity.

Other rare forms of metaphyseal chondrodysplasia to which eponyms have been applied include those of Maroteaux, Roy, Wiedemann and Spranger, and Rimoin. There is undoubtedly considerable overlap between early reports of ostensibly different conditions. In some instances, the same designation has been applied to disorders which may well be separate entities. The classification and nomenclature of the metaphyseal chondrodysplasias is therefore neither clear-cut nor complete.

11. MESOMELIC DWARFISM

The term 'mesomelic dwarfism' pertains to short stature with limb shortening which is most pronounced in the forearms and shins. Several distinct forms are known, and reports of atypical cases are probably indicative of further heterogeneity. The following types of mesomelic dwarfism are considered in this section:

(a) Nievergelt
(b) Langer
(c) Acromesomelic dwarfism
(d) Other types of mesomelic dwarfism.

Descriptive anatomical terms which are used in the context of short-limbed dwarfism include micromelia, rhizomelia, mesomelia and acromelia. Micromelia pertains to shortening of all segments of a limb, while rhizo-, meso- and acro- respectively imply maximal involvement of the proximal, middle and distal limb segments. Compound designations such as acromesomelic micromelia are sometimes employed. The mesomelic skeletal dysplasias have been reviewed by Kaitila, Leisti and Rimoin (1976).

By definition, dyschondrosteosis is a mesomelic dysplasia. However, affected individuals are not dwarfed, and this condition is therefore discussed elsewhere (see Chapter 3).

Nievergelt type of mesomelic dwarfism

The only reports of this extremely rare condition concern four members of a kindred studied by Nievergelt (1944) and affected boys described by Solonen and Sulamaa

(1958) and Young and Wood (1975). The major clinical features are gross shortening of the distal portions of the limbs, flexion deformities of the fingers and elbows, genu valgum, and club feet. Radiographically, the tibia has a pathognomonic rhomboidal configuration.

Good evidence for autosomal dominant inheritance was provided by a severely dwarfed male who transmitted the disorder to three sons, all by different mothers (Nievergelt, 1944). It is apparent that this individual's initiative amply compensated for his disability!

Langer type of mesomelic dwarfism

Using the designation 'mesomelic dwarfism of the hypoplastic ulna, fibula, mandible type', Langer (1967) reported two dwarfed individuals with severe shortening of the limbs and underdevelopment of the jaw. Radiographically, ossification was defective in the distal ulna and the proximal fibula. The mandible was hypoplastic.

It was originally thought that the disorder was inherited as an autosomal recessive. However, Espiritu, Chen and Wooley (1975) described two affected siblings whose parents had dyschondrosteosis (a similar but much milder condition — see Chapter 3) and advanced the fascinating proposition that these dwarfed children might be homozygous for the dyschondrosteosis gene. Silverman (1975) mentioned a similar situation and quoted other examples which could be recognised in the early case reports of Böök (1950) and Brailsford (1953).

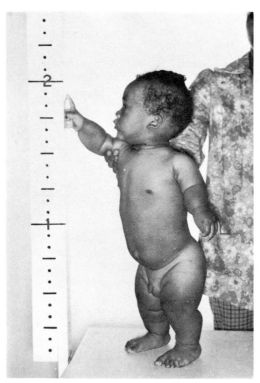

Fig. 1.27. Mesomelic dwarfism; a two-year-old boy with the Langer type of mesomelic dwarfism. Shortening is maximal in the middle segments of the limbs.

Acromesomelic dwarfism

Maroteaux, Martinelli and Campailla (1971) described a form of dwarfism in which shortening was present in the tubular bones of the hands and feet, as well as in the middle segments of the limbs. The term 'acromesomelic' is applicable to this distribution of abnormalities. Radiographically, shortening and hypoplasia was maximal in the fibula, proximal radius and distal ulna and in the tubular bones of the digits.

The Maroteaux type of acromesomelic dwarfism is inherited as an autosomal recessive. In a further case report, Campailla and Martinelli (1971) described an Italian brother and sister with acromesomelia. The status of the disorder in these sibs, as a distinct entity, is debateable. However, inheritance is certainly autosomal recessive, as indicated by the presence of the condition in five sisters in Cape Town, who had unaffected consanguineous parents (Beighton, 1974).

Other types of mesomelic dwarfism

Several other varieties of mesomelic dwarfism have been reported in individual patients or small kindreds. Each of these conditions has characteristic features which permit recognition.

In the Werner type of mesomelic dwarfism, which is inherited as an autosomal dominant, gross tibial hypoplasia is associated with polydactyly and absence of the thumbs (Werner, 1915; Eaton and McKusick, 1969; Pashayan et al., 1971).

The Robinow or 'fetal face' syndrome comprises limb shortening, hypoplasia of the genitals, an abnormal facies and anomalies of the vertebrae and ribs. A firm radiographic diagnosis can be made by recognition of the combination of mesomelia, hemivertebrae, rib fusion and bifid terminal phalanges. Familial cases have been reported by Robinow, Silverman and Smith (1973). Inheritance is apparently autosomal dominant.

The Reinhardt–Pfeiffer type of mesomelia is characterised by bowing and shortening of the bones of the forearms and shins, with variable synostosis in the carpus and tarsus (Reinhardt and Pfeiffer, 1967). In the single kindred which has been reported, 14 members of four generations were affected. The pattern of transmission was consistent with autosomal dominant inheritance (Reinhardt, 1976).

Leroy, de Vos and Timmermans (1975) described a form of mesomelic dwarfism in a father and his two sons. These individuals had severe shortening and bowing of the tibia, with milder changes in the radius, ulna and fibula. The skeleton was otherwise normal. It is probable that inheritance is autosomal dominant.

12. RHIZOMELIC DWARFISM

Proximal limb shortening or rhizomelia is a feature of several well-known forms of dwarfism. In addition, rhizomelia predominates in a number of rare malformation syndromes.

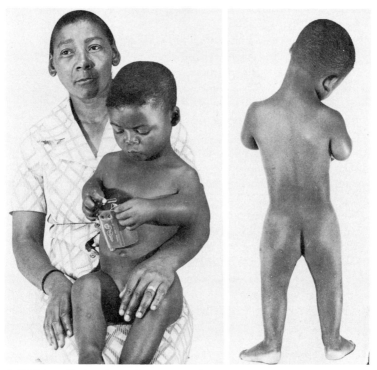

Fig. 1.28. (left) Rhizomelic dwarfism; a five-year-old boy with predominant shortening in the proximal segments of the limbs.

Fig. 1.29. (right) Rhizomelic dwarfism; the rhizomelia is particularly noticeable when the child is viewed from behind.

Under the designation 'humerospinal dysostosis with congenital heart disease', Kozlowski, Celermajer and Tink (1974) described two infants with rhizomelia, distal bifurcation of the humeri, subluxation of the elbow joints, coronal cleft vertebrae, talipes equinovarus and congenital heart disease. This brother and sister had the same mother but different fathers. As the parents were unaffected and non-consanguineous, it is likely that inheritance was autosomal dominant and that the gene was non-penetrant in the mother.

Patterson and Lowry (1975) reported a new dwarfing syndrome in which gross shortening of the humeri and coxa vara were the major features. As the patient in question was 86 years of age, it is apparent that the disorder is relatively innocuous!

13. CAMPOMELIC DWARFISM

More than 20 patients with campomelic dwarfism have been described, and reports contine to accumulate. The name of the condition, which is derived from the Greek, has the connotation 'bent limb'. The manifestations have been reviewed by Becker, Finegold and Genieser (1975), Eliachor *et al*. (1975) and Weiner, Benfield and Robinson (1976).

Clinical and radiographic features

Campomelic dwarfism is potentially lethal, and the majority of reports concern babies who have succumbed to respiratory obstruction and recurrent aspiration of food during the neonatal period. Bowing and shortening of the tubular bones is maximal in the legs. Hypertelorism, micrognathia and a cleft palate are additional features. The feet may have a calcaneovalgus or equinovarus deformity and a subcutaneous dimple is often present over the anterolateral aspect of the tibia. Radiographically, the scapula and fibula may be hypoplastic, while ossification of the vertebral bodies is sometimes defective.

Idiopathic bowing of the legs, in the absence of other stigmata of the campomelic syndrome, is not uncommon. However, this abnormality has probably been a source of confusion in some case reports. Congenital bowing of long bones is discussed in Chapter 3.

Genetics

The majority of reported cases of campomelic dwarfism have been sporadic. Spranger, Langer and Maroteaux (1970) mentioned that they had seen eight affected infants in a six-month period, and suggested that the condition might be increasing in frequency. There has been considerable speculation concerning the pathogenesis of the disorder and a variety of exogenous factors have been incriminated. Gardner, Assemany and Nue (1971) suggested that oral contraceptives might be involved. However, this contention has not been substantiated.

Some investigators have favoured a genetic aetiology, although there has been no general agreement. Bianchine *et al*. (1971) considered that new dominant mutation was a likely cause. Reports of a pair of sisters with the disorder (Stuve and Wiedemann, 1971) and an affected child with consanguineous parents (Cremin, Orsmond and Beighton, 1973) is evidence for autosomal recessive inheritance. Thurmon, DeFraites and Anderson (1973) encountered campomelic dwarfism in two sibs and their half-sister. The mother of all of these infants had slight tibial bowing. There was a suspicion of consanguinity in the kindred and the authors commented that the pedigree data were consistent with autosomal recessive inheritance, with minor manifestations in the heterozygous mother, or with autosomal dominant inheritance with very variable expression.

Khajavi *et al*. (1976) examined the radiographs of nine patients, reviewed the literature, and concluded that the campomelic syndrome was heterogeneous. These authors suggested that there might be a 'classic' or 'long bone' form and two 'short bone' types, one of which was associated with craniostenosis while the other was normocephalic. At the present time, there are no clear guide lines for genetic counselling in a sporadic case.

LETHAL SHORT-LIMBED DWARFISM

This problem, which has attracted considerable attention, has been reviewed and discussed by Housten, Awen and Kent (1972), Cremin and Beighton (1974), Curran, Sigmon and Optiz (1974), Martin (1975) and Maroteaux, Stanescu and Stanescu (1976). The following conditions enter into the differential diagnosis of the stillborn short-limbed dwarf:

	Eponym	Probable inheritance
Achondrogenesis type I (IA)	Parenti–Fraccaro	AR
Achondrogenesis type II (IB)	Saldino	AR
Achondrogenesis type III?	Verma	AR
Thanatophoric dwarfism		Polygenic?
Homozygous achondroplasia		
Asphyxiating thoracic dystrophy	Jeune syndrome	AR
Short rib-polydactyly syndrome I	Saldino–Noonan	AR
Short rib-polydactyly syndrome II	Majewski	?
Short rib-polydactyly syndrome III(?)	Naumoff	AR?
Chondroectodermal dysplasia	Ellis–van Creveld	AR
Chondrodysplasia punctata (severe rhizomelic type)		AR
Campomelic dwarfism		?
Osteogenesis imperfecta congenita (see Chapter 5)		AR/AD
Hypophosphatasia lethalis (see Chapter 5)		AR

Several of the disorders on this list are probably heterogeneous. Furthermore, there is little doubt that other forms of lethal short-limb dwarfism await delineation. At the present time, a spurious diagnosis of 'achondroplasia' often appears on the death certificate. In clinical practice, it is essential that radiographs are obtained of every stillborn dwarf. In this way, an accurate diagnosis can be made, thus facilitating meaningful genetic counselling.

REFERENCES

Preamble

Dorst, J. P., Scott, C. I. & Hall, J. G. (1972) The radiologic assessment of short stature — dwarfism. *Radiologic Clinics of North America*, 10/2, 393.

Kaufman, H. J. C. (1976) Classification of the skeletal dysplasias and the radiologic approach to their differentiation. *Clinical Orthopaedics and Related Research*, 114, 12.

Kopits, S. E. (1976) Orthopaedic complications of dwarfism. *Clinical Orthopaedics and related Research*, 114, 153.

Kozlowski, K. (1976) Bone dysplasias in radiographic diagnosis. *Pediatric Radiology*, 4/2, 66.

Lachman, R. S., Rimoin, D. L. & Hollister, D. W. (1974) Hip arthrography in the epiphyseal dysplasias. *Birth Defects: Original Article Series*, 10/12, 186.

Nelson, M. A. (1970) Orthopaedic aspects of the chondrodystrophies. The dwarf and his orthopaedic problems. *Annals of the Royal College of Surgeons of England*, 47, 185.

Rimoin, D. L., Silberberg, R. & Hollister, D. W. (1976) Chondro-osseous pathology in the chondrodystrophies. *Clinical Orthopaedics and Related Research*, 114, 137.

Achondroplasia

Bailey, J. A. (1970) Orthopaedic aspects of achondroplasia. *Journal of Bone and Joint Surgery*, 52A/7, 1285.

Bowen, P. (1974) Achondroplasia in two sisters with normal parents. *Birth Defects: Original Article Series*, **10/12**, 31.

Gardner, R. J. M. (1977) A new estimate of the achondroplasia mutation rate. *Clinical Genetics*, **11**, 31.

Globus, M. S. & Hall, B. D. (1974) Failure to diagnose achondroplasia *in utero*. *Lancet*, **i**, 624.

Hall, J. G., Dorst, J. P., Taybi, H., Langer, L. O., Scott, C. I. & McKusick, V. A. (1969) Two probable cases of homozygosity for the achondroplasia gene. *Birth Defects: Original Article Series*, **5(4)**, 24.

Harris, R. & Patton, J. T. (1971) Achondroplasia and thanatophoric dwarfism in the newborn. *Clinical Genetics*, **2**, 61.

Langer, L. O., Spranger, J. W., Greinacher, I. & Herdman, R. C. (1969) Thanatophoric dwarfism. *Radiology*, **92**, 285.

Langer, L. O., Jr, Baumann, P. A. & Gorlin, R. J. (1967) Achondroplasia. *American Journal of Roentgenology, Radium Therapy and Nuclear Medicine*, **100**, 12.

Murdoch, J. L., Walker, B. A., Hall, J. G., Abbey, H., Smith, K. K. & McKusick, V. A. (1970) Achondroplasia — a genetic and statistical survey. *Annals of Human Genetics*, **33**, 227.

Morch, E. T. (1941) Chondrodystrophic dwarfs in Denmark, Opera Ex Demo. *Biologiae Hereditariae Humanae, Vol. 3*. Copenhagen: Ejnar Munksgaard.

Opitz, J. M. (1969) Delayed mutation in achondroplasia? *Birth Defects: Original Article Series*, **5/4**, 20.

Potter, E. L. & Coverstone, V. A. (1948) Chondrodystrophy fetalis. *American Journal of Obstetrics and Gynecology*, **56**, 790.

Scott, C. I. (1976) Achondroplastic and hypochondroplastic dwarfism. *Clinical Orthopaedics and Related Research*, **114**, 18.

Stevenson, A. C. (1957) Achondroplasia: an account of the condition in Northern Ireland. *American Journal of Human Genetics*, **9**, 81.

Todorov, A. B. & Bolling, D. R. (1974) A computerised file for studying growth development in achondroplasia. *Birth Defects: Original Article Series*, **10/9**, 241.

Warkany, J. (1971) In *Congenital Malformations: Notes and Comments*. Chicago: Year Book Medical Publishers.

Hypochondroplasia

Beals, R. K. (1969) Hypochondroplasia: a report of five kindreds. *Journal of Bone and Joint Surgery*, **51A**, 728.

Frydman, M., Hertz, M. & Goodman, R. M. (1974) The genetic entity of hypochondroplasia. *Clinical Genetics*, **5**, 223.

McKusick, V. A., Kelley, T. E. & Dorst, J. P. (1973) Observations suggesting allelism of the achondroplasia and hypochondroplasia genes. *Journal of Medical Genetics*, **10**, 11.

Scott, C. I. (1976) Achondroplastic and hypochondroplastic dwarfism. *Clinical Orthopaedics and Related Research*, **114**, 18.

Specht, E. E. & Daentl, D. L. (1975) Hypochondroplasia. *Clinical Orthopaedics and Related Research*, **110**, 249.

Achondrogenesis

Harris, R., Patton, I. T. & Barson, A. J. (1972) Pseudoachondrogenesis with fractures. *Clinical Genetics*, **3**, 435.

Houston, C. S., Awen, C. F. & Kent, H. P. (1972) Fatal neonatal dwarfism. *Journal of the Canadian Association of Radiologists*, **23**, 45.

Laxova, R., O'Hara, P. T., Ridler, M. A. C. & Timothy, J. A. D. (1973) Family with probable achondrogenesis and lipid inclusions in fibroblasts. *Archives of Diseases in Childhood*, **48**, 212.

McKusick, B. A. (1975) Achondrogenesis, type IB. In *Mendelian Inheritance in Man*, 4th edn, p.333. Baltimore and London: Johns Hopkins University Press.

Ornoy, A., Sekeles, E. & Smith, (1976) Achondrogenesis type I in three sibling fetuses. Scanning and transmission electron microscopic studies. *American Journal of Pathology*, **82/1**, 71.

Scott, C. I. (1972) In *Progress in Medical Genetics, Vol. 8*, Ch. 7. Ed. Steinberg, A. G. & Bearn, A. G. New York: Grune and Stratton.

Verma, I. C., Bharagava, S. & Agarwal, A. (1975) An autosomal recessive form of lethal chondrodystrophy with severe thoracic narrowing, rhizoacromelic type of micromelia, polydactyly and genital anomalies. *Birth Defects: Original Article Series*, **11/6**, 167.

Wiedemann, H. R., Remagen, W. & Hienz, H. A. (1974) Achondrogenesis within the scope of connately manifested generalised skeletal dysplasias. *Zeitschrift für Kinderheilkunde*, **116/4**, *223*.

Yang, S. S., Brougn, A. J., Garewal, G. S. & Bernstein, J. (1974) Two types of inheritable lethal achondrogenesis. *Journal of Paediatrics*, **85/6**, 796.

Thanatophoric dwarfism

Bergstrom, K., Gustavon, K. H. & Jorulf, H. (1972) Thanatophoric dwarfism: diagnosis in utero. *Australasian Radiology*, **16**, 155.

Bouvet, J. P., Maroteaux, P. & Feingold, J. (1974) Genetic study of thanatophoric dwarfism. *Annals of Human Genetics*, 17/3, 181.

Chemke, J., Graff, G. & Lancet, M. (1971) Familial thanatophoric dwarfism (Letter). *Lancet*, i, 1358.

Cremin, B. J. & Shaff, M. I. (1977) Ultrasonic diagnosis of lethal dwarfism from the short limbs. *American Journal of Roentgenology* (in press).

Danes, B. S. (1974) Achondroplasia and thanatophoric dwarfism: a study in cell culture. *Birth Defects: Original Article Series*, 11/2,238.

Eaton, A. P., Sommer, A. & Sayers, M. P. (1975) The kleeblattschadel anomaly. *Birth Defects: Original Article Series*, 11/2, 238.

Feingold, M., Miller, D. & Bull, M. J. (1973) The demise of a syndrome? *Syndrome Identification*, 1, 21.

Harris, R. & Patton, J. T. (1971) Achondroplasia and thanatophoric dwarfism in the newborn. *Clinical Genetics*, 2, 61.

Hodach, R. J., Viseskul, C., Gilbert, E. F. *et al.* (1975) Studies of malformation syndrome in man. XXXVI: The Pfeiffer syndrome, association with kleeblattschadel and multiple visceral anomalies: case report and review. *Zeitschrift für Kinderheilkunde*, 119/2, 87.

Langer, L. O., Jr, Spranger, J. W., Greinacher, I. & Herdman, R. C. (1969) Thanatophoric dwarfism. A condition confused with achondroplasia in the neonate, with brief comments on achondrogenesis and homozygous achondroplasia. *Radiology*, 92, 285.

Maroteaux, P., Lamy, M. & Robert, J. M. (1967) Le nanisme thanatophore. *Presse Médicale*, 75, 2519.

Moir, D. H. & Kozlowski, K. (1976) Long survival in thanatophoric dwarfism. *Pediatric Radiology*, 5, 123.

O'Malley, B. P., Parker, R., Saphyakhafon, P. & Qizilbach, A. H. (1972) Thanatophoric dwarfism. *Journal of the Canadian Association of Radiologists*, 23, 62.

Partington, M. W., Gonzales-Crussi, F., Khakee, S. G. & Wollin, D. G. (1971) Cloverleaf skull and thanatophoric dwarfism. Report of four cases, two in the same sibship. *Archives of Diseases in Childhool*, 46, 656.

Pena, S. D. J. & Goodman, H. O. (1973) The genetics of thanatophoric dwarfism. *Pediatrics*, 51, 104.

Pilz, E. & Swoboda, W. (1975) Cloverleaf skull syndrome. *Pädiatrishe Fortbildungskurse für die Praxis*, 16/2, 275.

Sabry, A. (1974) Thanatophoric dwarfism in triplets (Letter). *Lancet*, ii, 533.

Temtamy, S. A., Shoukry, A. S., Fayad, I. & El Meligy, M. R. (1975) Limb malformations in the cloverleaf skull anomaly. *Birth Defects: Original Article Series*, 11/2, 247.

Thompson, B. H. & Parmley, T. H. (1971) Obstetric features of thanatophoric dwarfism. *American Journal of Obstetrics and Gynaecology*, 109, 396.

Widdig, K., Steinhoff, R. & Guenther, H. (1974) The cloverleaf skull syndrome. *Zentralblatt für allgemeine Pathologie und pathologische Anatomie*, 118/4, 358.

Young, R. S., Pocharzevsky, R., Leonicas, J. C., Wexley I. B. & Ratney, H. (1973) Thanatophoric dwarfism and cloverleaf skull (Kleeblattschädel). *Radiology*, 106, 401.

Asphyxiating thoracic dysplasia

Barnes, N. D., Hull, D., Milner, A. D. & Waterston, D. J. (1971) Chest reconstruction in thoracic dystrophy. *Archives of Diseases in Childhood*, 46, 833.

Friedman, J. M., Kaplan, H. G. & Hall, J. G. (1975) The Jeune syndrome (asphyxiating thoracic dystrophy) in an adult. *American Journal of Medicine*, 59/6, 857.

Hanissian, A. S., Riggs, W. W. & Thomas, D. A. (1967) Infantile thoracic dystrophy. A variant of the Ellis–van Creveld syndrome. *Journal of Pediatrics*, 71, 855.

Herdman, R. C. & Langer, L. O. (1968) The thoracic asphyxiant dystrophy and renal disease. *American Journal of Diseases of Children*, 116, 309.

Jeune, M., Beraud, C. & Carron, R. (1955) Dystrophie thoracique asphyxiante de caratère familial. *Archives françaises de Pédiatre*, 12, 886.

Kaufman, H. J. & Kirkpatrick, J. A., Jr. (1974) Jeune thoracic dysplasia — a spectrum of disorders? *Birth Defects: Original Article Series*, 10/9, 101.

Kohler, E. & Babbitt, D. P. (1970) Dystrophic thoraces and infantile asphyxia. *Radiology*, 94, 55.

Kozlowski, K. & Masel, J. (1976) Asphyxiating thoracic dystrophy without respiratory distress. *Pediatric Radiology*, 5, 30.

Majewski, F., Pfeiffer, R. A., Leng, W., Müller, R., Feil, G. & Seiler, R. (1971) Polysyndactylie, verkurzte Gliedmassen und Genitalfaltbildungen: Kennzeichen eines selbstandigen Syndroms? *Zeitschrift für Kinderheilkunde*, 111, 118.

Maroteaux, P. & Savart, P. (1964) La dystrophie thoracique asphyxiante. Etude radiologique et rapports avec le syndrôme d'Ellis–van Creveld. *Annals of Radiology*, 7, 332.

Naumoff, P., Young, L. W., Mazer, J. & Amortegui, A. J. (1977) Short rib-polydactyly syndrome type 3. *Radiology*, 122/2, 443.

Niemann, N., Mangiaux, M., Rauber, G., Pernot, C. & Bretagne-de Kersauson, M. C. (1963) Dystrophie thoracique asphyxiante du nouveau-né. *Pédiatrie*, **18**, 387.

Ozonoff, M. B. (1974) Asphyxiating thoracic dysplasia as a complication of metaphyseal chondrodysplasia (Jansen type). *Birth Defects: Original Article Series*, **10/12**, 72.

Saldino, R. M. & Noonan, C. D. (1972) Severe thoracic dystrophy with striking micromelia, abnormal osseous development, including the spine, and multiple visceral anomalies. *American Journal of Roentgenology, Radium Therapy and Nuclear Medicine*, **114**, 257.

Spranger, J., Langer, L. O., Weller, M. H. & Herrmann, J. (1974) Short rib-polydactyly syndromes and related conditions. *Birth Defects: Original Article Series*, **10/9**, 117.

Shokeir, M. H. K., Houston, C. S. & Awen, C. F. (1971) Asphyxiating thoracic chondrodystrophy. Association with renal disease and evidence for possible heterozygous expression. *Journal of Medical Genetics*, **8**, 107.

Chondroectodermal dysplasia

Ellis, R. W. B. & Van Creveld, S. (1940) A syndrome characterised by ectodermal dysplasia, polydactyly, chondroplasia and congenital morbus cordis. *Archives of Diseases in Childhood*, **15**, 65.

McKusick, V. A., Egeland, J. A., Eldridge, R. & Krusen, D. E. (1964) Dwarfism in the Amish. The Ellis–van Creveld syndrome. *Bulletin of the Johns Hopkins Hospital*, **115**, 306.

Metrakos, J. D. & Fraser, F. C. (1954) Evidence for a heredity factor in chondroectodermal dysplasia (Ellis–van Creveld syndrome). *American Journal of Human Genetics*, **6**, 260.

Murdoch, J. L. & Walker, B. A. (1969) Ellis–van Creveld syndrome. *Birth Defects: Original Article Series*, **5/4**, 279.

Multiple epiphyseal dysplasia

Barrie, H., Carter, C. & Sutcliffe, J. (1958) Multiple epiphyseal dysplasia. *British Medical Journal*, **ii**, 133.

Diamond, L. S. (1974) Pleomorphism of hip disease in a large kinship with spondylo- and multiple epiphyseal dysplasia. *Birth Defects: Original Article Series*, **10/12**, 406.

Fairbank, H. A. T. (1935) Generalised diseases of the skeleton. *Proceedings of the Royal Society of Medicine*, **28**, 1611.

Fairbank, H. A. T. (1951) *An Atlas of General Affections of the Skeleton*. Edinburgh, London: Churchill Livingstone.

Harper, P. S., Brotherton, B. J. & Cochlin, D. (1976) Genetic risks in Perthe's disease. *Clinical Genetics*, **10**, 178.

Hoefnagel, D., Sycamore, L. K., Russell, S. W. & Bucknall, W. E. (1967) Hereditary multiple epiphyseal dysplasia. *Annals of Human Genetics*, **30**, 201.

Juberg, R. C. & Holt, J. F. (1968) Inheritance of multiple epiphyseal dysplasia tarda. *American Journal of Human Genetics*, **20**, 549.

Lie, S. O., Siggers, D. C., Dorst, J. P. & Kopits, D. E. (1974) Unusual multiple epiphyseal dysplasias. *Birth Defects: Original Article Series*, **12**, 165.

Lowry, R. B. & Wood, B. J. (1975) Syndrome of epiphyseal dysplasia, short stature, microcephaly and nystagmus. *Clinical Genetics*, **8/4**, 269.

Monty, C. P. (1962) Familial Perthe's disease resembling multiple epiphyseal dysplasia. *Journal of Bone and Joint Surgery*, **44B**, 565.

Pfeiffer, R. A., Jünemann, G., Polster, J. & Bauer, H. (1973) Epiphyseal dysplasia of the femoral head, severe myopia and perceptive hearing loss in three brothers. *Clinical Genetics*, **4**, 141.

Stephens, F. E. & Kerby, J. P. (1946) Hereditary Legg–Calvé–Perthe's disease. *Journal of Heredity*, **37**, 153.

Wamoscher, Z. & Farhi, A. (1963) Hereditary Legg–Calvé–Perthe's disease. *American Journal of Diseases of Children*, **106**, 131.

Chondrodysplasia punctata AD

Bergstrom, K., Gustavson, K. H. & Jorulf, H. (1972) Chondrodystrophia calcificans congenita in a mother and her child. *Clinical Genetics*, **3**, 158.

Fairbank, H. A. T. (1927) Some general diseases of the skeleton. *British Journal of Surgery*, **15**, 120.

Happle, R., Matthiass, H. H. & Macher, E. (1977) Sex-linked chondrodysplasia punctata? *Clinical Genetics*, **11**, 73.

Hyndman, W. B., Alexander, D. S. & Mackie, K. W. (1976) Chondrodystrophia calcificans congenita (Conradi–Hünermann syndrome). Report of a case recognised antenatally. *Clinical Pediatrics*, **15/4**, 311.

Sheffield, L. J., Danks, D. M., Mayne, V. & Hutchinson, A. L. (1976) Chondrodysplasia punctata — 23 cases of a mild and relatively common variety. *Journal of Pediatrics*, **89/6**, 916.

Silverman, F. N. (1961) Dysplasie epiphysaires: entités proteiformes. *Annals of Radiology*, **4**, 833.

Silverman, F. N. (1969) Discussion on the relation between stippled epiphyses and the multiplex form of epiphyseal dysplasia. *Birth Defects: Original Article Series*, **5/4**, 68.

Spranger, J., Opitz, J. M. & Bidder, U. (1971) Heterogeneity of chondrodysplasia punctata. *Humangenetik*, **11**, 190.

Vinke, T. H. & Duffy, F. P. (1947) Chondrodystrophia calcificans congenita: report of two cases. *Journal of Bone and Joint Surgery*, **29**, 509.

Chondrodysplasia punctata AR

Fraser, F. C. & Scriver, J. B. (1954) A hereditary factor in chondrodystrophia calcificans congenita. *New England Journal of Medicine*, **250**, 272.

Gilbert, E. F., Opitz, J. M., Spranger, J. W., Langer, L. O., Wolfson, J. J. & Visekul, C. (1976) Chondrodysplasia punctata — rhizomelic form. Pathologic and radiologic studies of three infants. *European Journal of Pediatrics*, **123/2**, 89.

Kaufmann, H. J., Mahboubi, S., Spackman, T. J., Capitano, M. A. & Kirkpatrick, J. (1976) Tracheal stenosis as a complication of chondrodysplasia punctata. *Annals of Radiology*, **19/1**, 203.

Mason, R. C. & Kozlowski, K. (1973) Chondrodysplasia punctata. A report of 10 cases. *Radiology*, **109**, 145.

Melnick, J. C. (1965) Chondrodystrophia calcificans congenita. *American Journal of Disease of Children*, **110**, 218.

Mosekilde, E. (1952) Stippled epiphyses in the newborn. *Acta radiologica*, **37**, 291.

Pauli, M. P., Madden, J. D., Kranzler, K. J., Culpepper, W. & Port, R. (1976) Warfarin therapy initiated during pregnancy and phenotypic chondrodysplasia punctata. *Journal of Paediatrics*, **88/3**, 506.

Metaphyseal chondroplasia preamble

Spranger, J. W. (1976) Metaphyseal chondrodysplasias. *Birth Defects: Original Article Series*, **12/6**, 33.

Metaphyseal chondrodysplasia – Jansen type

Cameron, J. A. P., Young, W. B. & Sissons, H. A. (1954) Metaphyseal dysostosis — report of a case. *Journal of Bone and Joint Surgery*, **36B**, 622.

Gram, P. B., Fleming, J. L., Frame, B. & Fine, G. (1959) Metaphyseal chondrodysplasia of Jansen. *Journal of Bone and Joint Surgery*, **41A**, 951.

Haas, W. H. D., Boer, W. de & Griffioen, F. (1969) Metaphyseal dysostosis; a late follow-up of the first reported case. *Journal of Bone and Joint Surgery*, **51B**, 2-0.

Holthusen, W., Holt, J. F. & Stoeckenius, M. (1975) The skull in metaphyseal chondrodysplasia type Jansen. *Paediatric Radiology*, **3/3**, 137.

Jansen, M. (1934) Über atypische chondrodystrophie (achondroplasie) und über eine noch nicht beschriebene angeborene Wachstumsstörung des Knochensystems: metaphysäre Dysostosis. *Zeitschrift für Orthopädische Chirugie*, **61**, 253.

Kikuchi, S., Hasue, M., Watanabe, M. & Hasebe, K. (1976) Metaphyseal dysostosis (Jansen type). *Journal of Bone and Joint Surgery*, **58B**, 102.

Lenz, W. D. (1969) Discussion in first conference on the clinical delineation of birth defects. *Birth Defects: Original Article Series*, **5/4**, 71.

Weil, S. (1957) Die metaphysären Dysostoses. *Zeitschrift für Orthopaedie und ihre Grenzgebiete,* **89**, 1.

Metaphyseal chondrodysplasia – Schmid type

Debray, H., Poissonnier, M., Brault, J. & D'Angely, S. (1975) Metaphyseal chondrodysplasia. *Annals of Pediatrics*, **51/3**, 253.

Dent, C. E. & Normand, I. C. S. (1964) Metaphyseal dysostosis type Schmid. *Archives of Diseases in Childhood*, **39**, 444.

Maroteaux P. & Lamy, M. (1958) La dysostose métaphysaire. *Semaine de Hôpitaux de Paris*, **34**, 1729.

Rosenbloom, A. L. & Smith, D. W. (1965) The natural history of metaphyseal dysostosis. *Journal of Pediatrics*, **66**, 857.

Schmid, F. (1949) Beitrag zur Dysostosis enchondralis metaphysaria. *Monatsschrift für Kinderheilkunde*, **97**, 393.

Stephens, F. E. (1943) An achondroplastic mutation and the nature of its inheritance. *Journal of Heredity*, **34**, 229.

Metaphyseal chondrodysplasia – McKusick type

Lowry, R. B., Wood, B. J., Birkbeck, J. A. & Padwick, P. H. (1970) Cartilage-hair hypoplasia. A rare and recessive cause of dwarfism. *Clinical Pediatrics*, **9**, 44.

Lux, S. E., Johnston, R. B., Jr, August, C. S., Say, B., Penchaszadeh, V. B., Rosen, F. S. & McKusick, V. A. (1970) Neutropenia and abnormal cellular immunity in cartilage-hair hypoplasia. *New England Journal of Medicine*, **282**, 234.
Maroteaux, P., Savart, P., Lefebvre, J. & Royer, P. (1963) Les formes partielles de la dysostose metaphysaire. *Presse Medicale*, **71**, 1523.
McKusick, V. A., Eldridge, R., Hostetler, J. A., Ruangwit, J. A. & Egeland, J. A. (1965) Dwarfism in the Amish. II. Cartilage-hair hypoplasia. *Bulletin of the Johns Hopkins Hospital*, **116**, 285.

Metaphyseal chondrodysplasia – Schwachman type
Bodian, M., Sheldon, W. & Lightwood, R. (1964) Congenital hypoplasia of the exocrine pancreas. *Acta paediatrica*, **53**, 282.
Burke, V., Colebatch, J. H., Anderson, C. M. & Simions, M. J. (1967) Association of pancreatic insufficiency and chronic neutropenia in childhood. *Archives of Diseases in Childhood*, **42**, 147.
Schwachman, H. & Holsclaw, D. (1972) Some clinical observations on the Schwachman syndrome (pancreatic insufficiency and bone marrow hypoplasia). *Birth Defects: Original Article Series*, **8/3**, 46.
Stanley, P. & Sutcliffe, J. (1973) Metaphyseal chondrodysplasia with dwarfism, pancreatic insufficiency and neutropenia. *Pediatric Radiology*, **1**, 119.
Taybi, H., Mitchell, A. D. & Friedman, G. D. (1969) Metaphyseal dysostosis and the associated syndrome of pancreatic insufficiency and blood disorders. *Radiology*, **93**, 563.

Metaphyseal chondrodysplasia – Davis type
Alexander, W. J. & Dunbar, J. S. (1968) Unusual bone changes in thymic alymphoplasia. *Annals of Radiology*, **2**, 289.
Davis, J. A. (1966) A case of Swiss-type agammaglobulinaemia and achondroplasia. *British Medical Journal*, **ii**, 1371.
Fulginiti, V. A., Hathaway, W. E., Pearlman, D. S. & Kempe, C. H. (1967) Agammaglobulinaemia and achondroplasia (Letter). *British Medical Journal*, **ii**, 242.
Gatti, R. A., Platt, N., Pomerance, H. H., Hong, R., Langer, L. O., Kay, H. E. M. & Good, R. A. (1969) Hereditary lymphopenic agammaglobulinaemia associated with a distinctive form of short-limbed dwarfism and ectodermal dysplasia. *Journal of Pediatrics*, **75**, 675.

Metaphyseal chondrodysplasia – miscellaneous
Kozlowski, K. (1964) Metaphyseal dysostosis. Report of five familial and two sporadic cases of mild type. *American Journal of Roentgenology, Radium Therapy and Nuclear Medicine*, **91**, 602.
Kozlowski, K. & Sikorska, B. (1970) Dysplasia metaphysaria type Vaandrager–Pena. *Zeitschrift für Kinderheilkunde*, **108**, 165.
Pena, J. (1965) Disostosis metafisaria. Una revision. Con aportacion do una observacion familar. Una forma mieva de la enfermedael. *Radiologia*, **47**, 3.
Spahr, A. & Spahr-Hartmann, I. (1961) Dysostose metaphysaire familiale. Étude de 4 cas cans une fratrie. *Helvetica paediatrica acta*, **16**, 836.
Vaandrager, G. J. (1960) Metafysaire dysostosis. *Nederlands Tydschrift voor Geneeskunder*, **104**, 547.

Mesomelic dwarfism – preamble
Kaitila, I. I., Leisti, J. T. & Rimoin, D. L. (1976) Mesomelic skeletal dysplasia. *Clinical Orthopaedics and Related Research*, **114**, 94.

Mesomelic dwarfism – Nievergelt type
Nievergelt, K. (1944) Positiver Vaterschaftsnachweis auf Grund erblicher Missbildungen der Extremitaten. *Archiv der Julius Klaus-Stiftung für Vererbungsforschung Sozialanthropologie und Rassenhygiene*, **19**, 157.
Solonen, K. A. & Sulamaa, M. (1958) Nievergelt syndrome and its treatment. *Annales Chirugiae et Gynaecologiae Fenniae*, **47**, 142.
Young, L. W. & Wood, B. P. (1975) Nievergelt syndrome (mesomelic dwarfism type Nievergelt). *Birth Defects: Original Article Series*, **11/5**, 81.

Mesomelic dwarfism – Langer type
Brailsford, J. F. (1953) Dystrophies of the skeleton. *British Journal of Radiology*, **8**, 533.
Book, J. A. (1950) A clinical genetical study of disturbed skeletal growth (chondrohypoplasia). *Hereditas*, **36**, 161.
Espiritu, C., Chen, H. & Wooley, P. V. (1975) Mesomelic dwarfism as the homozygous expression of dyschondrosteosis. *American Journal of Diseases of Children*, **129**, 375.
Langer, L. O. (1967) Mesomelic dwarfism of the hypoplastic ulna, fibula and mandibular type. *Radiology*, **89**, 654.
Silverman, F. N. (1975) Intrinsic diseases of bones. In *Progress in Paediatric Radiology*, Vol. 4, p.546. Ed. Kaufman, H. J. Basel: Karger.

Acromesomelic dwarfism

Beighton, P. (1974) Autosomal recessive inheritance in the mesomelic dwarfism of Campailla and Martinelli. *Clinical Genetics*, 5, 363.

Campailla, E. & Martinelli, B. (1971) Deficit staturate con micromesomelia. Presentazione di due case familiari. *Minerva Orthopédica*, 22, 180.

Maroteaux, P., Martinelli, B. & Campailla, E. (1971) Le nanisme acromesomelique. *Presse Médicale*, 79, 1839.

Other types of mesomelic dwarfism

Eaton, G. O. & McKusick, V. A. (1969) A seemingly unique polydactyly-syndactyly syndrome in four persons in three generations. *Birth Defects: Original Article Series*, 5/3, 221.

Gidion, A., Mattaglia, G. F., Bellini, F. & Fancone, G. (1975) The radiological diagnosis of the fetal face (Robinow) syndrome (mesomelic dwarfism and small genitalia). Report of three cases. *Helvetica paediatrica acta*, 30/4-5, 409.

Leroy, J. G., De Vos, J. & Timmermans, J. (1975) Dominant mesomelic dwarfism of the hypoplastic tibia, radius type. *Clinical Genetics*, 7/4, 280.

Pashayan, H., Fraser, F. C., McIntyre, J. M. & Dunbar, J. S. (1971) Bilateral aplasia of the tibia, polydactyly and absent thumb in father and daughter. *Journal of Bone and Joint Surgery*, 53B, 495.

Reinhardt, K. (1976) A dominant-autosomal transmitted micromesomelia with dysplasia of radius and ulna (Reinhardt–Pfeiffer syndrome). *V International Congress of Human Genetics*, Mexico, D.F.

Reinhardt, K. & Pfeiffer, R. A. (1967) Ulno-fibulare dysplasie. Eine autosomal-dominant vererbte Mikromesomelie ähnlich dem Nievergeltsyndrom. *Fortschritte auf dem Gebiete der Röntgenstrahlen und der Nuklearmedizin*, 107, 379.

Robinow, M., Silverman, F. N. & Smith, H. D. (1969) A newly recognised dwarfing syndrome. *American Journal of Diseases of Children*, 117, 645.

Vera-Raman, J. M. (1973) Robinow dwarfing syndrome accompanied by penile agenesis and hemivertebrae. *American Journal of Diseases of Children*, 126, 202.

Wadlington, W. B., Tucker, V. L. & Schminke, R. N. (1973) Mesomelic dwarfism with hemivertebrae and small genitalia (the Robinow syndrome). *American Journal of Diseases of Children*, 126, 202.

Werner, P. (1915) Ueber einen seltenen Fall von Zwergwuchs. *Archiv für Gynaekologie*, 104, 278.

Rhizomelic dwarfism

Kozlowski, K. S., Celermajer, J. M. & Tink, A. R. (1974) Humero-spinal dysostosis with congenital heart disease. *American Journal of Diseases of Children*, 127, 407.

Patterson, C. & Lowry, R. B. (1975) A new dwarfing syndrome with extreme shortening of humeri and severe coxa vara. *Radiology*, 114/2, 341.

Camptomelic dwarfism

Becker, M. H., Finegold, M., Genieser, N. B. *et al.* (1975) Campomelic dwarfism. *Birth Defects: Original Article Series*, 11/6, 113.

Bianchine, J. W., Rismberg, H. M., Kanderian, S. S. & Harrison, H. E. (1971) Camptomelic dwarfism. *Lancet*, i, 1017.

Cremin, B. J., Orsmond, G. & Beighton, P. (1973) Autosomal recessive inheritance in campomelic dwarfism (Letter). *Lancet*, i, 488.

Eliachar, E., Baux, S., Maroteaux, P. *et al.* (1975) Congenital curvature of the long bones: a new case. *Semaine de Hôpitaux de Paris*, 51, 161.

Gardner, L. I., Assemany, S. R. & Neu, R. L. (1971) Syndrome of multiple osseous defects with pretibial dimples. *Lancet*, ii, 98.

Khajavi, A., Lachman, R. S., Rimoin, D. L., Shimke, R. N., Dorsrt, J. P., Ebbin, A. J., Handmaker, S. & Perreault (1976) Heterogeneity in the campomelic syndromes: long and short bone varieties. *Birth Defects: Original Article Series*, 10/6, 93.

Spranger, J., Langer, L. O. & Maroteaux, P. (1970) Increasing frequency of a syndrome of multiple osseous defects? *Lancet*, ii, 716.

Stuve, A. & Wiedemann, H. R. (1971) Congenital bowing of the long bones in two sisters (Letter). *Lancet*, i, 495.

Thurmon, T. F., Defraites, E. B. & Anderson, E. E. (1973) Familial camptomelic dwarfism. *Journal of Pediatrics*, 83/5, 841.

Weiner, D. S., Benfield, G. & Robinson, H. (1976) Camptomelic dwarfism. Report of a case and review of the salient features. *Clinical Orthopaedics and Related Research*, 116, 29.

Lethal short-limbed dwarfism

Cremin, B. J. & Beighton, P. (1974) Dwarfism in the newborn: the nomenclature, radiological features and genetic significance. *British Journal of Radiology*, 47, 77.

Curran, J. P., Sigmon, B. A. & Optiz, J. M. (1974) Lethal forms of chondrodysplastic dwarfism. *Pediatrics*, **53/1**, 76.
Houston, C. S., Awen, C. F. & Kent, H. P. (1972) Fatal neonatal dwarfism. *Journal of the Canadian Society of Radiologists*, **23**, 45.
Maroteaux, P., Stanescu, V. & Stanescu, R. (1976) The lethal chondrodysplasias. *Clinical Orthopaedics and Related Research*, **114**, 31.
Martin, C. (1975) Osteochondrodysplasias recognisable at birth. *Médicine Infantile*, **82/1**, 5.

2. Skeletal dysplasias with significant spinal involvement

The conditions considered in this section, which are of unknown pathogenesis, are all characterised by dwarfism and spinal abnormalities. Many of them are heterogeneous and there is little doubt that other entities in this general category await delineation.

1. Pseudoachondroplasia
2. Spondyloepiphyseal dysplasia
3. Spondylometaphyseal dysplasia
4. Schwartz syndrome
5. Metatropic dwarfism
6. Kniest syndrome
7. Diastrophic dwarfism
8. Dyggve-Melchior-Clausen syndrome
9. Parastremmatic dwarfism.

It is tempting to speculate that Richard Crookback, Duke of Gloucester, who became King Richard III, might have been afflicted with an osteochondrodysplasia of this type. His prematurity and the presence of clinically obvious malformations at the time of birth are clues which might lead to a more precise diagnosis.

> 'deformed, unfinished, sent before my time in this breathing world scarce half made-up.'
> Richard III. Act I, Scene 1. Shakespeare.

The major practical problems in this group of disorders are the consequence of spinal abnormality. Extensive and progressive deformity may lead to cardio-respiratory embarrassment or spinal cord compression. Hypoplasia of the odontoid process predisposes to atlantoaxial dislocation. In this respect, hyperextension of the neck during anaesthesia is particularly dangerous. As orthopaedic measures such as prosthetic joint replacement are playing an increasingly important part in the management of the osteochondrodysplasias, this hazard is of considerable practical significance.

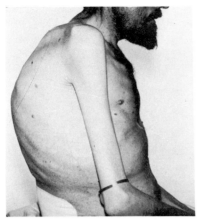

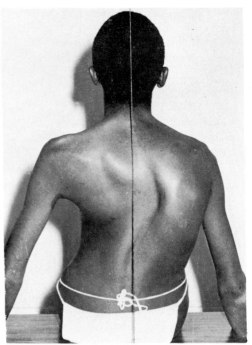

Fig. 2.1. (above) Kyphosis, backward curvature of the spine.

Fig. 2.2. (right) Scoliosis, sideways curvature of the spine.

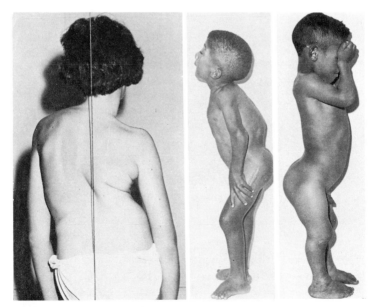

Fig. 2.3. (left) Kyphoscoliosis, backward and sideways curvature of the spine.

Fig. 2.4. (centre) Gibbus, localised backward angulation of the spine.

Fig. 2.5. (right) Lordosis, forwards curvature of the spine.

1. PSEUDOACHONDROPLASIA

Maroteaux and Lamy (1959) recognised that this disorder was distinct from true achondroplasia and the various types of spondyloepiphyseal dysplasia. Although less than 100 cases have been reported, pseudoachondroplasia is probably a relatively common osteochondrodystrophy.

Clinical and radiographic features
Growth retardation is apparent in early childhood. Body proportions resemble those of achondroplasia, with disproportionate shortening and deformity of the limbs. Lumbar lordosis and scoliosis are sometimes present and the joints may be hyper-mobile. There is no craniofacial involvement. Secondary osteo-arthritis supervenes in early adulthood, particularly in the weight bearing joints. Management centres upon orthopaedic correction of hip and knee problems (Kopits, Lindstrom and McKusick, 1974).

Fig. 2.6. Pseudoachondroplasia; a 22-year-old woman with her mother. The kindred are normal and it is likely that this patient represents a new mutation for the severe autosomal dominant form of the condition. (Heselson, N. G., Cremin, B. J. & Beighton, P. (1977) *British Journal of Radiology*, 50, 473.)

Radiographic abnormalities appear during late infancy and evolve throughout childhood. The epiphyses of the long bones are irregular, while the adjacent metaphyses are cup-shaped and widened. The diaphyses are relatively broad and may be mis-shapen. The tubular bones of the hands are shortened, and the phalanges have broad bases. During the developmental phase, the vertebrae are irregular and biconvex, with central projections. However, these changes are much less marked in the adult. In the pelvis, the acetabulae are flattened and the pubis and

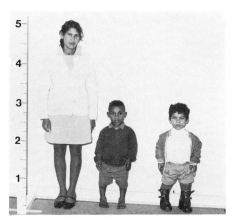 

Fig. 2.7. (left) Pseudoachondroplasia; two affected brothers, aged 14 and 19, with their mother. The parents and six other sibs were normal and it is probable that these young men had the severe autosomal recessive form of the disorder.

Fig. 2.8. (right) Pseudoachondroplasia; an 11-year-old boy with his unaffected younger brother. The limbs are short, but in distinction to achondroplasia, the head and face are normal.

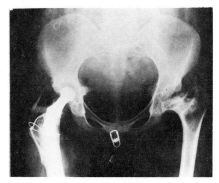

Fig. 2.9. Pseudoachondroplasia; prosthetic hip joint replacement is proving to be of value. (Courtesy of Mr F. J. Heddon, FRCS, Durban.)

ischium are hypoplastic. The radiographic features of 13 cases including individuals with autosomal dominant and recessive forms of pseudoachondroplasia have been discussed by Heselson, Cremin and Beighton (1977).

Genetics

Hall and Dorst (1969) reviewed the manifestations in 32 patients from 12 kindreds and delineated four forms of pseudoachondroplasia. According to this classification, types I and II were the autosomal dominant and recessive 'Kozlowski' types, while types III and IV were the autosomal dominant and recessive 'Maroteaux-Lamy' forms. However, the four subdivisions are not universally accepted, and delineation cannot be regarded as complete. McKusick (1975) suggested that the two dominant forms might be allelic.

Maynard, Cooper and Ponseti (1972) observed inclusion material in the endoplasmic reticulum of the chondrocytes from affected individuals. In further studies of five patients with various types of pseudoachondroplasia Cranley, Williams, Kopits and Dorst (1975) found consistent histological abnormalities in the cartilage and confirmed the previous reports of intracellular inclusion bodies. These changes are characteristic of the pseudoachondroplasias as a whole and they do not permit differentiation of the various types of the condition.

Dennis and Renton (1975) described the clinical and radiographic features of a kindred in which four out of seven sibs had the severe autosomal recessive form of pseudoachondroplasia. The authors noted that the parents were of short stature, without any signs of pseudoachondroplasia, and they suggested that there might be partial manifestation of the abnormal gene in the heterozygote.

Lachman *et al.* (1975) emphasised that in clinical practice, it is not always possible to assign a sporadic individual to a specific category. As the autosomal dominant and autosomal recessive forms cannot be recognised on clinical or radiological grounds alone, the genetic counsellor may be faced with a difficult situation when discussing recurrence risks for pseudoachondroplasia.

2. SPONDYLOEPIPHYSEAL DYSPLASIA

The predominant features of the spondyloepiphyseal dysplasias (SED) are dwarfism and spinal deformity. The changes are maximal in the vertebrae and in the epiphyses of the long bones, while involvement of the metaphyses is of lesser degree.

Rubin (1964) suggested that SED could be subdivided into 'congenita' and 'tarda' forms. In the 'congenita' type stigmata are present at birth, while in the 'tarda' type the disorder becomes evident in later childhood. These categories have met with general acceptance. In an alternative classification, Maroteaux (1969) described three forms of SED, on the basis of the anatomical distribution of the skeletal changes. In recent years, pseudoachondroplasia and spondylometaphyseal dysplasia have been split off from SED and recognised as entities in their own right. There is no doubt that considerable heterogeneity exists within the SED group of conditions. Indeed, in the author's own experience, 'atypical' or 'unclassifiable' forms of SED are encountered more frequently than the traditional types.

The term 'Morquio syndrome' is often loosely and incorrectly applied to any short-limbed dwarf with spinal abnormalities. In this way, conditions such as SED, pseudoachondroplasia, spondylometaphyseal dysplasia and the mucopolysaccharidoses have been erroneously grouped together. In the strict sense, the eponym 'Morquio syndrome' is applicable only to a distinct entity, mucopolysaccharidosis (MPS) type IV. Nevertheless, this nosological problem is still the cause of considerable confusion.

Spondyloepiphyseal dysplasia congenita

In spondyloepiphyseal dysplasia congenita (SEDC) spinal and epiphyseal changes are present in the neonate. Spranger and Weidemann (1966) recognised the distinction between this condition and the Morquio syndrome, and subsequently Spranger and Langer (1970) described a series of 29 cases.

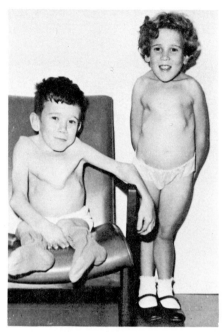

Fig. 2.10. SED; atypical varieties probably outnumber the classical types of the condition. These sibs have an undelineated autosomal recessive form of spondyloepiphyseal dysplasia.

Clinical and radiographic features

Short-limbed dwarfism and other features, including hypertelorism, cleft palate and talipes equinovarus are present at birth. Extension of the elbow joints is limited, and the hips are sometimes dislocated. Severe kyphoscoliosis and thoracic deformity develop in childhood. Leg length may be disproportionate, with genu valgum, genu recurvatum, lateral displacement of the patellae and metatarsus adductus. Ophthalmological problems occur in more than 50 per cent of the affected children, and retinal detachment in association with myopia represents a threat to vision. The adult with SED congenita is usually less than 140 cm in height and has a flat face, short neck and barrel chest. Secondary osteoarthritis develops in the weight bearing joints and backache is a frequent problem. Spinal and thoracic deformity lead to cardiorespiratory embarrassment, and cor pulmonale may supervene in middle age.

Radiographic changes in the newborn are particularly obvious in the spine and pelvis. Shortening of the tubular bones is not prominent but there is a generalised delay in development of ossification centres. In childhood the vertebrae are flattened and irregular, and progressive dorsal kyphoscoliosis develops. The odontoid process may be hypoplastic but the skull is usually normal. The epiphyses of the large joints are irregular, while the metaphyses are involved to a lesser extent. The hip joints become increasingly dysplastic and severe degenerative changes may be present by adulthood.

Genetics

SED congenita is inherited as an autosomal dominant. Kindreds with generation to generation transmission have been reported by Spranger and Wiedemann (1966) and

Spranger and Langer (1970). The inconsistency of clinical stigmata may be indicative of heterogeneity rather than variations in phenotypic expression of the gene.

Spondyloepiphyseal dysplasia tarda

The manifestations of spondyloepiphyseal dysplasia tarda (SEDT) appear in mid-childhood. Involvement is predominantly spinal, with shortening of the trunk relative to the limbs. The condition was delineated by Maroteaux, Lamy and Bernard (1957). However, a number of earlier reports under a variety of designations can be recognised in the literature.

Clinical and radiographic features

Clinical manifestations are variable, and at the mild end of the spectrum, affected individuals may be recognised only by demonstration of radiographic changes in the spine. Dorsal kyphoscoliosis usually develops in mid-childhood and the trunk becomes progressively shortened, although the limbs remain relatively uninvolved. The hamstrings tighten and pain in the legs and back is a common problem. Progressive degenerative osteoarthropathy of the spine and hip joints may cause severe disability in middle age.

Radiographically, the skeleton is virtually normal until the age of five. Later, generalised platyspondyly develops, with kyphoscoliosis and thoracic cage deformity. Heaping up of the posterior part of the upper surfaces of the bodies of the lumbar vertebrae is a pathognomonic feature. The articular surfaces of the large joints become flattened and dysplastic, the changes being maximal in the hip joints.

Genetics

Several extensive pedigrees have been published, indicating that SEDT is inherited as an X-linked recessive. The site of the abnormal gene on the X chromosome has not been identified. However, Bannerman, Ingall and Mohn (1971) have shown that the genes for SEDT and the Xg blood group are not closely linked. (The Xg blood group gene is situated on the X chromosome, and serves as a useful marker in linkage studies.)

SEDT is probably heterogeneous, as there is good evidence for autosomal recessive and autosomal dominant inheritance in some kindreds. Affected sibs with normal consanguineous parents were described by Klenerman (1961) and Martin et al. (1970). This form of SEDT seems to be inherited as an autosomal recessive. O'Brien et al. (1976) described a girl with SEDT in whom a deficiency of beta-galactosidase activity was demonstrated in cultured fibroblasts. The activity of this enzyme in fibroblasts from both unaffected parents was approximately 50 per cent of normal. On this evidence, it is likely that these parents were heterozygous for the abnormal gene. Generation to generation transmission of SEDT consistent with autosomal dominant inheritance was reported by Moldauer, Hanelin and Bauer (1962).

The clinical and radiographic features of these uncommon forms of SEDT are by no means clear-cut and diagnostic precision rests largely upon recognition of the pattern of transmission within a kindred. For this reason, genetic counselling in the case of a sporadic individual should be undertaken with caution.

3. SPONDYLOMETAPHYSEAL DYSPLASIA

Spondylometaphyseal dysplasia (SMD) was delineated by Kozlowski, Maroteaux and Spranger (1967) and since that time, about 30 cases have been reported (Kozlowski, 1976). SMD has been a source of great diagnostic and terminologic confusion and it is likely that there is considerable heterogeneity.

Clinical and radiographic features

In SMD, metaphyseal abnormalities are associated with spinal changes. The predominant features are short-trunked dwarfism, kyphoscoliosis, pectus carinatum, limited movements of the hips and elbows and knee deformity. In the form of the condition described by Murdoch and Walker (1969), a flat facies, cleft palate, limb bowing, talipes and marked joint laxity were also present.

The skeleton is radiographically normal at birth. Platyspondyly, which is maximal in the thoracic vertebrae, develops in early childhood. The metaphyses of the long bones become irregular at this time and in some instances show multiple radiolucent areas. Coxa vara may be associated with a short femoral neck. The epiphyses are uninvolved. The radiographic changes in SMD have been reviewed by Thomas and Nevin (1977).

Genetics

Controversy exists concerning the genetics of SMD. Kozlowski et al. (1967) proposed that inheritance was autosomal recessive, while Michel et al. (1970) favoured dominant transmission. Subsequently, Kozlowski (1973) has suggested that there might be several distinct forms of the disorder. This viewpoint almost certainly reflects the true situation.

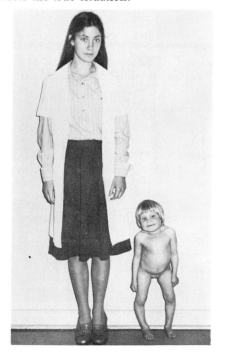

Fig. 2.11. SMD; a five-year-old girl with dwarfism, a short neck, thoracic deformity and marked genu varum.

4. SCHWARTZ SYNDROME

The Schwartz syndrome is a unique disorder in which myotonia coexists with dysplasia of the skeleton. The short eponym is preferable to the long descriptive designation, 'spondylo-epimetaphyseal dysplasia with myotonia.' The first recognisable cases were a pair of sisters investigated by Catel (1951). At the present time, reports concern about 20 patients, all of whom were in their childhood at the time of description.

Clinical and radiographic features

The main features are short stature, stiff joints, spinal malalignment and pectus carinatum. Blepharophimosis is present and the face is immobile and 'mask like'. Myotonia progresses until a plateau is reached in mid-childhood. Operative treatment for dislocation of the hips and talipes equinovarus has been successful. However, the anaesthetist may encounter problems on endotracheal intubation from the small size of the mouth, rigidity of the temporomandibular joint and shortness of the neck (Horan and Beighton, 1975).

Radiographically, the skeleton is undermineralised. The vertebrae are flattened and anterior wedging may develop. The epiphyses and metaphyses, particularly of the large joints, are dysplastic. The hips may be dislocated, with varus or valgus deformity of the femoral necks.

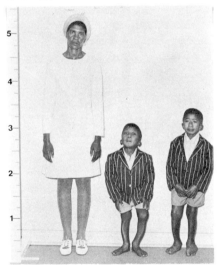

Fig. 2.12. Schwartz syndrome; brothers, aged 7 and 10, with short stature and limb deformity.

Genetics

Schwartz and Jampel (1962) described an affected brother and sister, and other cases have been reported by Huttenlocher *et al.*, 1969; Mereu, Porter and Hug, 1969; Kozlowski and Wise, 1974; Greze *et al.*, 1975 and Cadilhac *et al.*, 1975. All the parents were normal, but consanguinity was present in the kindreds reported by Saadat *et al.* (1972) from the Middle East and by Beighton (1973) from South Africa. There is little doubt that the Schwartz syndrome is inherited as an autosomal recessive.

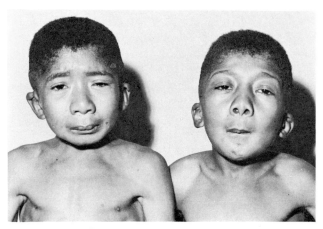

Fig. 2.13. Schwartz syndrome; the face is immobile and mask-like. (From Horan, F. T. & Beighton, P. (1975) *Journal of Bone and Joint Surgery*, **57**, 544.)

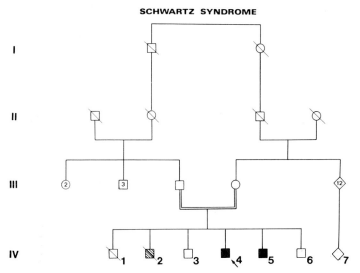

Fig. 2.14. Schwartz syndrome; the pedigree of the affected children. The consanguinity of their parents is evidence in favour of autosomal recessive inheritance.
Key to pedigree: □ normal male; ○ normal female; ▣ affected male; ● affected female; / deceased. (From Beighton, P. (1973) *Clinical Genetics*, **4**, 548.)

Simpson and Degnan (1975) described a boy with features resembling the Schwartz syndrome, but lacking myotonia, and with blepharophimosis of mild degree. The authors suggested that this child probably had the same condition as the infants reported by Marden and Walker (1966) and Fitch, Karpati and Pinsky (1971) and that this disorder was distinct from the true Schwartz syndrome. Temtamy *et al.* (1975), using the designation 'Marden-Walker syndrome' reported two affected cousins, both of whom were the offspring of consanguineous marriages, and suggested that inheritance was autosomal recessive.

5. METATROPIC DWARFISM

Metatropic dwarfism was described by Maroteaux, Spranger and Weidemann (1966). The term 'metatropic' pertains to the reversal of bodily proportions which occurs during early childhood, and it is derived from the Greek 'metatropos', meaning 'changing pattern'. The manifestations have been reviewed by Jenkins, Smith and McKinnell (1970), Gefferth (1973) and Rimoin *et al.* (1976).

Clinical and radiographic features
Affected neonates have a relatively long trunk, a narrow cylindrical thorax and short limbs. The face is normal but the palate may be cleft. The joints are knobbly and stiff. A coccygeal cutaneous fold or 'tail', which is sometimes present, represents a valuable diagnostic feature. In early childhood, growth of the spine is retarded, and kyphoscoliosis becomes pronounced when the child begins to walk. In this way, the trunk becomes short in comparison with the limbs. Spinal deformity may be very marked and patients are severely dwarfed and crippled.

Gross widening of the intervertebral disc spaces is apparent at birth, in conjunction with marked platyspondyly. The metaphyses are very broad. The iliac crests are crescentic and the acetabular roofs are horizontal and irregular. In later childhood, the vertebrae become wedged and the proximal end of the femur takes on a configuration which has been likened to a 'halberd' or 'battle-axe.'

Genetics
The majority of case reports have concerned sporadic individuals of either sex, although affected brothers with normal parents have been encountered by Michael *et al.* (1956) and Crowle, Astley and Insley (1970). It is likely that inheritance is autosomal recessive.

Pseudometatropic dwarfism, in which joint laxity and dermal extensibility are prominent features, might be a separate entity (Bailey, 1971). Patients have all been males, and it is possible that this condition is X-linked. However, definitive delineation will not be accomplished until further cases have been reported, and at present, the mode of genetic transmission remains uncertain.

6. KNIEST SYNDROME

This rare disorder was delineated by Kniest (1952) and a further eight cases were reported by Siggers *et al.* (1974). McKusick (1975) proposed the designation 'metatropic dwarfism type II'.

Clinical and radiographic features
The face is round and flattened, and the eyes are prominent. Inguinal hernia, cleft palate and club feet are inconsistent components of the syndrome. In early childhood, short stature, kyphoscoliosis and stiff joints are evident. The thorax is broad and the trunk is short. Deafness, myopia and retinal detachment are important complications.

The skeleton is generally osteoporotic. In infancy, the vertebral bodies are flattened, with coronal clefts and anterior wedging. The epiphyses and metaphyses of the tubular bones are bulky and flared. The femoral necks are broad and short, and the femoral capital epiphyses remain unossified throughout childhood.

Specific histological and ultrastructural changes have been identified in the collagen of the cartilage. Excess urinary excretion of keratan sulphate has been demonstrated in an affected mother and daughter (Brill *et al.*, 1975).

Genetics
In a review of the Kniest syndrome, Siggers *et al.* (1974) pointed out that apart from a pair of male indentical twins, all reported patients have been sporadic. Subsequently, Kim *et al.* (1975) and Gnamey, Farriaux and Fontaine (1976) described affected mothers and daughters. So far, there have been no reports of male to male transmission, and the autosomal status of the abnormal gene has not yet been confirmed.

7. DIASTROPHIC DWARFISM

Diastrophic dwarfism was delineated by Lamy and Maroteaux (1960). The term 'diastrophic', adapted from the Greek word meaning 'twisted', aptly fits the disorder. Diastrophic dwarfism is not uncommon and over 120 cases have been reported, including a series of 51 described by Walker *et al.* (1972).

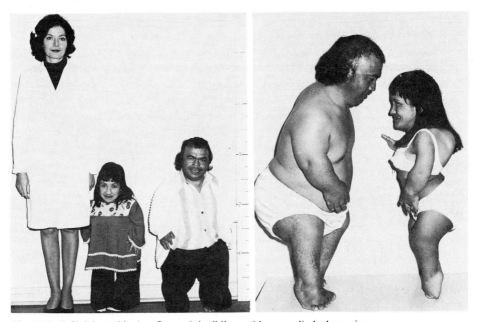

Fig. 2.15. (left) Diatrophic dwarfism; adult siblings with severe limb shortening.

Fig. 2.16. (right) Diastrophic dwarfism; the sister has a marked dorsal kyphoscoliosis although her brother's spine is virtually normal. Both have gross rigid talipes equinovarus.

Clinical and radiographic features

Shortness of stature and micromelia are obvious at birth. Severe talipes equinovarus is a universal finding, while cleft palate is present in about 50 per cent of patients. The hands are broad with rigidity of the interphalangeal joints. The first metacarpal is short and the thumb is subluxed into the 'hitch-hiker' position. Curious episodes of spontaneous swelling and inflammation of the pinna of the ear during early childhood lead to a 'cauliflower' appearance. Although the face is normal, affected individuals bear a close resemblance to each other. Life expectancy is reasonably good but dwarfing is extreme and deformity is severe. Kyphoscoliosis may predispose to cardiopulmonary complications and spinal cord compression. The orthopaedic management of diastrophic dwarfism has been discussed by Hollister and Lachman (1976).

The spine is relatively normal at birth, but lumbar lordosis and kyphoscoliosis develop during childhood. The femoral heads are flattened and the hip joints are usually dislocated. The long bones are reduced in length, the epiphyses are irregular and the metaphyses are flared. Mineralisation of the skeleton may be delayed. The radiographic criteria for diagnosis in the newborn have been reviewed by Saule (1975).

Genetics

A considerable body of evidence indicates that diastrophic dwarfism is inherited as an autosomal recessive. There have been several descriptions of multiple affected sibs, all with normal parents (Jackson, 1951; Lamy and Maroteaux, 1960; Paul *et al.*, 1965). Parental consanguinity was mentioned by Taybi (1963). The only reports of affected females who have reproduced concern two women who gave birth to normal children (Walker *et al.*, 1972).

Rimoin (1975) reviewed the features of a 'diastrophic variant' or 'pseudo-diastrophic' dwarfism, an uncommon entity in which clinical and radiographic stigmata resemble those of mild diastrophic dwarfism. From their study of 20 patients, including three pairs of sibs, Norton *et al.* (1976) concluded that the disorder was probably inherited as an autosomal recessive. As the histological appearances of cartilage are identical to those of diastrophic dwarfism, these authors suggested that the genes which determine these conditions might be allelic. Alternatively, these disorders could represent variations in phenotypic expression of the same basic genetic defect.

8. DYGGVE-MELCHIOR-CLAUSEN SYNDROME

Dyggve, Melchior and Clausen (1962) reported three mentally retarded dwarfed sibs from Greenland. The clinical features resembled those of the Morquio syndrome and it was suggested that the condition might be a mucopolysaccharidosis. However, initial reports of mucopolysacchariduria could not be confirmed on repeated urinary testing. Later, radioactive sulphate uptake by fibroblasts was shown to be normal and Spranger, Maroteaux and Kaloustian (1975) concluded that the syndrome was not a mucopolysaccharidosis.

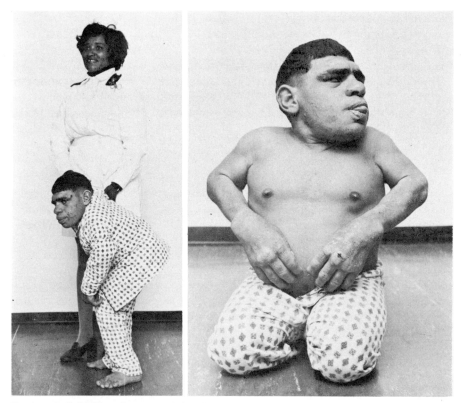

Fig. 2.17. (left) DMC syndrome; an adult male with short stature and mental deficiency. His brother and sister also had the condition. The unaffected parents were consanguineous.

Fig. 2.18. (right) DMC syndrome; the neck is short, and the chest is barrel-shaped.

Clinical and radiographic features

Major characteristics are short-limbed dwarfism with a short neck, barrel chest, lumbar lordosis, genu valgum and a crouching stance. The majority of patients have been mentally defective. Important radiographic features are platyspondyly, in association with metaphyseal and epiphyseal changes in the long bones. The acetabulae are dysplastic and the hips may be dislocated. During childhood, the margin of the iliac crests have a pathognomonic lace-like configuration. This appearance does not persist into adult life.

Genetics

The patients mentioned in the initial report were the product of an uncle–neice relationship (Dyggve, Melchior and Clausen, 1962), and affected sibs with consanguineous parents have also been observed by Spranger, Maroteaux and Kaloustian (1975).

In South African institutions for the mentally defective, the author has encountered two brothers and a sister with the condition. Their parents, who were of Lebanese extraction, were first cousins. At least three other descriptions have concerned Lebanese kindreds, and the gene evidently reaches a relatively high frequency in this particular population (Naffah, 1976). At the present time, about 30 case reports can be found in the literature. Sex distribution is approximately equal and there has been no instance of generation to generation transmission. There is little doubt that inheritance is autosomal recessive.

Three affected sibs of Japanese stock, including a pair of non-identical twins, were reported by Smith and McCort (1959). These children were mentally normal, but otherwise they had the typical features of the syndrome. This anomalous observation might indicate heterogeneity.

9. PARASTREMMATIC DWARFISM

The designation of this disorder is derived from the Greek 'parastremma', meaning 'distorted limb' (Langer, Peterson and Spranger, 1970). In this rare entity, dwarfism and kyphoscoliosis are associated with severe malformations of the extremities.

Clinical and radiographic features
The forehead is high, with brachycephaly and a temporal bulge. Scoliosis appears in early infancy and becomes increasingly severe. The extremities are short, with bilateral genu valgum, bowing of the shins, osseous enlargement of the knees, and contractures of the hip joints.

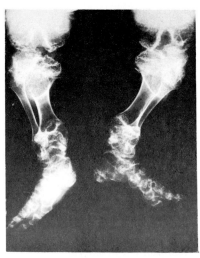

Fig. 2.19. (left) Parastremmatic dwarfism; a five-year-old girl with dwarfism and severe limb deformity.

Fig. 2.20. (right) Parastremmatic dwarfism; radiographically, the skeleton has a pathognomonic 'flocky' appearance. (From Horan, F. T. & Beighton, P. (1976) *Journal of Bone and Joint Surgery*, **58**, 343.)

The skeleton is grossly undermineralised. The zones of endochondral bone are lucent, widened and coarsely trabeculated, and contain areas of irregular stippling. The bone has a pathognomonic 'flocky' appearance. The vertebral bodies are flattened and irregular and the pelvic bones are very dysplastic. The metaphyses and epiphyses of the tubular bones are grossly deformed. The femoral necks are short and the femoral heads are distorted.

Genetics

Seven cases have been reported, including three unrelated females (Langer, Peterson and Spranger, 1970), a father and daughter (Rask, 1963) and a girl of Cape Coloured stock (Horan and Beighton, 1976). Neither affected sibs nor parental consanguinity have been recorded. It is generally assumed that inheritance is autosomal dominant, and that sporadic individuals represent new mutations.

NON-LETHAL SHORT-LIMBED DWARFISM IN THE NEWBORN

Apart from the potentially lethal conditions listed in Chapter 1, short-limbed dwarfism may be apparent at birth in a number of disorders in which survival is usual:

	Inheritance
Achondroplasia	AD
Mesomelic dwarfism (various types)	AD/AR
Rhizomelic dwarfism (various types)	AD/AR
Spondyloepiphyseal dysplasia congenita	AD
Spondylometaphyseal dysplasia	AD/AR
Metatropic dwarfism	AR?
Kniest syndrome	?
Diastrophic dwarfism	AR
Parastremmatic dwarfism	AD?

Other forms of short-limbed dwarfism, such as hypochondroplasia and pseudoachondroplasia, are not usually recognised in the newborn. However, if a parent or sibling is affected, the presence of minor changes may permit diagnosis in the neonatal period.

REFERENCES

Pseudochondroplasia
Cranley, R. E., Williams, B. R., Kopits, S. E. & Dorst, J. P. (1975) Pseudoachondroplastic dysplasia: five cases representing clinical, roentgenographic and histologic heterogeneity. *Birth Defects: Original Article Series*, **11/6**, 205.
Dennis, N. R. & Renton, P. (1975) The severe recessive form of pseudoachondroplastic dysplasia. *Pediatric Radiology*, **3/3**, 169.
Hall, J. G. & Dorst, J. P. (1969) Four types of pseudoachondroplastic spondyloepiphyseal dysplasia (SED). *Birth Defects: Original Article Series*, **5/4**, 242.
Heselson, N. G., Cremin, B. J. & Beighton, P. (1977) Pseudoachondroplasia; a report of 13 cases. *British Journal of Radiology*, **50**, 473.
Kopits, S. E., Lindstrom, J. A. & McKusick, V. A. (1974) Pseudoachondroplastic dysplasia: pathodynamics and management. *Birth Defects: Original Article Series*, **10/12**, 341.
Lachman, R. S., Rimoin, D. L. & Hall, J. G. (1975) Difficulties in the classification of the epiphyseal dysplasias. *Birth Defects: Original Article Series*, **11/6**, 231.

Maroteaux, P. & Lamy, M. (1959) Les formes pseudoachondroplastiques des dysplasies spondyloepiphysaires. *Presse Médicale*, **67**, 383.
Maynard, J. A., Cooper, R. R. & Ponseti, I. V. (1972) A unique rough surfaced endoplasmic reticulum inclusion in pseudoachondroplasia. *Laboratory Investigations*, **26**, 40.
McKusick, V. A. (1975) Pseudoachondroplastic dysplasia I (formerly pseudoachondroplastic spondyloepiphyseal dysplasia). In *Mendelian Inheritance in Man*, p.279. Baltimore and London: The Johns Hopkins University Press.

SED

Maroteaux, P. (1969) Spondyloepiphyseal dysplasias and metatropic dwarfism. *Birth Defects: Original Article Series*, **5/4**, 35.
Rubin, P. (1964) In *Dynamic classification of Bone Dysplasias*. Chicago: Year Book Publishers.

SEDC

Spranger, J. & Wiedemann, H. R. (1966) Dysplasia spondyloepiphysaria congenita. *Helvetica Paediatrica Acta*, **21**, 598.
Spranger, J. & Langer, L. O. (1970) Spondyloepiphyseal dysplasia congenita. *Radiology*, **94**, 313.

SEDT

Bannerman, R. M., Ingall, A. B. & Mohn, J. F. (1971) X-linked spondyloepiphyseal dysplasia tarda: clinical and linkage data. *Journal of Medical Genetics*, **8**, 291.
Klenerman, L. (1961) An adult case of chondro-osteodystrophy. *Proceedings of the Royal Society of Medicine*, **54**, 71.
Maroteaux, P., Lamy, M. & Bernard, J. (1957) La dysplasie spondylo-epiphysaire tardive. *Presse Médicale*, **65**, 1205.
Martin, J. R., Macewan, D. W., Blais, J. A., Metrakos, J., Gold, P., Langer, F. & Hill, R. O. (1970) Platyspondyly, polyarticular osteoarthritis, and absent beta-2-globulin in two brothers. *Arthritis and Rheumatism*, **13**, 53.
Moldauer, M., Hanelin, J. & Bauer, W. (1962) Familial precocious degenerative arthritis and the natural history of osteochondrodystrophy. In *Medical and Clinical Aspects of Aging*, p.226. Ed. Blumenthal, H. T. New York: Columbia University Press.
O'Brien, J. S., Gugler, E., Giedion, A., Weissman, U., Herschkowitz, N., Meier, C. & Leroy, J. (1976) Spondyloepiphyseal dysplasia, corneal clouding, normal intelligence and acid beta-galactosidase deficiency. *Clinical Genetics*, **9**, 495.

SMD

Kozlowski, K., Maroteaux, P. & Spranger, J. (1967) La dysostose spondylometaphysaire. *Presse Médicale*, **75**, 2769.
Kozlowski, K. (1973) Spondylometaphyseal dysplasia. *Progress in Pediatric Radiology*, **4**, 299.
Kozlowski, K. (1976) Metaphyseal and spondylometaphyseal chondrodysplasias. *Clinical Orthopaedics and Related Research*, **114**, 83.
Michel, J., Grenier, B., Castaing, J., Augier, J. L. & Desbuquois, G. (1970) Deux cas familiaux de dysplasie spondylometaphysaire. *Annals of Radiology*, **13**, 251.
Murdock, J. L. & Walker, B. A. (1969) A 'new' form of spondylometaphyseal dysplasia. *Birth Defects: Original Article Series*, **5**, 368.
Thomas, P. S. & Nevin, N. C. (1977) Spondylometaphyseal dysplasia. *American Journal of Roentgenology*, **128**, 89.

Schwartz syndrome

Beighton, P. (1973) The Schwartz syndrome in Southern Africa. *Clinical Genetics*, **4**, 548.
Cadilhac, J., Baldet, P., Greze, J. & Duday, H. (1975) E.M.G. studies of two familial cases of the Schwartz and Jampel syndrome (osteo-chondro-muscular dystrophy with myotonia). *Electromyography and Clinical Neurophysiology*, **15/1**, 5.
Catel, W. (1951) Diffentialdiagnostische syptomatologie von krankheiten des kindesalters. In *Klinische Vorlesungen*, p.48. Stuttgart: G. Thieme.
Fitch, N., Karpati, G. & Pinsky, L. (1971) Congenital blepharophimosis, joint contractures and muscular hypotonia. *Neurology*, **21**, 1214.
Greze, J., Baldet, P. & Dumas, R. (1975) Schwartz Jampel's osteo-chondro-muscular dystrophy. Two familial cases. *Archives Francaises de Pédiatrie*, **32/1**, 59.
Horan, F. & Beighton, P. (1975) Orthopaedic aspects of the Schwartz syndrome. *Journal of Bone and Joint Surgery*, **57A/4**, 542.

Huttenlocher, P. R., Landwirth, J., Hanson, V., Gallagher, B. B. & Bench, K. (1969) Osteo-chondro-muscular dystrophy. A disorder manifested by multiple skeletal deformities, and dystrophic changes in muscle. *Pediatrics,* **44,** 945.

Kozlowski, K. & Wise, G. (1974) Spondylo-epi-metaphyseal dysplasia with myotonia. A radiographic study. (Catel–Jampel syndrome, Schwartz–Jampel syndrome, Aberfeld syndrome, chondrodystrophic myotonia). *Radiologia Diagnostica,* **6,** 817.

Marden, P. M. & Walker, W. A. (1966) A new generalised connective tissue syndrome. *American Journal of Diseases of Children,* **112,** 225.

Mereu, T. R., Porter, I. H. & Hug, G. (1969) Myotonia, shortness of stature and hip dysplasia: Schwartz-Jampel syndrome. *American Journal of Diseases of Children,* **117,** 470.

Saadat, M., Mokfi, H., Vakil, H. & Ziai, M. (1972) Schwartz syndrome: myotonia with blepharophimosis and limitation of joints. *Journal of Pediatrics,* **81,** 348.

Schwartz, O. & Jampel, R. S. (1962) Congenital blepharophimosis associated with a unique generalised myopathy. *Archives of Ophthalmology,* **68,** 52.

Simpson, J. L. & Degnon, M. (1975) A child with facial and skeletal dysmorphism reminiscent of Schwartz syndrome. *Birth Defects: Original Article Series,* **11/2,** 456.

Temtamy, S. A., Shoukry, A. S., Raafat, M. & Mihareb, S. (1975) Probable Marden–Walker syndrome: evidence for autosomal recessive inheritance. *Birth Defects: Original Article Series,* **11!2,** 104.

Metatropic dwarfism

Bailey, J. A. (1971) Forms of dwarfism recognisable at birth. *Clinical Orthopaedics and Related Research,* **76,** 150.

Crowle, P., Astley, R. & Insley, J. (1976) A form of metatropic dwarfism in two brothers. *Pediatric Radiology,* **4/3,** 172.

Gefferth, K. (1973) Metatropic dwarfism. In *Progess in Pediatric Radiology,* Vol. 4, p.137. Ed. Kaufman H. J. Basel: Karger.

Jenkins, P., Smith, M. B. & McKinnel, J. S. (1970) Metatropic dwarfism. *British Journal of Radiology,* **43,** 561.

Maroteaux, P., Spranger, J. & Wiedemann, H. (1966) Metatropischer Zwergwuchs. *Archiv für Kinderheilkunde,* **173,** 211.

Michael, J., Matsovkas, J., Theodorou, S. & Houliaras, K. (1956) Maladie de Morquio (osteochondrodystrophie polyepiphysaire deformante) chez deux frères. *Helvetica Paediatrica Acta,* **2,** 403.

Rimoin, D. L., Siggers, D. C., Lachman, R. S. & Silberberg, R. (1976) Metatropic dwarfism, the Kniest syndrome and the pseudoachondroplastic dysplasias. *Clinical Orthopaedics and Related Research,* **114,** 70.

Kniest syndrome

Brill, P. W., Kim, H. J., Beratis, N. G. & Hirschhorn, K. (1975) Skeletal abnormalities in the Kniest syndrome with mucopolysacchariduria. *American Journal of Roentgenology, Radium Therapy and Nuclear Medicine,* **125/3,** 731.

Gnamey, D., Farriaux, J. P. & Fontaine, G. (1976) Kniest's disease. A familial case report. *Archives Francaises de Pédiatrie,* **33/2,** 143.

Kim, H. J., Beratis, N. G. & Brill, P. (1975) Kniest syndrome with dominant inheritance and mucopolysacchariduria. *American Journal of Human Genetics,* **27/6,** 755.

Kniest, W. (1952) Zur Abgrenzung der Dysostosis enchondralis von der Chondrodystrophie. *Zeitschrift für Kinderheilkunde,* **70,** 633.

McKusick, V. A. (1975) Metatropic dwarfism. In *Mendelian Inheritance in Man,* 4th Edition, p.496. Baltimore–London: The Johns Hopkins University Press.

Siggers, D., Rimoin, D., Dorst, J., Doty, S., Williams, B., Hollister, D., Silberberg, R., Cranley, R., Kaufman, R. & McKusick, V. (1974) The Kniest syndrome. *Birth Defects: Original Article Series,* **10/9,** 193.

Diastrophic dwarfism

Hollister, D. W. & Lachman, R. S. (1976) Diastrophic dwarfism. *Clinical Orthopaedics and Related Research,* **114,** 61.

Horton, W. A., Rimoin, D. L., Lachman, R. S., Hollister, D. W., Dorst, J. P., Skovby, F., Scott, C. I. & Hall, J. G. (1976) The diastrophic variant. *Fifth International Congress of Human Genetics,* Mexico, D.F.

Jackson, W. P. U. (1951) Irregular familial chondro-osseous defect. *Journal of Bone and Joint Surgery,* **33B,** 420.

Lamy, M. & Maroteaux, P. (1960) Le nanisme diastrophique. *Presse Médicale,* **68,** 1977.

Paul, S. S., Rao, P. L., Mullick, P. & Saigal, S. (1965) Diastrophic dwarfism. A little known disease entity. *Clinical Pediatrics*, **4**, 95.

Rimoin, D. L. (1975) The chondrodystrophies. *Advances in Genetics*, **5**, 1.

Saule, H. (1975) Diastrophic dwarfism. *Radiolge*, **15/2**, 50.

Taybi, H. (1963) Diastrophic dwarfism. *Radiology*, **80**, 1.

Walker, B. A., Scott, C. I., Hall, J. G., Murdoch, J. L. & McKusick V. (1972) Diastrophic dwarfism. *Medicine*, **51**, 41.

D.M.C. sundrome

Dyggve, H. V., Melchior, J. C. & Clausen, J. (1962) Morquio–Ulrich's disease. An inborn error of metabolism? *Archives of Diseases in Childhood*, **37**, 525.

Naffah, J. (1976) The Dyggve–Melchior–Clausen syndrome. *American Journal of Human Genetics*, **28/6**, 607.

Smith, R. & McCort, J. J. (1959) Osteochondrodystrophy (Morquio–Brailsford type); occurrence in three siblings. *California Medicine*, **88**, 53.

Spranger, J., Maroteaux, P. & Der Kaloustian, V. M. (1975) The Dyggve–Melchior–Clausen syndrome. *Radiology*, **114/2**, 415.

Parastremmatic dwarfism

Horan, F. & Beighton, P. (1976) Parastremmatic dwarfism. *Journal of Bone and Joint Surgery*, **58B**, 343.

Langer, L. O., Petersen, D. & Spranger, J. (1970) An unusual bone dysplasia: parastremmatic dwarfism. *American Journal of Roentgenology, Radium Therapy and Nuclear Medicine*, **110**, 550.

Rask, M. R. (1963) Morquio-Brailsford osteochondrodystrophy and osteogenesis imperfecta: report of a patient with both conditions. *Journal of Bone and Joint Surgery*, **45A**, 561.

3. Miscellaneous skeletal dysplasias

Dysplasia of the skeleton is a feature of a number of disorders of unknown pathogenesis which do not fit neatly into any nosologic category. The following conditions are reviewed in this chapter:

1. Cleidocranial dysplasia
2. Dyschondrosteosis
3. Hereditary arthro-ophthalmopathy (Stickler)
4. Larsen syndrome
5. Acrodysplasia (Brailsford, Thiemann)
6. Trichorhinophalangeal syndrome (Giedion)
7. Coffin-Lowry syndrome
8. Freeman-Sheldon syndrome
9. Nail-patella syndrome
10. Congenital bowing of long bones.

1. CLEIDOCRANIAL DYSPLASIA

Cleidocranial dysplasia, formerly known as cleidocranial dysostosis, is a well defined condition in which maldevelopment of the clavicles is associated with mild shortness of stature and a characteristic facies. More than 500 cases have been reported. As many individuals with the disorder are totally asymptomatic, it is probable that cleidocranial dysplasia is even commoner than this figure suggests.

Clinical and radiographic features
The forehead is broad and the parietal region is wide. Clavicular hypoplasia, which may be asymmetrical, permits undue mobility of the shoulder girdle. Many photographs have been published depicting patients in a classic pose, with their arms wrapped around their chests and their shoulders almost touching anteriorly. The only consistent complication is recurrent dislocation, chiefly of the shoulder, elbow and hip joints. Vertebral malalignment is a less frequent but more serious problem.

In skull radiographs, patency of the fontanelles, widening of the sutures and multiple Wormian bones may be evident. The sphenoid is short, the foramen magnum is enlarged and the paranasal sinuses are absent. The clavicles may be totally absent, although the medial portions are often spared. The pubic and iliac bones are hypoplastic. The long bones are gracile and the distal phalanges are shortened.

Fig. 3.1. Cleidocranial dysplasia; two brothers demonstrating the unusual mobility of their shoulder girdles. These boys are members of the Arnold kindred of Cape Town, a large extended family in which more than 500 individuals have the disorder.

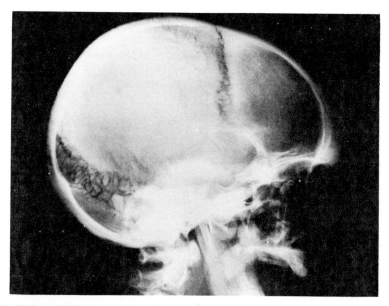

Fig. 3.2. Cleidocranial dysplasia; skull radiograph of an affected adult showing widening of the sutures and multiple Wormian bones. (Wormian bones, which are found in a few other skeletal dysplasias, notably osteogenesis imperfecta, are named after Oluff Worm (1588–1654), professor of anatomy at the University of Copenhagen. Worm was known for his opposition to William Harvey's concepts concerning the circulation of the blood in the human body.)

Genetics

Autosomal dominant inheritance is well established. Large series were reported by Lasker (1946) who studied 73 kindreds, and Jarvis and Keats (1974), who reviewed the radiological features of 40 patients. A family with concordant monozygous twins and discordant dizygous twins was investigated by de Weerdt and Wildervanck (1973). Cleidocranial dysplasia is well known in Cape Town, where several hundred affected individuals had a common ancestor in an energetic polygamous immigrant of Chinese stock (Jackson, 1951). This extended kindred are aware of the configuration of their skulls and use the family name to describe this feature, terming it the 'Arnold head'.

Herndon (1951) estimated that about 16 per cent of patients are apparently sporadic. On this basis, it appears that the mutation rate is high, but variability of phenotypic expression and underdiagnosis could also account for this observation. Goodman et al. (1973) described three severely affected individuals in two Iraqi–Jewish kindreds in Israel. As both sets of parents were consanguineous these authors postulated that there might be an autosomal recessive form of the disorder. It is therefore possible that a proportion of the apparently sporadic patients have inherited the condition as an autosomal recessive. Nevertheless, for the purposes of genetic counselling, it is reasonable to assume that the vast majority of patients have an autosomal dominant disorder. As cleidocranial dysplasia is comparatively innocuous, it is unlikely that many parents would wish to limit the size of their families because of their possession of the gene.

2. DYSCHONDROSTEOSIS

Dyschondrosteosis was delineated by Léri and Weill (1929), when they recognised the association of mild mesomelic dysplasia with a bilateral Madelung deformity of the forearm. This deformity, which carries the name of Otto Madelung, a German surgeon of 1846–1926, consists of shortening and dorsilateral bowing of the shaft of the radius. The ulna is subluxed dorsally at the wrist and the bones of the carpus are wedged between the inclined articular surfaces of the deformed radius and protruding ulna. Seen from the side, the forearm and wrist have a 'dinner fork' configuration. The Madelung deformity may be the result of trauma or infection, or part of a generalised skeletal dysplasia such as diaphyseal achalasis or multiple enchondromatosis. The deformity may also be present as a non-genetic isolated congenital abnormality, which seems to be confined to females (Golding and Blackburne, 1976).

Clinical and radiographic features

In dyschondrosteosis, defective growth becomes evident in mid-childhood. The ultimate height of affected adults is usually between 135 and 170 cm (Kaitila, Leisti and Rimoin, 1976). This stunting is largely the result of symmetrical shortening of the tibia and fibula. The Madelung deformity of the forearm is often asymmetrical and the range of movement at the elbow joints is usually limited. In the lower limbs, genu valgum is a common feature. The skeleton is otherwise normal.

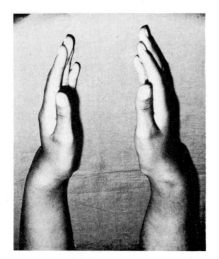 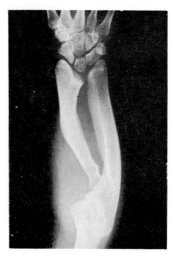

Fig. 3.3. (left) Dyschondrosteosis; the Madelung or dinner fork deformity of the forearm.

Fig. 3.4. (right) Dyschondrosteosis; in the Madelung deformity, the radius is shortened and bowed, and the bones of carpus are wedged between inclined articular surfaces.

Genetics

There has been considerable controversy as to whether all individuals with a primary Madelung deformity have dyschondrosteosis (Langer, 1965) or whether the disorders exist as separate entities (Felman and Kirkpatrick, 1969; Kozlowski and Zychowicz, 1971). Evidence in support of the latter contention was provided by Golding and Blackburne (1976) when they studied 26 individuals with primary Madelung deformity of the wrist and demonstrated that none had any additional radiographic stigmata of dyschondrosteosis. As no bony abnormalities were detected in 65 first degree relatives of seven of these patients, it is probable that the isolated primary Madelung deformity is non-genetic.

Dyschondrosteosis is transmitted as a dominant trait (Henry and Thornburn, 1967; Carter and Currey, 1974; Beals and Lovrien, 1976). The preponderance of females has led some authorities to believe that the condition may be an X-linked dominant. However, as females are often more severely affected than their male relatives, this discrepancy is also explicable in terms of bias of ascertainment. Nevertheless, clearcut male to male transmission has not yet been reported. As mentioned in Chapter 1, the Langer form of mesomelic dwarfism may be the result of homozygosity of the dyschondrosteosis gene (Silverman, 1975; Espirutu, Chen and Wooley, 1975).

Funderburk et al. (1976) described a kindred in which members of four generations had dyschondrosteosis and chronic nephritis. The authors speculated that two closely linked dominant genes might be responsible, although simple X-linked dominant inheritance could not be ruled out.

3. HEREDITARY ARTHRO-OPHTHALMOPATHY (STICKLER SYNDROME)

At the turn of the century, Dr C. H. Mayo, of the Mayo clinic, examined a middle aged woman with serious eye complications and swollen joints. In the ensuing years other members of the family presented with the same stigmata and ultimately Stickler *et al*. (1965) documented the results of a comprehensive investigation of the kindred. Stickler and Pugh (1967) expanded the diagnostic features of the syndrome, mentioning that deafness could be a component.

Clinical and radiographic features

Enlargement of large joints, particularly the wrists, knees and ankles, may be present at birth. Repeated episodes of acute arthritis precede degenerative arthropathy and physical activity may be considerably impaired by middle age. Myopia and choroidoretinal abnormalities constitute the ophthalmological facets of the syndrome. Retinal detachment sometimes leads to blindness in childhood and painful secondary glaucoma may necessitate eventual enucleation of the eyeball. Conductive deafness, cleft palate and structural abnormalities of the vertebrae are inconsistent features. Radiographically, the epiphyses are dysplastic and in adulthood severe degenerative changes may be evident. The vertebral bodies may show some degree of flattening and irregularity.

Genetics

The sex distribution of patients and the transmission of the disorder through five generations of the Mayo clinic kindred is indicative of autosomal dominant inheritance. This genetic mechanism was confirmed when Popkin and Polomeno (1974) described 22 cases in two large Canadian kindreds. The clinical stigmata are very variable in the same family (Turner, 1974; Kozlowski and Turner, 1975). Hall and Herrod (1975) emphasised this point when they compared the ocular, skeletal and orofacial features in members of three generations of a kindred.

O'Donnell, Sirkin and Hall (1976) described a father and two sons with a spondyloepiphyseal type of skeletal dysplasia which was complicated by cataracts and neural deafness. The authors pointed out that although this disorder resembled Stickler syndrome, it was probably a separate entity. They speculated that their patients might have the Marshall syndrome, a condition in which involvement of the skeleton had not been previously recognised.

4. LARSEN SYNDROME

The Larsen syndrome is an unusual disorder in which marked articular hypermobility and a flattened facies are associated with various skeletal abnormalities (Larsen, Schottstaedt and Bost, 1950). It is likely that the condition exists in mild autosomal dominant and severe autosomal recessive forms. More than 40 cases have now been reported.

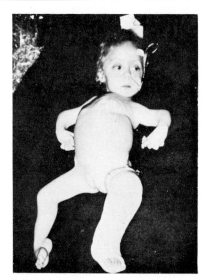 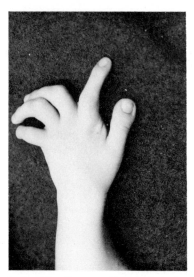

Fig. 3.5. (left) Larsen syndrome; a girl with gross joint laxity. The elbows are dislocated, the knees are unstable and the feet and thorax are deformed. Her affected brother died in infancy from a structural cardiac defect. No other members of the family had the condition, but as the parents were consanguineous, it is likely that these sibs had the autosomal recessive form of the disorder.

Fig. 3.6. (right) Larsen syndrome; the thumb is spatulate and the fingers are hypermobile.

Clinical and radiographic features

Laxity of the joints is the cardinal feature of the Larsen syndrome. The knee joint is frequently unstable, and dislocations may be recurrent. Equinovarus deformities of the feet are common and spinal malalignment may develop. The depressed nasal bridge and widely spaced eyes produce a 'dish face'. The fingers have a cylindrical configuration and the thumbs are spatulate. The manifestations are variable in degree and mildly affected individuals have few problems. However, those with severe involvement may have a disturbed gait, spinal deformity and cord compression. Structural cardiac malformation has been observed in a few patients. The orthopaedic complications in the Larsen syndrome have been discussed by Oki *et al*. (1976), Michel, Hall and Watts (1976) and Habermann, Sterling and Dennis (1976).

The radiographic stigmata of the Larsen syndrome have been reviewed by Kozlowski, Robertson and Middleton (1974). The presence of an extra ossification centre in the calcaneum is an important diagnostic feature. This centre appears in infancy and fuses by the end of the first decade. The radial heads are often dislocated and the sequelae of repeated dislocation or subluxation may be evident in other joints. Vertebral anomalies are sometimes encountered in the cervical and dorsal regions. Supernumerary bones may be present in the carpus, and the phalanges are usually mis-shapen.

Genetics

The heterogeneity of the Larsen syndrome has been emphasised by Maroteaux (1975). In keeping with autosomal recessive transmission, Curtis and Fisher (1970) and Steel and Kohl (1972) reported affected sibs with normal parents.

There have been several case descriptions which are consistent with autosomal dominant inheritance. Harris and Cullen (1971) reported an affected mother and daughter. The maternal grandfather had a similar facies and probably also had the condition. Latta *et al.* (1971) and Sugarman (1975) also described generation to generation transmission. Retrospectively, it is possible that McFarlane (1947) had encountered the dominant form of the Larsen syndrome when he reported a woman who, in three marriages, had produced children with bilateral dislocation of the knees.

Piussan *et al.* (1975) described six patients with short stature, hypermobility and diffuse skeletal sclerosis. Payet (1975) reported five sporadic children with the typical stigmata of the Larsen syndrome, in whom advanced bone age and osteoporosis were also present. The relationship of these disorders to the classical Larsen syndrome is uncertain, but they are probably separate entities.

In the absence of affected kin, genetic counselling in the Larsen syndrome is not easy. In the present state of knowledge, the severity of the stigmata and the presence of cardiac and vertebral malformations are probably indicative of the autosomal recessive type, while mild manifestions are suggestive of the autosomal dominant form.

5. ACRODYSPLASIAS

The acrodysplasias are a group of disorders in which the peripheral tubular bones are maldeveloped, so that the digits are shortened. Brailsford (1948) drew attention to these conditions and subsequently his name has been used in conjunction with the designation 'peripheral dysostosis' or 'epiphysometaphyseal acrodysplasia'. Giedion (1976) emphasised that acrodysplasia may be part of several major syndromes, in which generalised skeletal changes overshadow the digital manifestations. Peripheral dysostosis of clinically significant degree, in the absence of other associated features, has been inherited as an autosomal dominant in kindreds reported by Singleton, Daeschner and Teng (1960) and Bachman and Norman (1967). In the acrodysplasias, the phalangeal epiphyses have a cone-shaped configuration. However, epiphyses of this type are non-specific, and they can also occur as unimportant isolated minor variants (Newcombe and Keats, 1969).

The Thiemann form of acrodysplasia is a rare disease in which soft tissue swelling of the proximal interphalangeal joints develops at puberty. Rubinstein (1975) reviewed the topic and concluded that Thiemann disease was inherited as an autosomal dominant. Giedion (1976) disagreed, commenting that although the designation is frequently encountered in the early literature, there is considerable doubt as to whether the condition exists as a specific entity.

6. TRICHORHINOPHALANGEAL DYSPLASIA

In the decade since Giedion (1966) delineated this syndrome, more than 80 cases have been reported. The disorder causes little disability and it is probably under-diagnosed.

Clinical and radiographic features

Individuals with the condition are of short stature, and have a bulbous pear-shaped nose and sparse hair. Expansion of the interphalangeal joints may lead to an erroneous diagnosis of rheumatoid arthritis. Prior to skeletal maturation, cone-shaped epiphyses are radiographically evident in the digits. In the adult, the tubular bones of the hands are shortened and in some instances, the articular surfaces may be indented. The femoral capital epiphyses are small, and Perthe-like changes may supervene in the hip joints.

Fig. 3.7. Trichorhinophalangeal dysplasia; a girl with short stature, expansion of the interphalangeal joints, sparse hair and a bulbous nose.

Genetics

There have been several reports of generation to generation transmission and autosomal dominant inheritance is well established (Murdoch, 1969; Beals, 1973; Giedion et al., 1973). In a recent review of the genetics of the condition, Weaver, Cohen and Smith (1974) described three sibs who had inherited the disorder from their father. Giedion (1976) pointed out that in four kindreds, a total of nine sibs with the condition had normal parents. Parental consanguinity was present in one instance. This author postulated that in these families inheritance might be recessive or conversely, that the dominant gene may be non-penetrant.

Trichorhinophalangeal dysplasia type II, or the Langer-Giedion syndrome, is a separate entity in which the facial appearance is reminiscent of the classical form of the disorder. Additional features are microcephaly, mental retardation, loose joints and exostoses. All seven reported patients have been sporadic, and of these, six have been males (Giedion, 1976). The mode of genetic transmission is unknown.

7. COFFIN–LOWRY SYNDROME

The Coffin–Lowry syndrome is a rare disorder in which facial and digital abnormalities are associated with mental deficiency and skeletal deformities. In a review of 28 cases in eight kindreds, including eight patients in three unreported families, Temtamy, Miller and Maumenee (1975) demonstrated that the conditions reported independently by Coffin, Siris and Wegienka (1966) and Lowry, Miller and Fraser (1971) were the same entity.

Clinical and radiographic features

Intellectual impairment is the major problem in the syndrome. The face becomes progressively coarsened and the eyes have an antimongoloid slant. The hands are stubby, the fingers are hypermobile and the skin is extensible. The cervical vertebrae may be fused. Radiographically, minor dysplastic changes are evident throughout the skeleton.

Genetics

The Coffin–Lowry syndrome has been transmitted from generation to generation in several families. Males are more severely affected than females, in whom the stigmata are inconsistent. It is possible that the condition is X-linked, with variable manifestations in the female heterozygote. However, as no affected male has reproduced, the issue remains unresolved. Temtamy, Miller and Maumenee (1975) have suggested that the syndrome might be transmitted as an autosomal dominant sex-influenced trait. Membrane-limited intracytoplasmic inclusions have been identified in the skin and conjunctiva of one patient with the Coffin–Lowry syndrome. These findings may be of eventual significance in the elucidation of the mode of genetic transmission and in the development of techniques for antenatal diagnosis of the disorder.

Christian *et al*. (1977) have described four male cousins with short stature, mental retardation and skeletal anomalies which were reminiscent of the Coffin–Lowry syndrome. The condition was X-linked and of five female obligatory carriers, three had fusions of the cervical vertebrae.

8. FREEMAN–SHELDON SYNDROME

Since the first report by Freeman and Sheldon (1938), more than 50 cases have been described. The descriptive designation 'whistling face syndrome' pertains to the shape of the small puckered mouth. The title 'craniocarpotarsal dystrophy', which is sometimes employed, is neither specific nor accurate, and could well be discarded.

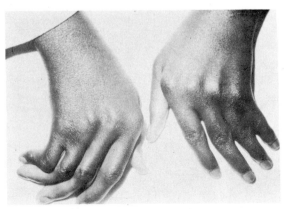

Fig. 3.8. Freeman–Sheldon syndrome; ulnar deviation of the elongated fingers produces a windmill-vane configuration.

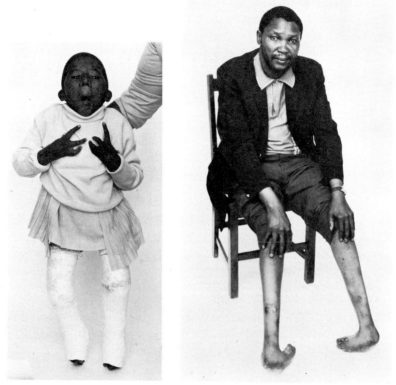

Fig. 3.9. (left) Freeman–Sheldon syndrome; microstomia and myotonia of the facial muscles produce the typical whistling face. Abnormalities of the lower limbs have been corrected surgically.

Fig. 3.10. (right) Freeman–Sheldon syndrome; the father of the patient shown in Fig. 3.9. Although the feet are severely deformed, the face is comparatively normal. Phenotypic expression of this autosomal dominant condition is notoriously variable.

Clinical and radiographic features

The face is immobile, with ptosis, strabismus, a long philtrum, microstomia and a dimpled chin. The elongated fingers have ulnar deviation and a 'windmill vane' configuration. Skeletal changes are very variable. Talipes equinovarus and scoliosis may be severe and a misdiagnosis of arthrogryposis is not unusual. Intelligence is normal, but stature is sometimes reduced. The orthopaedic features of 28 patients have been analysed by Rinsky and Bleck (1976). Sauk *et al.* (1974) have shown that the facial musculature is electromyographically and histologically abnormal.

Genetics

The Freeman–Sheldon syndrome was identified in seven individuals in four generations of a kindred by Jacquemain (1966) and in a father and son by Fraser, Pashayan and Kadish (1970). Pedigree data is consistent with autosomal dominant inheritance. In a review of 24 cases, MacLeod and Patriquin (1974) emphasised the phenotypic variability of the syndrome. For this reason, diagnosis may be difficult in the sporadic individual and genetic counselling can be a complex matter.

9. NAIL–PATELLA SYNDROME

The nail–patella syndrome or osteo-onychondysostosis is relatively common, occurring in about one in 50,000 newborn infants (Renwick and Izatt, 1965). More than 250 cases have been described.

Clinical and radiographic features

Dystrophy of the nails and hypoplasia or absence of the patellae are the most obvious manifestations. Extension of the elbow joints is sometimes limited. Many patients develop renal lesions, which are now recognised as an integral component of the syndrome. Structural abnormalities of the elbow joints and horn-shaped protuberances on the lateral aspects of the ilia may be radiographically evident.

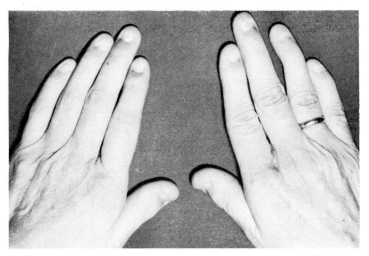

Fig. 3.11. Nail–patella syndrome; hypoplasia of the finger nails.

Genetics

Large kindreds have been reported by many authors, including Beals and Eckhardt (1969), Aggarwal and Mittal (1970) and Bennet *et al.*, (1973) and autosomal dominant inheritance is well established. The nail–patella syndrome is notable in that it is one of the few clinically important conditions in which genetic linkage has been demonstrated (Renwick and Lawler, 1955). The gene is linked to the ABO blood group locus, with a recombination fraction of approximately 10 per cent (Renwick and Schulz, 1965). Further linkage studies, in which these findings were confirmed, were undertaken in a large Corsican kindred by Serville, Verger and Astruc (1974).

The renal lesions seem to aggregate in families and it is possible that there are nephropathic and non-nephropathic forms of the condition. McKusick (1975) pointed out that as there are no obvious discrepancies in the linkage data, these two forms of the condition, if they indeed exist, may be allelic.

An infant with a 48XXXY chromosome constitution, together with the stigmata of the nail–patella syndrome, was reported by Jansen *et al.* (1976). The child's mother and grandfather also had the nail–patella syndrome, and the authors concluded that the occurrence of the two syndromes in the child was fortuitous. In the

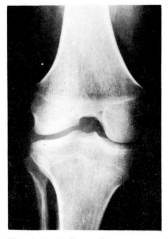

Fig. 3.12. Nail–patella syndrome; absence of the patella.

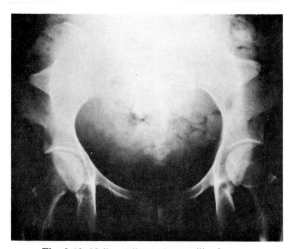

Fig. 3.13. Nail–patella syndrome; iliac horns.

same way, Gilula and Kantor (1975) encountered a family in which carcinoma of the colon coexisted with the nail–patella syndrome. These authors speculated that predisposition to colonic carcinoma might represent a previously undetected component of the nail–patella syndrome. However, as there have been no similar reports, this association is almost certainly a chance event.

Absence of the patella and aniridia have been observed in members of three generations of a kindred by Mirkinson and Mirkinson (1975). No other features of osteo-onychodysostosis were present and this condition is apparently a distinct autosomal dominant entity.

10. CONGENITAL BOWING OF LONG BONES

Bowing of the long bones, particularly those of the lower limbs, may occur in the absence of any obvious metabolic disturbance. Thompson, Oliphant and Grossman (1976) reviewed the problem of bowed limbs in the neonate and emphasised that this abnormality may be a self-limiting physiological condition or the presenting feature of a variety of disorders. Campomelic dwarfism, a potentially lethal condition in the newborn in which the long bones are bowed, has been discussed in Chapter 1.

Bowlegs due to tibial torsion was recognised in eight individuals in four generations of a kindred by Blumel, Eggers and Evans (1957). A second family was reported by Fitch (1974). Inheritance is apparently autosomal dominant and it is probable that this disorder is more common than is generally recognised. A kindred with autosomal dominant inheritance of congenital tibial bowing in association with pseudoarthrosis and pectus excavatum was reported by Beals and Fraser (1975). These authors reviewed the causes of tibial bowing and concluded that this particular syndrome had not been previously reported. Newell and Durbin (1976) discussed the pathogenesis of congenital angulation and the relationship of this deformity to pseudoarthrosis.

In Blount disease or tibia vara, bowing of the legs develops during the second year of life. The medial aspect of the upper tibial epiphysis is primarily involved, but

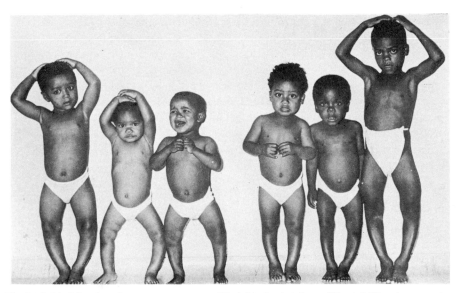

Fig. 3.14. Limb bowing; there are many causes of limb bowing in infancy. These children have a variety of conditions, including dietary and metabolic rickets, metaphyseal dysplasia and Blount disease. (Courtesy of Mr R. A. de Méneaud, Cape Town.)

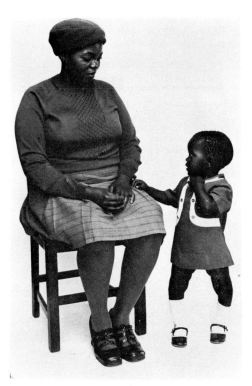

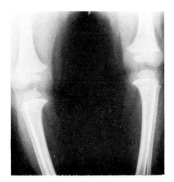

Fig. 3.15. (left) Blount disease; bowing of the legs develops during the second year of life. The aetiology is unknown.

Fig. 3.16. (above) Blount disease; beaking of the upper medial tibial metaphysis and buttressing of the cortex of the medial side of the tibial shaft.

changes are sometimes present in the adjacent tibial metaphysis and at the lower end of the femur (Blount, 1937). Blount disease is particularly common in the African negro population of Southern Africa. However, the aetiology is unknown. In a survey of over 100 affected children, the author has been unable to identify any pathogenic factor. Blount disease of late onset, which appears at puberty, is almost certainly a separate condition from the infantile form of Blount disease. An isolated report of late-onset Blount disease in a father and his two sons is suggestive of autosomal dominant inheritance (Tobin, 1957).

REFERENCES

Cleidocranial dysplasia

Goodman, R. M., Tadmor, R., Zaritsky, A. & Becker, S. A. (1975) Evidence for an autosomal recessive form of cleidocranial dysostosis. *Clinical Genetics*, **8**/1, 20.

Herndon, C. N. (1951) Cleidocranial dysostosis. *American Journal of Human Genetics*, **3**, 314.

Jackson, W. P. U. (1951) Osteo-dental dysplasia (cleido-cranial dysostosis) the 'Arnold head'. *Acta Medica Scandinavica*, **139**, 292.

Jarvis, L. J. & Keats, T. E. (1974) Cleidocranial dysostosis, a review of 40 new cases. *American Journal of Roentgenology, Radium Therapy and Nuclear Medicine*, **121**, 5.

Lasker, G. W. (1946) The inheritance of cleidocranial dysostosis. *Human Biology*, **18**, 103.

Weerdt, C. J. de & Wildervanck, L. S. (1973) A family with dysostosis cleido-cranialis in twins, with rare or never mentioned aspects in the relatives. *Clinical Genetics*, **4**, 490.

Dyschondrosteosis

Beals, R. D. & Lovrien, E. W. (1976) Dyschondrosteosis and Madelung's deformity: report of three kindreds and review of the literature. *Clinical Orthopaedics and Related Research*, **116**, 24.

Carter, A. R. & Currey, H. L. F. (1974) Dyschondrosteosis (mesomelic dwarfism); A family study. *British Journal of Radiology*, **47**/562, 634.

Espiritu, C., Chen, H. & Wooley, P. V. (1975) Mesomelic dwarfism as the homozygous expression of dyschondrosteosis. *American Journal of Diseases of Children*, **129**, 375.

Felman, A. H. & Kirkpatrick, J. A. (1969) Madelung's deformity; observations in 17 patients. *Radiology*, **93**, 1037.

Funderburk, S. J., Smith, L., Falk, R. E., Bergstein, J. M. & Winter, H. (1976) A family with concurrent mesomelic shortening and hereditary nephritis. *Birth Defects: Original Article Series*, **12**/6, 47.

Golding, J. S. R. & Blackburne, J. S. (1976) Madelung's disease of the wrist and dyschondrosteosis. *Journal of Bone and Joint Surgery*, **58B**, 350.

Kaitila, I. I., Leisti, J. T. & Rimoin, D. L. (1976) Mesomelic skeletal dysplasias. *Clinical Orthopaedics and Related Research*, **114**, 94.

Henry, A. & Thornburn, M. J. (1967) Madelung's deformity. *Journal of Bone and Joint Surgery*, **49B**, 66.

Kozlowski, K. & Zychowicz, D. (1971) Dyschondrosteosis. *Acta Radiologica*, **11**, 459.

Langer, L. O. (1965) Dyschondrosteosis: a heritable bone dysplasia with characteristic roentgenographic features. *American Journal of Roentgenology, Radium Therapy and Nuclear Medicine*, **95**, 178.

Léri, A. & Weill, J. (1929) Une affection congénitale et symétrique du développement osseux; la dyschondrostéose. *Bulletins et Mémoires de la Société Médicale des Hôpitaux de Paris*, **53**, 1491.

Silverman, F. N. (1975) Mesomelic dwarfism. In *Progress in Pediatric Radiology*, Vol. 4, p.456. Ed. Kaufman, H. J. Basel: Karger.

Hereditary artho-ophthalmopathy

Hall, J. G. & Herrod, H. (1975) The Stickler syndrome presenting as a dominantly inherited cleft palate and blindness. *Journal of Medical Genetics*, **12**/4, 397.

Kozlowski, K. & Turner, G. (1975) Stickler syndrome — report of a second Australian family. *Paediatric Radiology*, **3**, 230.

O'Donnell, J. J., Sirkin, S. & Hall, B. D. (1976) Generalised osseous abnormalities in the Marshall syndrome. *Birth Defects: Original Article Series*, **12**/5, 299.

Popkin, J. S. & Polemeno, R. C. (1974) Stickler's syndrome (hereditary progressive artho-ophthalmopathy). *Canadian Medicine Association Journal*, **111**/10, 107.

Stickler, G. B., Belau, P. G., Farrell, F. J., Jones, J. D., Pugh, D. G., Steinberg, A. G. & Ward, L. E. (1965) Hereditary progressive arthro-ophthalmopathy. *Mayo Clinic Proceedings*, **40**, 433.
Stickler, G. B. & Pugh, D. G. (1967) Hereditary progressive arthro-ophthalmopathy II. Additional observation on vertebral abnormalities, a hearing defect, and a report of a similar case. *Mayo Clinic Proceedings*, **42**, 495.
Turner, G. (1974) The Stickler syndrome in a family with the Pierre Robin syndrome and severe myopia. *Australian Paediatric Journal*, **10**/2, 103.

Larsen syndrome
Curtis, B. H. & Fisher, R. L. (1970) Heritable congenital tibio-femoral subluxation. *Journal of Bone and Joint Surgery*, **52A**, 1104.
Habermann, E. T., Sterling, A. & Dennis, R. I. (1976) Larsen syndrome: a heritable disorder. *Journal of Bone and Joint Surgery*, **58**/4, 558.
Harris, R. & Cullen, C. H. (1971) Autosomal dominant inheritance in Larsen's syndrome. *Clinical Genetics*, **2**, 87.
Kozlowski, K., Robertson, F. & Middleton, R. (1974) Radiographic findings in Larsen's syndrome. *Australasian Radiology*, **18**/3, 336.
Larsen, L. J., Schottstaedt, E. R. & Bost, F. D. (1950) Multiple congenital dislocations associated with a characteristic facial abnormality. *Journal of Paediatrics*, **37**, 574.
Latta, R. J., Graham, C. B., Aase, J. M., Scham, S. M. & Smith, D. W. (1971) Larsen's syndrome: a skeletal dysplasia with multiple joint dislocations and unusual facies. *Journal of Paediatrics*, **78**, 291.
Maroteaux, P. (1975) Heterogeneity of Larsen's syndrome. *Archives Francaises de Pédiatrie*, **32**/7, 597.
McFarlane, A. L. (1947) A report of four cases of congenital genu recurvatum occurring in one family. *British Journal of Surgery*, **34**, 388.
Michel, L. J., Hall, J. E. & Watts, H. G. (1976) Spinal instability in Larsen's syndrome: report of three cases. *Journal of Bone and Joint Surgery*, **58**/4, 562.
Oki, T., Terashima, Y., Murachi, S. & Nogami, H. (1976) Clinical features and treatment of joint dislocations in Larsen's syndrome. Report of three cases in one family. *Clinical Orthopaedics and Related Research*, **119**, 206.
Payet, G. (1975) Dwarfism with hyperlaxity, facial malformations and multiple dislocations. Larsen's syndrome. *Archives Francaises de Pédiatrie*, **32**/7, 601.
Piussan, C., Maroteaux, P., Castroviejo, I. & Risbourg, B. (1975) Bone dysplasia with dwarfism and diffuse skeletal abnormalities. *Archives Francaises de Pédiatrie*, **32**/6, 541.
Steel, H. H. & Kohl, E. J. (1972) Multiple Congenital dislocations associated with other skeletal anomalies. (Larsen's syndrome) in three siblings. *Journal of Bone and Joint Surgery*, **54A**, 75.
Sugarman, G. I. (1975) The Larsen syndrome. Autosomal dominant form. *Birth Defects: Original Article Series*, **11**/2, 121.
Ventrutoin, Festa, B., Sebastio, L., Sebastio, G. & Catani, L. (1976) Larsen syndrome in two generations of an Italian family. *Journal of Medical Genetics*, **13**/6, 538.

Acrodysplasias
Bachman, R. K. & Norman, A. P. (1967) Hereditary peripheral dysostosis (three cases). *Proceedings of the Royal Society of Medicine*, **60**, 21.
Brailsford, J. F. (1948) *The Radiology of Bones and Joints*. London: Churchill.
Giedion, A. (1976) Acrodysplasias: peripheral dysostosis, acrodysostosis and Thiemann's disease. *Clinical Orthopaedics and Related Research*, **114**, 107.
Newcombe, D. S. & Keats, T. E. (1969) Roentgenographic manifestations of hereditary peripheral dysostosis. *American Journal of Roentgenology, Radium Therapy and Nuclear Medicine*, **106**, 178.
Rubinstein, H. M. (1975) Thiemann's disease: a brief reminder. *Arthritis and Rheumatism*, **18**/4, 357.
Singleton, E. B., Daeschner, C. W. & Teng, C. T. (1960) Peripheral dysostosis. *American Journal of Roentgenology, Radium Therapy and Nuclear Medicine*, **84**, 499.

Trichorhinophalangeal dysplasia
Beals, R. K. (1973) Tricho-rhino-phalangeal dysplasia. *Journal of Bone and Joint Surgery*, **55**, 821.
Giedion, A. (1966 Das tricho-rhino-phalangeale syndrome. *Helvetica Paediatrica Acta*, **21**, 475.
Giedion, A., Burdea, M., Fruchter, Z., Meloni, T. & Trosc, V. (1973) Autosomal dominant transmission of the tricho-rhino-phalangeal syndrome. Report of four unrelated families and a review of 60 cases. *Helvetica Paediatrica Acta*, **28**, 249.
Giedion, A. (1976) Acrodysplasias: peripheral dysostosis, acrodysostosis and Thiemann's disease. *Clinical Orthopaedics and Related Research*, **114**, 107.
Murdoch, J. L. (1969) Tricho-rhino-phalangeal dysplasia with possible autosomal dominant transmission. In *The Clinical Delineation of Birth Defects*. Chapter 2, p.218. New York: National Foundation.

Weaver, D. D., Cohen, M. M. & Smith, D. W. (1974) The tricho-rhino-phalangeal syndrome. *Journal of Medical Genetics*, **11**/3, 312.

Coffin–Lowry syndrome

Christian, J. C., Demyer, W., Franken, E. A., Huff, J. S., Khairi, S. & Reed, T. (1977) X-linked skeletal dysplasia with mental retardation. *Clinical Genetics*, **11**/2, 128.

Coffin, G. S., Siris, E. & Wegienka, L. C. (1966) Mental retardation with osteocartilagenous anomalies. *American Journal of Diseases of Children*, **112**, 205.

Lowry, R. B., Miller, J. R. & Fraser, F. C. (1971) A new dominant gene mental retardation syndrome: associated with small stature, tapering fingers, characteristic facies, and possible hydrocephalus. *American Journal of Diseases of Children*, **121**, 496.

Temtamy, S. A., Miller, J. D. & Maumenee, I. (1975) The Coffin–Lowry syndrome: An inherited faciodigital mental retardation syndrome. *Journal of Pediatrics*, **86**/5, 724.

The Freeman–Sheldon syndrome

Fraser, F. C., Pashayan, H. & Kadish, M. E. (1970) Cranio-carpo-tarsal dysplasia. Report of a case in father and son. *Journal of the American Medical Association*, **211**, 1374.

Freeman, E. A. & Sheldon, J. H. (1938) Cranio-carpo-tarsal dystrophy. An undescribed congenital malformation. *Archives of Diseases in Childhood*, **13**, 277.

Jacquemain, B. (1966) Die angeborene Windmuehlenfluegelstellung als erbliche Kombinations-missbildung. *Zeitschrift für Orthopaedie und ihre Grenzgebiete*, **102**, 146.

MacLeod, P. & Patriquin, H. (1974) The whistling face syndrome: cranio-carpo-tarsal dysplasia. Report of a case and a survey with the literature. *Clinical Pediatrics*, **13**/2, 184.

Rinsky, L. A. & Bleck, E. E. (1976) Freeman–Sheldon ('whistling face') syndrome. *Journal of Bone and Joint Surgery*, **58A**, 148.

Sauk, J. J., Delaney, J. R., Reaume, C., Brandjord, R. & Witkop, C. J. (1974) Electromyography of oral-facial musculature in craniocarpotarsal dysplasia (Freeman–Sheldon syndrome). *Clinical Genetics*, **6**, 132.

The nail–patella syndrome

Aggarwal, N. D. & Mittal, R. L. (1970) Nail–patella syndrome. *Journal of Bone and Joint Surgery*, **52B**, 29.

Beals, R. K. & Eckhardt, A. L. (1969) Hereditary onycho-osteodysplasia (nail–patella syndrome). *Journal of Bone and Joint Surgery*, **51A**, 505.

Bennett, W. M., Musgrave, J. E., Campbell, R. A., Elliot, D., Cox, R., Brooks, R. E., Lovrien, E. W., Beals, R. K. & Porter, G. A. (1973) The nephropathy of the nail–patella syndrome: clinico-pathologic analysis of 11 kindreds. *American Journal of Medicine*, **54**, 304.

Gilula, L. A. & Kantor, O. S. (1975) Familial colon carcinoma in nail–patella syndrome. *American Journal of Roentgenology, Radium Therapy and Nuclear Medicine*, **123**/4, 783.

Jansen, J., Hansen, E., Holbolth, N., Jacobson, P. & Mikkelsen (1976) 48,XXXY Klinefelter syndrome and nail–patella syndrome in the same child. *Clinical Genetics*, **9**, 163.

McKusic, V. A. (1975) Nail–patella syndrome. In *Mendelian Inheritance in Man*, 4th Edition, p.225. Baltimore London: The Johns Hopkins Press.

Mirkinson, A. E. & Mirkinson, N. K. (1975) A familial syndrome of aniridia and absence of the patella. *Birth Defects: Original Article Series*, **11**/5, 129.

Renwick, J. H. & Lawler, S. D. (1955) Genetic linkage between the ABO and nail–patella loci. *Annals of Human Genetics*, **28**, 312.

Renwick, J. H. & Schulze, J. (1965) Male and female recombination fractions for the nail–patella ABO linkage in man. *Annals of Human Genetics*, **28**, 379.

Renwick, J. H. & Izatt, M. M. (1965) Some genetical parameters of the nail–patella locus. *Annals of Human Genetics*, **28**, 369.

Serville, F., Verger, P. & Astruc, J. (1974) Osteo-onychodysostosis: a new family. *Humangenetika*, **24**/4, 333.

Congenital bowing of long bones

Beals, R. K. & Fraser, W. (1975) Familial congenital bowing of the tibia with pseudoarthrosis and pectus excavatum. *Birth Defects: Original Article Series*, **11**/6, 87.

Blount, W. P. (1937) Tibia vara: osteochondrosis deformans tibiae. *Journal of Bone and Joint Surgery*, **19**, 1.

Blumel, J., Eggers, G. W. & Evans, E. B. (1957) Eight cases of hereditary bilateral tibial torsion in four generations. *Journal of Bone and Joint Surgery*, **39A**, 1198.

Fitch, N. (1974) Male-to-male transmission of tibial torsion. *American Journal of Human Genetics*, **26**, 662.

Newell, R. L. M. & Durbin, F. C. (1976) The aetiology of congenital angulation of tubular bones with constriction of the medullary canal, and its relationship to congenital pseudoarthrosis. *Journal of Bone and Joint Surgery*, **58B/4**, 444.

Thompson, W., Oliphant, M. & Grossman, H. (1976) Bowed limbs in the neonate: significance and approach to diagnosis. *Annales de Pédiatrie*, **5/1**, 50.

Tobin, W. J. (1957) Familial osteochondritis dissecans with associated tibia vara. *Journal of Bone and Joint Surgery*, **39A**, 1091.

4. Disorders with disorganised development of cartilage and fibrous tissue

The majority of the conditions in this category are very uncommon and with a few exceptions, no pattern of Mendelian inheritance has been demonstrated. However, for the sake of clarity and completion, the following have been included in this chapter:

1. Dysplasia epiphysealis hemimelica
2. Multiple cartilagenous exostoses (diaphyseal achalasis)
3. Enchondromatosis (Ollier)
4. Enchondromatosis with haemangiomata (Maffucci)
5. Neurofibromatosis
6. Fibrous dysplasia (monostotic and polyostotic).

The distinction between the non-genetic disorders in this group is not always clearcut, and it is not unusual to encounter patients who do not fit into any precise diagnostic category. In view of the embryological relationships of these conditions, it is possible that a number of them are fundamentally the same entity, differing only in their clinical manifestations.

1. DYSPLASIA EPIPHYSEALIS HEMIMELICA

Dysplasia epiphysealis hemimelica is a rare localised disorder of cartilage which was described by Trevor (1950), and discussed in detail by Fairbank (1956). Kettelkamp, Campbell and Bonfiglio (1966) presented 15 cases and reviewed the literature.

Pain and swelling call attention to the condition and diagnostic confirmation is obtained by recognition of the characteristic radiographic changes. Areas of sclerosis may be evident in the epiphyses of a single segment of a limb. The lateral aspect of the lower end of the tibia and the corresponding bones of the tarsus and metatarsus are the sites of predeliction, although the knee joint may also be involved. Onset usually occurs during childhood and growth of the affected bones may be disturbed. The disorder becomes quiescent following epiphyseal fusion but residual deformity persists. The orthopaedic implications of dysplasia epiphysealis hemimelica have been discussed by Wolfgang and Heath (1976) and Fasting and Bjerkreim (1976).

There have been no reports of more than one patient within a kindred and it is usually accepted that the disorder is non-genetic.

 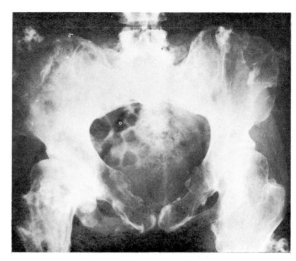

Fig. 4.1. (left) Diaphyseal achalasis; the exostoses, which are usually found at the ends of the long bones, may interfere with growth and lead to deformity.

Fig. 4.2. (right) Diaphyseal achalasis; the exostoses may be very numerous. Involvement of the pelvis may prevent normal childbearing.

2. MULTIPLE CARTILAGENOUS EXOSTOSES (Diaphyseal Achalasis)

Diaphyseal achalasis is one of the most common inherited skeletal disorders. Indeed, a collection of 1100 cases was reported more than 50 years ago (Stocks and Barrington, 1925), and since that time other large series have been published (Krooth, Macklin and Hillbish, 1961; Murken, 1963; Solomon, 1963).

Clinical and radiographic features

Multiple bony swellings make their appearance in infancy and increase both in number and in size until growth ceases. The ends of the long bones, the pelvis and shoulder girdle are most commonly involved, while the skull and spine are usually spared. The bone lesions may be painful or 'silent'. Infrequently, there is severe skeletal deformity and stature may be diminished. Problems arise from pressure upon tendons, nerves and blood vessels. Pelvic distortion may preclude normal parturition. The most important complication is malignant degeneration, as sarcoma develops in adulthood in about 10 per cent of patients. There is considerable variation in the degree to which members of the same kindred are affected. In general, the disorder is more severe in males than in females.

Radiological examination invariably reveals many more bony lesions than are clinically evident. The juxtaepiphyseal regions are the site of predeliction in childhood. Later in life, the exostoses migrate towards the diaphyses. Marked irregular expansion of the metaphyses often interferes with normal bone development, resulting in malalignment of joints and limb deformity. A 'Madelung' configuration of the forearm is a common consequence of this disturbance of growth.

Genetics

Autosomal dominant inheritance is well documented. The excess of affected males which has been reported in several series led Harris (1948) to postulate that phenotypic expression might be influenced by a sex-limited modifying gene. However, when full radiographic studies have been carried out in kindreds, asymptomatic affected females have been detected, and the equal sex ratio has been re-established (Solomon, 1963).

About 30 per cent of affected individuals apparently represent new mutations of the gene. In his investigation in Germany, Murken (1963) calculated that the mutation rate lay between six and nine per million. Diaphyseal achalasis has a wide geographic distribution, but it is particularly common on the Pacific island of Guam. It has been established that the prevalence is about one in 1000 in the islanders (Krooth, Macklin and Hillbish, 1961).

As diaphyseal achalasis is relatively common and readily recognisable, it would lend itself to linkage studies. Investigations of this type yielded negative results a decade ago (Scholz and Murken, 1963), but as new gene markers are now available, repetition and augmentation of these studies might be worthwhile. Diaphyseal achalasis is one of the few dominant disorders in which a possible homozygote has been encountered. Giedion, Kesztler and Muggiasca (1975) studied two brothers who were the offspring of affected parents. These authors suggested that the severe and precocious manifestations of the condition in these boys might be indicative of homozygosity.

Solomon (1963) identified a kindred in which eight individuals in three generations had multiple exostoses which were virtually confined to the bones of the hands. As the intrafamilial manifestations of diaphyseal achalasis are usually very variable, it is possible that this particular condition is a unique genetic entity, differing from the common form of the disorder.

3. ENCHONDROMATOSIS (OLLIER)

Enchondromatosis (Ollier disease) is much less common than diaphyseal achalasis, with which it is sometimes confused. However, there is little similarity in the manifestations of these disorders, and diagnostic distinction is not difficult.

Clinical and radiographic features

Multiple bony swellings, particularly of the digits, are the presenting feature of Ollier disease. These enchondromatous lesions become static or regress after puberty. Localised disturbance of growth by enchondromata in the metaphyseal regions of long bones may produce deformity and limb asymmetry. Pathological fracture is a well recognised complication and the development of malignancy has been reported. Precise figures are not available, but this latter problem seems to be uncommon. Radiographically, the enchondromata appear as multiple, oval, lucent defects. The digits and tubular bones are predominantly involved, while the spine and skull are spared.

Fig. 4.3. Enchondromatosis; the bony swellings are often most evident in the digits.

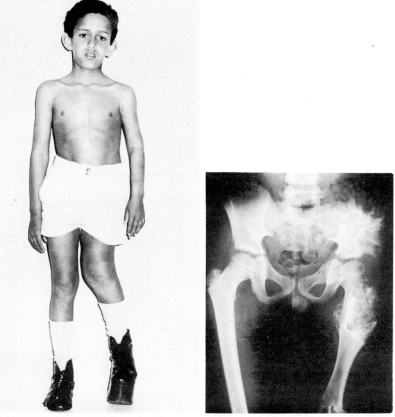

Fig. 4.4. (left) Enchondromatosis; the enchondromata may interfere with growth, producing marked disparity in limb length.

Fig. 4.5. (right) Enchondromatosis; radiograph of the boy depicted in Fig. 4.4. In the pelvis, the fan-like lesions radiate towards the iliac crest. The upper part of the shaft of the femur is expanded and distorted.

Genetics

Enchondromatosis is usually considered to be non-genetic and the majority of reports have been concerned with sporadic individuals. The presence of the disorder in a male member of a pair of dizygotic twins was described by Boron, Danilewicz, Wytrychowska and Zajac (1975). Sibs have been reported by Lamy *et al*. (1954) while Rossberg (1959) described a brother and sister with the condition, whose paternal grandfather had also been affected. McKusick (1975) suggested that these observations were compatible with autosomal dominant inheritance with reduced penetrance.

Other enchondromatoses

(a) Metachondromatosis is a condition in which multiple exostoses and enchondromata occur together. Maroteaux (1971) reported two kindreds in which metachondromatosis was inherited as an irregular dominant. Kozlowski and Scougall (1975) suggested that metachondromatosis was probably often misdiagnosed as diaphyseal achalasis.

(b) A distinct disorder in which multiple enchondromata and eccondromata were associated with Perthe-like changes in the femoral heads has been observed in 13 members of three generations of an Afrikaner kindred (Schweitzer, Jones and Timme, 1971). The authors reported the condition under the title 'Upington disease, a familial dyschondrodysplasia'. This geographic appellation pertained to the patients' origin in a rural area in South Africa. In this kindred the pattern of transmission was consistent with autosomal dominant inheritance.

(c) Using the designation 'spondyloenchondrodysplasia', Schorr, Legum and Ochshorn (1976) reported two Israeli brothers who had enchondromatosis together with platyspondyly. These individuals were of short stature. It is likely that this disorder is yet another distinct genetic entity.

4. ENCHONDROMATOSIS WITH HAEMANGIOMA (MAFFUCCI)

A patient with enchondromatosis and haemangiomata was described by Maffucci in the last century and the syndromic relationship of these anomalies was firmly established by Carleton *et al*. (1942). At that time, about 20 cases have been reported. This total had risen to 62 when the literature was reviewed two decades later (Anderson, 1965).

Clinical and radiographic features

The components of the Maffucci syndrome are single or multiple enchondromata and dermal angiomatous lesions. The radiographic changes resemble those of the Ollier syndrome although bone lesions do not have the same predilection for the digits. The skeletal anomalies appear during childhood and progress until growth ceases. Limb deformity of significant degree is present in at least 50 per cent of cases. Cavernous or capillary haemangiomata of the skin are often evident at birth, although their appearance is sometimes delayed until infancy. These lesions may be widespread, with angiomatous involvement of internal organs. The most important consequence of the Maffucci syndrome is chondrosarcoma, which arises in about 10 per cent of patients. It has been suggested that the predisposition to neoplasia is not limited to the skeletal system and that there is an increased risk of malignancy at other sites (Anderson, 1965).

Genetics

The Maffucci syndrome must be distinguished from the Klipper–Trenaunay–Weber syndrome, in which telangectasia is associated with bone hypertrophy, in the absence of enchondromata. Uncomplicated hemihypertrophy also has features in common with these disorders and in the individual patient, accurate categorisation can be difficult. Although the interrelationship of these conditions is sufficiently controversial to inspire argument, their distinction is largely academic, as there is no evidence to indicate that any of them has a genetic basis.

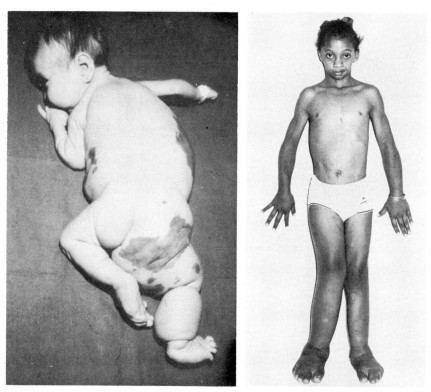

Fig. 4.6. (left) Klippel–Trenaunay–Weber syndrome; haemangiomata and hypertrophy are striking features.

Fig. 4.7. (right) Hemihypertrophy; excess growth of all components of the right leg, without localised bone or vascular lesions.

The essential features of this group of disorders can be summarised as follows:

Disorder	Manifestations	Inheritance
Diaphyseal achalasis	Multiple exostoses	AD
Ollier disease	Multiple enchondromata	Non-genetic
Metachondromatosis	Multiple exostoses and echondromata	AD
Upington disease	Enchondromata, eccondromata and Perthe-like changes in the femoral heads	AD
Maffucci syndrome	Enchondromata and haemangiomata	Non-genetic
Klippel–Trenaunay– Weber syndrome	Bone hypertrophy and haemangiomata	Non-genetic
Hemihypertrophy	Unilateral skeletal and soft tissue overgrowth	Non-genetic

5. NEUROFIBROMATOSIS

Neurofibromatosis or von Recklinghausen disease is a well known disorder which is inevitably encountered at some stage in every medical career — frequently during viva voce examination for a specialist qualification! The clinical manifestations can be very severe. Sir Frederick Treves, Surgeon Extraordinary to Queen Victoria, gave a poignant account of the life of John Merrick, who was grotesquely disfigured by the condition. This unfortunate individual was exhibited in a peep-show as the 'Elephant Man'. Sir Frederick rescued him from his predicament in 1886, and he spent his remaining years in seclusion at the London hospital.

Clinical and radiographic features

Multiple pedunculated and sessile dermal tumours may be very numerous, while café-au-lait pigmented macules are invariably present. (The smooth outline outline of these patches has been likened to the 'coast of California', in distinction to the irregular 'coast of Maine' configuration of the macules in polyostotic fibrous dysplasia. As these conditions bear a resemblance to each other, this geographic concept is of practical value.) Plexiform neuromata occasionally develop into pendulous lesions, while neuromata on peripheral nerves or in the spine sometimes cause neural compression. An acoustic neuroma may result in deafness. Phaeochromocytoma and other endocrine tumous occur and malignant degeneration is not uncommon. Skeletal manifestations include vertebral abnormalities, rib fusion, pseudoarthrosis of the tibia and hypertrophy of a limb or digit. In a Scandinavian series of 3209 patients with scoliosis, 3 per cent had neurofibromatosis (Rezaian 1976). Apter et al. (1975) have emphasised that the rate of development and the extent of the abnormalities is very variable, even in members of the same kindred.

Genetics

Neurofibromatosis is one of the commonest autosomal dominant disorders. The genetic basis was extensively reviewed by Crowe, Schull and Neel (1956). These authors estimated that about 50 per cent of patients represented new gene mutations, and that the mutation rate was approximately 1×10^{-4}. If this figure is accurate,

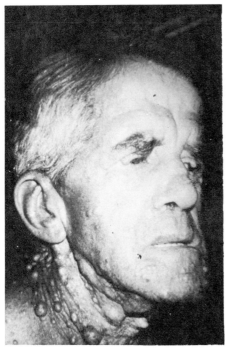

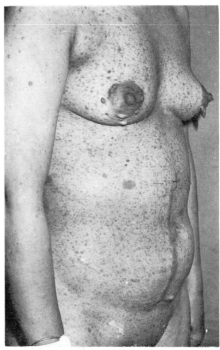

Fig. 4.8. (left) Neurofibromatosis; dermal tumours may cover the whole body. Phenotypic expression of this autosomal dominant gene is extremely variable.

Fig. 4.9. (right) Neurofibromatosis; the café-au-lait macule which is present over the right costal margin has the typical smooth 'coast of California' configuration.

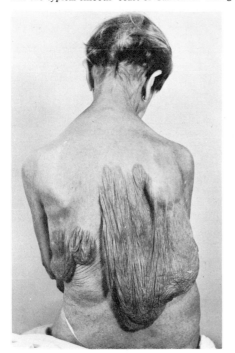

Fig. 4.10. Neurofibromatosis; pendulous dermal lesions.

then neurofibromatosis bears the dubious distinction of having the highest mutation rate of any clinically important genetic condition. However, in view of the notorious variability of expression of the gene, it is likely that neurofibromatosis is under-diagnosed in parents of sporadic cases and that the quoted mutation rate is errone-ously high. In family studies, skipped generations are not unusual. Nevertheless, minor stigmata are often recognisable on close examination of ostensibly normal obligatory carriers of the gene. In this context, it is of practical importance that the dermal manifestations of neurofibromatosis are progressive and that they may not become evident until mid-childhood.

Ansari and Nagamani (1976) described a woman in whom neurofibromata increased in size in late pregnancy. The authors considered that the risk of this complication, together with the dominant mode of transmission of the gene, would justify abortion. Steimly *et al*. (1975) reported a similar instance in which swelling of a spinal tumour during pregnancy resulted in quadriplegia.

Atypical cases of neurofibromatosis are not uncommon, and it is sometimes difficult to reach a firm diagnosis. The familial nature of acoustic neuromata and the association of café-au-lait macules with scoliosis are examples of this situation. In addition, there may well be poorly delineated entities which resemble neurofib-romatosis. For instance, Bradley, Richardson and Frew (1974) reported a mother and two daughters who shared the features of neurofibromatosis, congenital deaf-ness due to neuronal degeneration, partial albinism and a defect of the iris. The condition in this kindred could well be a 'private' syndrome.

For the purposes of genetic counselling and prognostication, the individual with minimal clinical manifestations poses considerable problems. It is generally assumed that the diagnosis can be confirmed or refuted on a basis of criteria such as the presence of a minimum number of café-au-lait macules which must be of a certain size. However, as the phenotype is very variable, a rigid approach will inevitably lead to diagnostic errors and inappropriate counselling.

6. FIBROUS DYSPLASIA

The association of fibrosis with cystic lesions of bone is well recognised. The term 'fibrous dysplasia' is used loosely to include several conditions which have similar features but which are separate pathological entities.

If the fibrocystic changes are limited to one bone, the term 'monostotic fibrous dysplasia' is used, while 'polyostotic fibrous dysplasia' implies widespread skeletal involvement. In either instance, irregular patches of light brown macular (café-au-lait' pigmentation may be present, not necessarily in anatomical relationship with the affected bone. The condition may remain unrecognised until pathological frac-ture occurs. However, if bone involvement is extensive, deformity may be severe. The cystic radiolucent lesions may be small and discreet, or may occupy and distort the whole shaft of the affected bone. Sarcomatous changes are uncommon.

The condition in this group which has attracted the most attention is the McCune–Albright syndrome. In this disorder, polyostotic fibrous dysplasia and café-au-lait dermal pigmentation are associated with sexual precocity (Albright *et al*., 1937). Females are predominantly affected, and secondary sexual charac-teristics, including pubic hair, breast enlargement and menstruation appear during

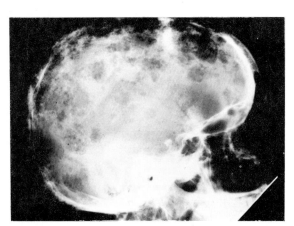
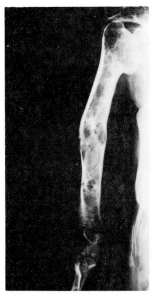

Fig. 4.11. (left) Polyostotic fibrous dysplasia; skull radiograph of a severely affected girl.

Fig. 4.12. (right) Polyostotic fibrous dysplasia; this girl had multiple bony lesions and areas of irregular café-au-lait dermal pigmentation. However, as sexual development was not precocious, an initial diagnosis of the McCune–Albright could not be substantiated.

the first decade. Lightner, Penny and Frasier (1975) have suggested that the endocrinopathies in the McCune–Albright syndrome are the result of hypothalamic dysfunction. In many patients, the epiphyses close prematurely, resulting in failure to achieve normal adult height. About 3 per cent of individuals in the polyostotic dysplasia group have the full McCune–Albright syndrome. There is no evidence to suggest that any of these monostotic or polyostotic fibrous dysplasia syndromes have a genetic basis. A simple classification is shown below:

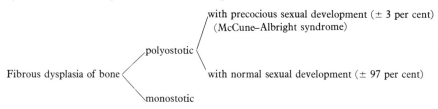

Genetic forms of fibrous dysplasia
(a) Fibrous dysplasia of the tibia was a feature in about 30 per cent of a series of 23 patients with Mulibrey nanism (Perheentupa *et al.*, 1975). The major stigmata of this unusual autosomal recessive disorder are low birth weight, proportionate short stature, muscular hypotonicity and pericardial constriction.

(b) A boy with craniofacial polyostotic fibrous dysplasia was described by Reitzik and Lownie (1975). Although there are no previous reports of genetic transmission of this disorder, the condition was familial in this particular kindred.

(c) In multiple fibromatosis, metaphyseal lesions are associated with fibrous soft tissue nodules (Stout, 1954). The bony changes resemble those of Ollier disease (Schlangen, 1976). Using the designation 'congenital generalised fibromatosis' Heiple, Perrin and Aikawa (1972) reviewed the features of 24 patients. In some, bone involvement was minimal or absent. It is not clear if this variability is indicative of heterogeneity. Baird and Worth (1976), who reported two sets of affected sibs in a consanguineous kindred, pointed out that the disorder had never been recorded in the parent of a child with the condition, and postulated that inheritance was autosomal recessive.

(d) Léri pleonosteosis is a rare disorder in which moderate shortening of stature and broadening of the thumbs is associated with flexion contractures of the digits and extremities. Excessive fibrosis is present in the fascia, joint capsules, tendons and ligaments (Watson-Jones, 1949). There have been several familial instances and inheritance is probably autosomal dominant (Léri, 1922; Rukavina et al., 1959).

(e) The descriptive designation 'cherubism' is applied to a rare disorder in which expansion of the maxilla and mandible produce an 'angelic' or 'cherubic' facies. This appearance is the result of multiple cystic changes in the jaw bones, and is accompanied by hyperplasia of the submandibular lymph nodes. Multilocular cysts may also be present in the anterior ends of the ribs. Cherubism appears during the first year of life and progresses until the skeleton is mature. The hereditary nature of cherubism was discussed by Anderson and McClendon (1962), who concluded that the disorder was inherited as an autosomal dominant with varying expression. Jones (1965) concurred with this opinion. Subsequently, generation to generation transmission was reported by Salzano and Ebling (1966) and Khosla and Korobkin (1970).

Using the term 'fibro-osseous dysplasia of the jaw' Chatterjee and Mazumder (1967) depicted a father and two sons who had gross facial abnormalities. These unfortunate individuals hardly resembled cherubs, and in view of the severity of the stigmata, the condition in this kindred is probably a distinct entity.

REFERENCES

Dysplasia epiphysealis hemimelica

Fairbank, T. J. (1956) Dysplasia epiphysialis hemimelica (tarso-epiphysial aclasis). *Journal of Bone and Joint Surgery*, **38B**, 257.
Fasting, O. J. & Bjerkreim, I. (1976) Dysplasia epiphysealis hemimelica. *Acta Medica Scandinavica*, **47/2**, 217.
Kettlekamp, D. B., Campbell, C. J. & Bonfiglio, M. (1966) Dysplasia epiphysialis hemimelica. A report of fifteen cases and a review of the literature. *Journal of Bone and Joint Surgery*, **48A**, 746.
Trevor, D. (1950) Tarso-epiphyseal aclasis: a congenital error of epiphysial development. *Journal of Bone and Joint Surgery*, **32B**, 204.
Wolfgang, G. L. & Heath, R. D. (1976) Dysplasia epiphysealis hemimelica: a case report. *Clinical Orthopaedics and Related Research*, **116**, 32.

Cartilagenous exostosis

Giedion, A., Kesztler, R. & Muggiasca, F. (1975) The widening spectrum of multiple cartilagenous exostosis (MCE). *Paediatric Radiology*, **3/2**, 93.
Harris, H. (1948) A sex-limiting modifying gene in diaphyseal aclasis. *Annals of Eugenics*, **14**, 165.
Krooth, R. S., Macklin, M. A. P. & Hillbish, T. F. (1961) Diaphyseal aclasis (multiple exostoses) on Guam. *American Journal of Human Genetics*, **13**, 340.
Murken, J. D. (1963) Über multiple cartilaginare Exostosen. *Zeitschrift für menschliche Vererbungs- und Konstitutionslehre*, **36**, 469.

Scholz, W. & Murken, J. D. (1963) Koppelungsuntersuchungen bei Familien mit multiplen cartilaeginaeren Exostosen. *Zeitschrift für menschliche Vererbungs- und Konstitutionslehre*, **37**, 178.
Solomon, L. (1963) Hereditary multiple exostosis. *Journal of Bone and Joint Surgery*, **45B**, 292.
Solomon, L. (1964) Hereditary multiple exostosis. *American Journal of Human Genetics*, **16**, 351.
Stocks, P. & Barrington, A. (1925) Hereditary disorders of bone development. In *Treasury of Human Inheritance*, Volume 3, part 1. London: Cambridge University Press.

Enchondromatosis

Boron, Z., Danilewicz Wytrychowska, T. & Zajac, J. (1975) Ollier's disease in a dizygotic twin. *Monatsschrift für Kinderheilkunde*, **123/2**, 91.
Kozlowski, K. & Scougall, J. S. (1975) Metachondromatosis: report of a case in a 6-year-old boy. *Australian Paediatric Journal*, **11/1**, 42.
Lamy, M., Aussannaire, M., Jammet, M. L. & Nezelop, C. (1954) Trois cas de maladie d'Ollier dan une fratrie. *Bulletins et Mémoires de la Societé Médicale de Hôpitaux de Paris*, **53**, 1491.
Maroteaux, P. (1971) La metachondromatose. *Zeitschrift für Kinderheilkunde*, **109**, 246.
McKusick, V. A. (1975) In *Mendelian Inheritance in Man*, 4th Edition, p.242. Baltimore–London: The Johns Hopkins University Press.
Rossberg, A. (1959) Zur Erblichkeit der Knochenchondromatose. *Fortschritte auf dem Gebiete der Rötgenstrahlen und der Nuklearmedizin*, **90/1**, 138.
Schweitzer, G., Jones, B. & Timme, A. (1971) Upington disease: a familial dyschondroplasia. *South African Medical Journal*, **45**, 994.
Schorr, S., Legum, C. & Ochshorn, M. (1976) Spondyloenchondrudysplasia. Enchondromatosis with severe platyspondyly in two brothers. *Radiology*, **118/1**, 133.

Maffucci syndrome

Anderson, I. F. (1965) Maffucci's syndrome: report of a case, with a review of the literature. *South African Medical Journal*, **39**, 1066.
Carleton, A., Elkington, J. S. C., Breenfield, J. G. & Robb-Smith, A. H. T. (1942) Maffucci's syndrome (dyschondroplasia with haemangiomata). *Quarterly Journal of Medicine*, **2**, 203.

Neurofibromatosis

Ansari, A. H. & Nagamani, M. (1976) Pregnancy and neurofibromatosis (von Recklinghausen's disease). *Obstetrics and Gynecology*, **47/1**, 25.
Apter, N., Chemke, J., Hurwitz, N. & Levin, S. (1975) Neonatal neurofibromatosis: unusual manifestations with malignant clinical course. *Clinical Genetics*, **7/5**, 388.
Bradley, W. G., Richardson, J. & Frew, I. J. C. (1974) The familial association of neurofibromatosis, peroneal muscular atrophy, congenital deafness, partial albinism, and Axenfeld's defect. *Brain*, **97/3**, 521.
Crowe, F. W., Schull, W. J. & Neel, J. V. (1956) In *A Clinical, Pathological and Genetic Study of Multiple Neurofibromatosis*. Springfield, Illinois: Charles C. Thomas.
Rezaian, S. M. (1976) The incidence of scoliosis due to neurofibromatosis. *Acta Orthopaedica Scandinavica*, **47**, 534.
Steimle, R., Jacquet, G. & Gillet, J. Y. (1975) Subacute quadriplegia: pregnancy and von Recklinghausen's disease. *Journal of Medicine of Besançon*, **11/2**, 55.

Fibrous dysplasia

Albright, F., Bulter, A. M., Hampton, A. O. & Smith, P. (1937) Syndrome characterised by osteitis fibrosa disseminata, areas of pigmentation and endocrine dysfunction, with precocious puberty in females; report of five cases. *New England Journal of Medicine*, **216**, 727.
Baird, P. A. & Worth, A. J. (1976) Congenital generalised fibromatosis: an autosomal recessive condition? *Clinical Genetics*, **9**, 488.
Heiple, K. G., Perrin, E. & Aikawa, M. (1972) Congenital generalised fibromatosis. A case limited to osseous lesions. *Journal of Bone and Joint Surgery*, **54A**, 663.
Léri, A. (1922) Dystrophie osseuse generalisée congenitale et hereditaire: la pleonosteose familiale. *Presse Médicale*, **30**, 13.
Lightner, E. S., Penny, R. & Fraser, S. D. (1975) Growth hormone excess and sexual precocity in polyostotic fibrous dysplasia (McCune–Albright syndrome): evidence for abnormal hypothalamic function. *Journal of Pediatrics*, **87/61**, 922.
Perheentupa, J., Autio, S. & Leisti, S. (1975) Mulibrey nanism: review of a new autosomal recessive syndrome. *Birth Defects: Original Article Series*, **11/2**, 3.
Reitzik, M. & Lownie, J. F. (1975) Familial polyostotic fibrous dysplasia. *Oral Surgery*, **40/6**, 769.

Rukavina, J. G., Falls, H. F., Holt, J. F. & Block, W. G. (1959) Léri's pleonosteosis: a study of a family with a review of the literature. *Journal of Bone and Joint Surgery*, **41A**, 397.
Schlangen, J. T. (1976) Congenital generalised fibromatosis. A case report with roentgen manifestations of the skeleton. *Clinical Radiology*, **45/1**, 18.
Stout, A. P. (1954) Juvenile fibromatoses. *Cancer*, **7**, 953.
Watson-Jones, R. (1949) Léri's pleonosteosis, carpal tunnel compression of the medial nerves and Morton's metatarsalgia. *Journal of Bone and Joint Surgery*, **31**, 397.

Cherubism

Anderson, D. E. & McClendon, J. L. (1962) Cherubism — hereditary fibrous dysplasia of the jaws. 1. Genetic considerations. *Oral Surgery*, **15**, Supl. 2.
Chatterjee, S. K. & Mazumder, J. K. (1967) Massive fibro-osseous dysplasia of the jaws in two generations. *British Journal of Surgery*, **54**, 335.
Jones, W. A. (1965) Cherubism: a thumbnail sketch of its diagnosis and a conservative method of treatment. *Oral Surgery*, **20**, 648.
Khosla, V. M. & Korobkin, M. (1970) Cherubism. *American Journal of Diseases of Children*, **120**, 458.
Salzano, F. M. & Ebling, H. (1966) Cherubism in a Brazilian kindred. *Acta geneticae medicae et gemellologiae*, **15**, 296.

5. Disorders with diminished bone density

The conditions in this category share the common feature of increased radiolucency of the skeleton. It is likely that the majority of them are heterogeneous, and further subdivision can be foreseen.

1. Osteogenesis imperfecta
2. Juvenile idiopathic osteoporosis
3. Idiopathic osteolysis
   (a) acro-osteolysis; phalangeal type
   (b) acro-osteolysis; tarsocarpal type
   (c) multicentric osteolysis
4. Hypophosphatasia
5. Vitamin D-resistant rickets
6. Pseudohypoparathyroidism.

1. OSTEOGENESIS IMPERFECTA

Osteogenesis imperfecta (OI) is one of the most common and best known of the inherited disorders of the skeleton. OI has been described and discussed by many authors during the last 150 years and the eponyms 'Lobstein', 'Vrolik' and 'van der Hoeve' are still associated with the condition. A tentative retrospective diagnosis of OI has been made in a variety of historic circumstances. Perhaps the earliest and best documented case is that of an affected Egyptian mummy (Gray, 1970).

OI has been subdivided on a clinical basis into the severe 'congenita' form, where stigmata are present at birth, and a mild 'tarda' form, where problems arise in childhood. It must be emphasised that these are not necessarily distinct genetic entities (*vide infra*).

OI congenita is categorized radiographically into 'thick bone' and 'thin bone' types. In the present state of knowledge, it is thought that the former type is a rare autosomal recessive, while the latter represents early manifestations of the common autosomal dominant form of OI. However, OI is probably very heterogeneous, and this concept is almost certainly an over-simplification of the true state of affairs.

Clinical and radiographic features

Bone fragility is the cardinal feature of OI. This predisposition is variable and some patients suffer many fractures, while others have few problems. The fractures heal rapidly but there may be residual deformity. Abundant callus at a fracture site may simulate osteosarcoma, and there have been erroneous diagnoses and needless amputations. A wide bi-temporal diameter, blueness of the sclera and pearly grey

Fig. 5.1. Osteogenesis imperfecta; a kindred in which several members have autosomal dominant osteogenesis imperfecta. The orthopaedic sequelae of multiple fractures are very obvious.

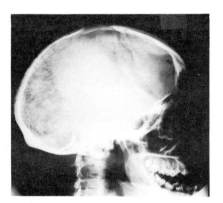

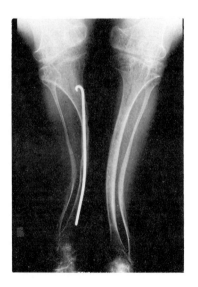

Fig. 5.2. (above) Osteogenesis imperfecta; skull radiograph showing multiple Wormian bones.

Fig. 5.3. (right) Osteogenesis imperfecta; the skeleton is gracile and undermineralised. In this patient, deformity of the tibia has been corrected by means of an intramedullary nail.

discolouration of the teeth are variable concomitants. Deafness supervenes in adult-hood in about 20 per cent of cases. The severity of OI is very variable, some patients being little troubled, while others are dwarfed, crippled and deformed. Multiple fractures and intracranial bleeding may cause stillbirth or death in the neonatal period and for this reason, Caesarian section is the preferential method of delivery for a potentially affected infant (Roberts and Solomons, 1975).

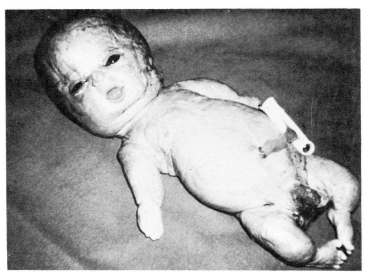

Fig. 5.4. Osteogenesis imperfecta; a stillborn premature infant. Blue sclerae, a poorly ossified cranial vault (caput membranaceum) and limb deformities were indicative of the diagnosis, which was confirmed radiographically.

In OI congenita, limb shortening and deformity are obvious at birth. The calvarium may be poorly ossified (caput membranaceum) and fractures of the ribs and long bones are radiographically evident. In the 'thick bone' type of OI, the shafts of the long bones are short and wide, while in the 'thin bone' type, the diaphyses are gracile. In later life, undermineralisation of the skeleton is a prominent feature.

The tubular bones are slender, although in the small proportion of survivors with the 'thick bone' form of OI, the shafts may be expanded and cystic. Wormian bones are present in the occipital sutures and the base of the skull is flattened.

Genetics

Autosomal dominant inheritance of OI was firmly established after Bell (1928) and Fuss (1935) analysed numerous pedigrees. OI is notoriously variable in expression, and complete non-penetrance has been reported. This variability is intrafamilial as well as interfamilial, and, although several subtypes have been proposed, accurate clinical delineation has been impossible. Francis, Bauze and Smith (1975) studied 58 affected individuals and divided them into 'mild', 'moderate' and 'severe' groups, on a basis of long bone deformity. These authors emphasised that white sclerae may be associated with severe bone disease, a finding previously noted by Ibsen (1967) and King and Bobechko (1971). Abnormalities of collagen and non-collagenous proteins of the organic matrix of bone in OI have recently been described (Lancaster *et al.*, 1975; Dickson, Millar and Veis, 1975). In the light of these developments, it is likely that subcategorisation of OI will be achieved on a biochemical rather than a clinical or radiographic basis. In view of the complexity of the collagen molecule, and in the light of precedents established for other connective tissue disorders. OI could turn out to be very heterogeneous indeed.

Numerous 'sporadic' individuals are encountered with either the 'tarda' or the 'congenita' forms of OI. This situation is explicable, in part, by non-penetrance, new

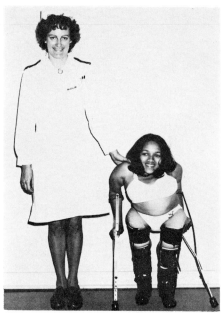

Fig. 5.5. Osteogenesis imperfecta; a young woman who suffered numerous fractures and now has severe dwarfing and deformity. Several sibs are affected, but her parents are normal. (From Horan, F. T. & Beighton, P. (1975) *Clinical Genetics*, **8**, 107.)

mutation or recessive inheritance. The possibility of the existence of an autosomal recessive type of OI has been the subject of considerable debate. Affected sibs with normal parents have been reported by several authors including Goldfarb and Ford (1954), Chawla (1964), Ibsen (1967), Wilson (1974) and Horan and Beighton (1975). Similarly, patients with unaffected consanguineous parents have been described by Freda, Vosburgh and DiLiberti (1961). The recessive hypothesis is strengthened by the observations of Kaplan and Baldino (1953), who encountered OI in many members of an inbred religious isolate, the Mosabites of Ghardia, in Southern Algeria. There is little doubt that there is at least one autosomal recessive form of OI. However, it must be emphasised that the great majority of affected individuals have an autosomal dominant disorder.

Tentative classification of OI

Clinical type	Genetics	Relative prevalence
OI congenita		
thick bone	AR?	Rare
thin bone	AD	Rare
OI tarda		
blue sclerae, few fractures	AD (heterogeneous)	Common
white sclerae, many fractures	AD	Common
white sclerae, severe dwarfing	AR	Rare

Genetic counselling in OI is usually fairly straightforward. However, the great variability of the manifestations should certainly be taken into consideration. A problem arises when young non-consanguineous normal parents produce a severely affected infant. In the present state of knowledge it is difficult to give them guidance concerning risks of recurrence. Hopefully, in the foreseeable future, advances in collagen chemistry will permit recognition of parental heterozygosity or determination of the precise genetic status of the affected child. Organisations such as the Brittle Bone Society of Britain and the National Osteogenesis Imperfecta Foundation of the USA can play an important role in the overall management of the disorder.

An affected fetus has been recognised radiographically during the third trimester (Heller, Winn and Heller, 1975). These authors stressed that their finding cannot necessarily be extrapolated to the earlier stages of pregnancy, and that routine antenatal diagnosis by radiological examination is not yet feasible. Smith, Francis and Bauze (1975) identified abnormalities in polymeric collagen in the skin of two neonates with lethal OI congenita. If it could be demonstrated that these changes are expressed in cultured amniotic fluid cells, it would be possible to monitor an 'at risk' pregnancy.

2. JUVENILE IDIOPATHIC OSTEOPOROSIS

Juvenile idiopathic osteoporosis is a rare disease of uncertain aetiology. The condition was delineated by Dent and Friedman (1965) and reviewed by Dent (1969). The diagnosis is usually reached after the exclusion of renal, gastrointestinal, endocrine or other metabolic disorders which lead to decreased skeletal density. At a clinical level, confusion with osteogenesis imperfecta is a common problem.

Clinical and radiographic features
Onset is in the prepubescent period. Bone pain, or fracture on minor trauma are the presenting features. The course is self-limiting, with remission within five years. However, there may be residual stunting and skeletal deformity. During the active phase, serum calcium concentrations are persistently low and intestinal absorption of calcium is impaired.

Radiographically, the skeleton is demineralised, and bowing of the tubular bones and concavity of the vertebral bodies may be seen. Unlike osteogenesis imperfecta there are no Wormian bones in the skull.

Genetics
All the case descriptions of juvenile idiopathic osteoporosis have concerned sporadic individuals and there is no evidence to indicate a genetic aetiology. The condition described by Jackson (1958) under the title 'Osteoporosis of unknown cause in younger people', affects adults in the third decade. It is probably a separate entity.

Bianchine and Murdoch (1969) reported a boy with osteoporosis, in whom both eyes had been enucleated in infancy for pseudoglioma. Subsequently, Bianchine et al. (1972) described two more kindreds and concluded that the osteoporosis-pseudoglioma syndrome was a specific disorder which was inherited as an autosomal recessive. In a further review, Neuhauser, Kaveggia and Opitz (1976) mentioned that mental deficiency may be a component of this syndrome. These authors pointed out that the osteoporotic process becomes quiescent in adulthood and confirmed that transmission is autosomal recessive.

3. IDIOPATHIC OSTEOLYSES

The idiopathic osteolyses are a group of unusual disorders, in which spontaneous skeletal rarefaction progresses to partial or complete disappearance of the affected bones. These conditions are classified according to the anatomical distribution of the areas which are involved. Although rare, they are very heterogeneous in terms of their clinical manifestations and genetic background. The idiopathic osteolyses can be listed under the following headings:

(a) acro-osteolysis; phalyngeal type
(b) acro-osteolysis; tarsocarpal type, with or without nephropathy
(c) multicentric osteolysis.

Osteolysis may be a component of a large number of genetic and acquired disorders, such as pycnodysostosis, hyperparathyroidism, ainhum, rheumatoid arthritis, Raynaud's disease, Sudeck arthropathy and polyvinylchloride poisoning. Similarly, bone reabsorption due to neurological damage is found in many conditions, including diabetes mellitus, leprosy and meningomyelocele. By definition, these secondary forms of osteolysis are excluded from the idiopathic category. In a recent review, Zugibe et al. (1974) have emphasised that 'osteolysis' is a radiological concept and that the underlying process may be either abnormal bone destruction or defective primary bone formation. The biochemical basis of acro-osteolysis has been discussed by Brown et al. (1976).

(a) Acro-osteolysis — phalangeal type

The majority of case reports have mentioned onset in childhood, with pain and swelling in the affected fingers. Osteolysis develops when the inflammatory process becomes quiescent. Bony collapse follows, with subsequent deformity and contracture of the digits.

A kindred in which the disorder was transmitted in a dominant pattern through four generations was described by Lamy and Maroteaux (1961). Families in which inheritance was probably autosomal recessive have been reported by Giaccai (1952) and Hozay (1953).

A syndrome in which phalangeal osteolysis is associated with generalised osteoporosis, absence of frontal sinuses, deafness, hypoplasia of the ramus of the mandible, articular laxity and Wormian bones in the skull was recognised in a mother and her four children by Cheney (1965). Further cases were reported by Dorst and McKusick (1969) and Herrmann et al. (1973). These latter authors reviewed the features of the disorder, mentioned an earlier report by Hajdu and Kauntze (1948) and proposed the term 'arthro-dento-osteodysplasia (Hajdu-Cheney syndrome)'. The evolution of the disease process was discussed in detail by Silverman, Dorst and Hajdu (1974), when they described a long-term follow-up on the cases of Hajdu and Kauntze (1948) and Herrmann et al. (1973). The phenotypic features of an affected boy and 13 other previously reported patients were tabulated by Zugibe et al. (1974). Apart from the members of the kindred mentioned by

Cheney (1965), these individuals have all been sporadic, and several have had normal children. The condition is probably inherited as an autosomal dominant, although this is not yet certain.

(b) Acro-osteolysis — tarsocarpal type

Acro-osteolysis which primarily involved the carpus and tarsus was recognised in three generations of a kindred by Gluck and Miller (1972). Dominant inheritance was also a feature of the disorder in the kindreds reported by Thieffry and Sorrel-Dejerine (1958) and Kohler et al. (1973). The condition in this latter family is probably a separate entity, as the affected individuals had the additional stigmata of frontal bossing, pes cavus, micrognathia and a Marfanoid habitus.

Autosomal dominant inheritance of tarsocarpal acro-osteolysis in association with nephropathy was reported by Shurtleff et al. (1974). Subsequently, sporadic patients were described by Torg and Steel (1968) and Macpherson, Walker and Kowall (1973). In a review of the literature, Beals and Bird (1975) identified 14 patients with tarsocarpal acro-osteolysis. Of these, nine were sporadic and five had affected kin.

The eponymous designation 'Winchester disease' is applied to a syndrome in which osteolysis of the tarsus and carpus is associated with short stature, severe joint contractures, corneal opacities, a coarse facies and generalised osteoporosis.

This disorder was first recognised in two sisters in a consanguineous Puerto Rican kindred (Winchester et al., 1969). By means of electron microscopic studies of corneal biopsy material, Brown and Kuwabara (1970) demonstrated that the condition was a mucopolysaccharide storage disease. The autosomal recessive mode of inheritance was confirmed by Hollister et al. (1974), when they investigated three patients in two consanguineous sibships.

(c) Multicentric osteolysis

Gorham and Stout (1955) reviewed the features of 24 children in whom widespread osteolysis had developed. No genetic basis was established. Subsequently, Torg et al. (1969) reported three affected sibs who had consanguineous parents. It is likely that the condition in this kindred was inherited as an autosomal recessive. Evidence for genetic heterogeneity was provided by Canún et al. (1976) following their studies of a mother and her two children, in whom generalised osteolysis was apparently transmitted as a dominant. The daughter experienced severe episodes of joint pain during childhood, and later developed generalised osteolysis and contractures. As the mother and son were less severely affected, it is evident that phenotypic expression of the abnormal gene is very variable. Following a review of the literature, Sage and Allen (1974) concluded that there are at least four separate conditions in the 'generalised osteolysis' group of disorders.

4. HYPOPHOSPHATASIA

The hypophosphatasias are a heterogeneous group of conditions with clinical and radiographic stigmata which resemble those of dietary rickets, in conjunction with a low level of serum alkaline phosphatase. These disorders are classified according to the age of onset and the severity of manifestations into varieties such as 'congenita', 'juvenile', 'tarda' and 'adult'. However, there is considerable overlap and inter-mediate types have been reported. The congenita form is lethal while in the tarda variety survival is usual.

Clinical and radiographic features

In hypophosphatasia congenita grossly defective calvarial ossification produces a caput membranaceum and the affected neonate usually dies as a result of respiratory distress or intracranial bleeding. The limbs are shortened and deformed, and clinical differentiation form the other lethal skeletal dysplasias may be difficult. Radiographically, the skeleton is poorly ossified, with marked metaphyseal irregularities. Changes in the tarda form are similar but less severe. Multiple fractures and spontaneous bowing distort the shafts of the long bones. The manifestations ameliorate during early childhood, but the patient may be left with short stature and bone deformities. Premature fusion of the sutures leads to craniostenosis and the teeth are often lost prematurely. In all forms of hypophosphatasia, serum levels of alkaline phosphatase are reduced and the urinary excretion of phosphoethanolamine is increased. Hypercalcaemia is sometimes present. The current status of knowledge concerning the biochemical abnormalities has been discussed by Gorodischer *et al.* (1976).

Genetics

Hypophosphatasia is one of the few conditions in which the heterozygote can be identified with ease and certainty (Rathbun *et al.*, 1961). Recognition is based upon the demonstration of diminished serum concentrations of alkaline phosphatase and increased urinary excretion of phosphoethanolamine. These heterozygotes often lose their permanent teeth at an early age (Pimstone, Eisenberg and Silverman, 1966). Parental consanguinity has been a feature of several cases (Svejcar and Walther, 1975), and there is little doubt that inheritance is autosomal recessive.

Controversy exists as to whether the various types of hypophosphatasia are the result of great variation in clinical expression of the same underlying genetic defect, or whether they are truly separate genetic entities. MacPherson, Kroeker and Houston (1972) have reported mild and severe cases within the same kindred, thus lending support to the former contention. On the other hand, Mehes *et al.* (1972) found consistent clinical manifestations in affected children in an inbred Hungarian community. This observation seems to confirm the existence of a distinct 'juvenile' form of hypophosphatasia. In further studies in the same kindred, Rubecz *et al.* (1974) found four homozygotes and 25 heterozygotes amongst the 53 individuals in whom biochemical studies were undertaken.

Benzie *et al.* (1976) reported the antenatal detection of hypophosphatasia in the 19th week of pregnancy by ultrasonographic demonstration of defective cranial ossification in one instance, and by radiographic identification of generalised skeletal undermineralisation in a second. These authors suggested that assessment of the activity of alkaline phosphatase in cultured fibroblasts might be of diagnostic value.

A 'rickety' form of hypophosphatasia, which was present in a man and his two sons, led Silverman (1962) to postulate that inheritance might be autosomal domin-ant. The possibility of pseudodominance in this kindred, due to unusually marked expression in the heterozygote, was discounted by the fact that the patient's wife had a normal serum alkaline phosphatase level. A very mild dominant type of hypophosphatasia, characterised principally by premature loss of primary teeth, was identified in female members of three generations of a kindred by Danovitch, Baer and Laster (1968).

Some idea of the relative prevelance of the various clinical categories of hypophosphatasia can be obtained from a multicentre radiographic analysis of 24 cases which was undertaken by Kozlowski *et al*. (1976). Three were considered to have the neonatal lethal form. The authors emphasised that the diagnostic situation was not clear-cut and that in a sporadic case it is not always possible to distinguish between the dominant and recessive forms of the condition.

5. VITAMIN D-RESISTANT RICKETS

Vitamin D-resistant rickets or familial hypophosphataemia is clinically and radiograhically very similar to other forms of dietary and metabolic rickets. Following delineation of the condition by Albright, Butler and Bloomberg (1937), there have been many reports of affected kindreds.

Clinical and radiographic features

Bony changes become evident during infancy, usually when walking commences. Bowing of the legs is the commonest presentation but knock-knees or a combined genu valgum and genu varum 'windswept' configuration is sometimes encountered. Expansion of the bone ends and chostochondral beading may develop. Hypophosphataemia differs from dietary rickets in that tetany and muscular weakness do not occur. The clinical course is very variable, but the stature of adults is usually diminished, with some residual bone deformity. The serum concentration of phosphorous is consistently low and in the active phase the serum alkaline phosphatase level may be moderately raised. The nature of the basic abnormality is not fully understood, and opinion is divided as to whether calcium, phosphorus or vitamin D absorption, excretion or metabolism are at fault.

Radiographically, the skeleton is demineralised, particularly in the juxtametaphyseal regions. The metaphyses are cupped, irregular and uneven. The shafts of the long bones of the legs become bowed, with secondary cortical buttressing in their concavities. Ligamentous calcification may develop in adulthood.

Genetics

Vitamin D-resistant rickets is one of the few clinically important X-linked dominant conditions. In this mode of inheritance, a female with the disorder would be expected to produce equal numbers of affected and unaffected sons and daughters, while an affected male would transmit the condition to all his daughters but none of his sons. Clinical expression is very variable within any kindred and in some instances, the only phenotypic manifestation is a persistently lowered serum phosphorus concentration.

The X-linked dominant mode of transmission was confirmed by Graham, McFalls and Winters (1959) following a genetic analysis of five affected kindreds. Subsequently, Burnett *et al*. (1964) described the pedigrees of 24 patients.

There have been a few reports of kindreds with autosomal dominant inheritance of

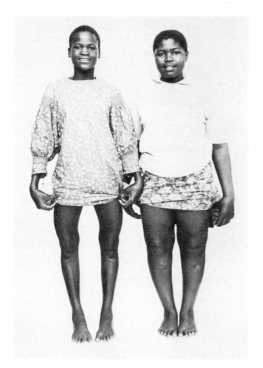

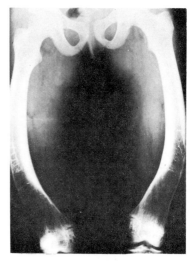

Fig. 5.6. (left) Vitamin D-resistant rickets; affected sisters with femoral bowing. Bilateral tibial osteotomy has been employed to straighten the shins. These girls are of Zulu stock, a group in whom the condition reaches a relatively high prevalence.

Fig. 5.7. (right) Vitamin D-resistant rickets; femoral bowing and demineralisation. Secondary cortical buttressing is evident on the medial side of the femora.

X-LINKED DOMINANT INHERITANCE

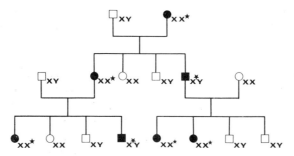

Fig. 5.8. Vitamin D-resistant rickets; a hypothetical pedigree showing X-linked dominant inheritance. A female with the condition would expect to have equal numbers of affected and unaffected sons and daughters. Conversely, all the daughters and none of the sons of an affected male will inherit the condition.

Key to pedigree: □ normal male; ○ normal female; ■ affected male; ● affected female; / deceased. X, normal X chromosome; X*, X chromosome bearing the abnormal gene; Y, Y chromosome.

hypophosphataemia (Harrison *et al.*, 1966; Pak *et al.*, 1972). The fact that expression of the abnormal gene may be limited to a low level of serum phosphorus complicates the issue. However in these particular families autosomal dominant inheritance seems to have been confirmed.

Pseudo vitamin D-resistant rickets or vitamin D-dependent rickets is a severe disorder in which gross skeletal dysplasia is associated with hypophosphataemia, marked hypocalcaemia and mild aminoaciduria. This condition differs from the common form of hypophosphataemic rickets in that tetany and myopathy occur, while the response to vitamin D therapy is usually good. Patients with normal consanguineous parents have been reported by Dent *et al.* (1968) and Arnaud *et al.* (1970), and affected sibs have been described by Birtwell *et al.* (1970). There is little doubt that pseudo vitamin D-resistant rickets is inherited as an autosomal recessive.

A non-hyperaminoaciduric form of vitamin D-dependent rickets was identified in two Mexican siblings by Cantu (1974). In a similar report, Stamp and Baker (1976) described a brother and sister with severe bowing of the legs, craniostenosis and nerve deafness, in association with persistent hypophosphataemia. As the unaffected parents were consanguineous, it is likely that inheritance of this entity was autosomal recessive.

6. PSEUDOHYPOPARATHYROIDISM

In a report of three patients, Albright *et al.* (1942) used the term 'pseudohypoparathyroidism' for the syndrome in which hypocalcaemia was associated with a characteristic facies and habitus in the absence of other evidence of hypoparathyroidism. Descriptions then followed of individuals with identical clinical stigmata, in whom serum calcium concentrations were normal. This latter condition, which was considered to be a separate entity, was designated 'pseudopseudohypoparathyroidism'. It was later recognised that the hypocalcaemia may fluctuate, and it is now accepted that these disorders are manifestations of the same abnormal gene. For this reason, to the great relief of the semantic purists, the repetitious term 'pseudo-pseudohypoparathyroidism' has been discarded. In view of the confusion which still persists, 'Albright's hereditary osteodystrophy' is a useful alternative common designation.

Clinical and radiographic features

Moderate shortness of stature, obesity and a 'moon face' are associated with some degree of shortening of the tubular bones of the extremities. The bone changes are very variable but the fourth metacarpal is most often affected. Hypocalcaemia, which is resistant to treatment with parathormone, may lead to cataracts, mental deficiency, tetany and ectopic calcification.

Radiographically, cone-shaped epiphyses may be identified in the phalanges. Other changes include thickening of the calvarium and calcification in the basal ganglia and subcutaneous tissues.

Genetics

The problems concerning the relationship of pseudohypoparathyroidism and pseudo-pseudohypoparathyroidism were finally settled when both of these disorders were recognised within the same kindred (Mann, Alterman and Hill, 1962; Brito Suarenz, Hernandez and de la Rosa, 1975).

There are several recorded instances of generation to generation transmission, and there is no doubt that pseudohypoparathyroidism is inherited as a dominant. An excess of affected females has led to a contention that the disorder may be an X-linked dominant. Against this is the fact that descriptions of affected fathers producing affected sons were given by Weinberg and Stone (1972) and Brito Suarez, Hernandez and de la Rosa, (1975). The situation remains confused but it is possible that the disorder is heterogeneous, one form being an X-linked dominant and the other an autosomal dominant trait. Alternative explanations would be incomplete ascertainment or partial sex limitation.

Cederbaum and Lippe (1973) reviewed the genetics of pseudohypoparathyroidism and reported an anomalous kindred in which inheritance was apparently autosomal recessive. Using the designation pseudohypoparathyroidism type II, Drezner, Neelon and Lebovitz (1973) reported a male infant with hypocalcaemia and other endocrine abnormalities. The genetic status of this condition and its relationship to the usual form of pseudohypothyroidism is uncertain.

REFERENCES

Osteogenesis imperfecta

Bell, J. (1928) Blue sclerotics and fragility of bone. In *Treasury of Human Inheritance*, Volume 2, part III. Cambridge — London: University of London, Cambridge University Press.

Chawla, S. (1964) Intrauterine osteogenesis imperfecta in four siblings. *British Medical Journal*, 1, 99.

Dickson, I. R., Millar, E. A. & Veis, A. (1975) Evidence for abnormality of bone matrix proteins in osteogenesis imperfecta. *Lancet*, 2/7935, 586.

Francis, M. J. O., Bauze, R. J. & Smith, R. (1975) Osteogenesis imperfecta: A new classification. *Birth Defects: Original Article Series*, 11/6, 99.

Freda, V. J., Vosburgh, G. J. & Di Liberti, C. (1961) Osteogenesis imperfecta congenita: a presentation of 16 cases and a review of the literature. *Obstetrics and Gynaecology*, 18, 535.

Fuss, H. (1935) Die erbliche Osteopsathyrosis. *Deutsche Zeitschrift für Chirurgie*, 245, 279.

Goldfarb, A. A. & Ford, D. (1954) Osteogenesis imperfecta congenita in consecutive siblings. *Journal of Paediatrics*, 44, 264.

Gray, P. H. K. (1970) A case of osteogenesis imperfecta, associated with dentinogenesis imperfecta, dating from antiquity. *Clinical Radiology*, 21, 106.

Heller, R. H., Winn, K. J. & Heller, R. M. (1975) The prenatal diagnosis of osteogenesis imperfecta congenita. *American Journal of Obstetrics and Gynaecology*, 121/4, 572.

Horan, F. & Beighton, P. (1975) Autosomal recessive inheritance of osteogenesis imperfecta. *Clinical Genetics*, 8/2, 107.

Ibsen, K. H. (1967) Distinct varieties of osteogenesis imperfecta. *Clinical Orthopaedics and Related Research*, 50, 279.

Kaplan, M. & Baldino, C. (1953) Dysplasie periostale paraissant familiale et transmise suivant le mode mendélien recessif. *Archives Francaises de Pédiatrie*, 10, 943.

King, J. D. & Bobechko, W. P. (1971) Osteogenesis imperfecta. An orthopaedic description and surgical review. *Journal of Bone and Joint Surgery*, 53B, 72.

Lancaster, G., Goldman, H. & Scriver, C. R. (1975) Dominantly inherited osteogenesis imperfecta in man: an examination of collagen biosynthesis. *Pediatric Research*, 9/2, 83.

Roberts, J. M. & Solomons, C. C. (1975) Management of pregnancy in osteogenesis imperfecta: new perspectives. *Obstetrics and Gynecology*, 45/2, 168.

Smith, R., Francis, M. J. O. & Bauze, R. J. (1975) Osteogenesis imperfecta. A clinical and biochemical study of a generalized connective tissue disorder. *Quarterly Journal of Medicine*, 44/176, 555.

Wilson, M. G. (1974) Congenital osteogenesis imperfecta. *Birth Defects: Original Article Series*, 10/12, 296.

Juvenile idiopathic osteoporosis

Bianchine, J. W. & Murdoch, J. L. (1969) Juvenile osteoporosis in a boy with bilateral enucleation of the eyes for pseudoglioma. In *The Clinical Delineation of Birth Defects*, p.225. New York: National Foundation.

Bianchine, J. W., Briard-Gullemot, Maroteaux, P., Frezal, J. & Harrison, H. E. (1972) Generalised osteoporosis with bilateral pseudoglioma — an autosomal recessive disorder of connective tissue: report of three families — review of the literature. *American Journal of Human Genetics*, **24**, 34A.

Dent, C. E. & Friedman, M. (1965) Idiopathic juvenile osteoporosis. *Quarterly Journal of Medicine*, **34**, 177.

Dent, C. E. (1969) Idiopathic juvenile osteoporosis (IJO). *Birth Defects: Original Article Series*, 5/**4**, 134.

Jackson, W. P. U. (1958) Osteoporosis of unknown cause in younger people. *Journal of Bone and Joint Surgery*, **40B**, 420.

Neuhauser, G., Kaveggia, E. G. & Opitz, J. M. (1976) Autosomal recessive syndrome of pseudogliomatous blindness, osteoporosis and mild mental retardation. *Clinical Genetics*, **9**, 324.

Idiopathic osteolysis—preamble

Brown, D. M., Bradford, D. S. & Gorlin, R. J. (1976) The acro-osteolysis syndrome: morphologic and biochemical studies. *Journal of Pediatrics*, 88/**41**, 573.

Zugibe, F. T., Herrmann, J., Opitz, J. M., Gilbert, E. F. & McMillan, G. (1974) Arthro-dento-osteodysplasia: A genetic 'acro-osteolysis' syndrome. *Birth Defects: Original Article Series*, **10**/5, 145.

Acro-osteolysis (phalangeal type)

Cheney, W. D. (1965) Acro-osteolysis. *American Journal of Roentgenology, Radium Therapy and Nuclear Medicine*, **94**, 595.

Dorst, J. P. & McKusick, V. A. (1969) Acro-osteolysis (Cheney syndrome). In *The Clinical Delineation of Birth Defects*, p.215. New York: National Foundation.

Giaccai, L. (1952) Familial and sporadic neurogenic acro-osteolysis. *Acta radiologica*, **38**, 17.

Herrmann, J., Zugibe, F. T., Gilbert, E. F. & Opitz, J. M. (1973) Arthro-dento-osteo dysplasia (Hajdu–Cheney syndrome). *Zeitschrift für Kinderheilkunde*, **11**, 1.

Hajdu, N. & Kauntze, R. (1948) Cranioskeletal dysplasia. *British Journal of Radiology*, **21**, 42.

Hozay, J. (1953) Sur une dystrophie familiale particulière (inhibition precoce de la croissance et osteolyse non mutilante acrales avec dysmorphie faciale). *Revista de Neurologia Clínica*, **89**, 245.

Lamy, M. & Maroteaux, P. (1961) Acro-osteolyse dominante. *Archives Francaises de pediatrie*, **18**, 693.

Silverman, F. N., Dorst, J. P. & Hajdu, N. (1974) Acro-osteolysis (Hajdu–Cheney syndrome). *Birth Defects: Original Article Series*, **10**/12, 106.

Zugibe, F. T., Herrmann, J., Optiz, J. M., Gilbert, E. F. & McMillan, G. (1974) Arthro-dento-osteodysplasia: a genetic 'acro-osteolysis' syndrome. *Birth Defects: Original Article Series*, **10**/6, 145.

Acro-osteolysis (tarsocarpal type)

Beals, R. K. & Bird, C. B. (1975) Carpal and tarsal osteolysis. *Birth Defects: Original Article Series*, **11**/6, 107.

Brown, S. I. & Kuwabara, T. (1970) Peripheral corneal opacification and skeletal deformities: a newly recognized acid mucopolysaccharidosis simulating rheumatoid arthritis. *Archives of Ophthalmology*, **83**, 667.

Gluck, J. & Miller, J. J. (1972) Familial osteolysis of the carpal and tarsal bones. *Pediatrics*, **81**, 506.

Hollister, D. W., Rimoin, D. L., Lachman, R. S., Cohen, A. H., Reed, W. B. & Westin, G. W. (1974) The Winchester syndrome: a nonlysosomal connective tissue disease. *Journal of Pediatrics*, **84**, 701.

Kohler, E., Babbitt, D., Huizenga, B. & Good, T. A. (1973) Hereditary osteolysis. *Radiology*, **108**, 99.

Macpherson, R. I., Walker, R. D. & Kowall, M. H. (1973) Essential osteolysis with nephropathy. *Journal of the Canadian Association of Radiologists*, **24**, 98.

Shurtleff, D. B., Sprakes, R. S., Clawson, K., Guntheroth, W. G. & Mottet, N. K. (1964) Hereditary osteolysis with hypertension and nephropathy. *Journal of American Medical Association*, **188**, 363.

Thieffry, S. & Sorrel-Dejerine, J. (1958) Forme speciale d'osteolyse essentielle hereditaire et familiale à stabilisation spontanée, survenant dans l'enfance. *Presse Médicale*, **66**, 1858.

Torg, J. S. & Steel, H. H. (1968) Essential osteolysis with nephropathy: a review of the literature and case report of an unusual syndrome. *Journal of Bone and Joint Surgery*, **50**, 1629.

Winchester, P., Grossman, H., Lim, W. N. & Danes, B. S. (1969) A new acid mucopolysaccharidosis with skeletal deformities simulating rheumatoid arthritis. *American Journal of Roentgenology, Radium Therapy and Nuclear Medicine*, **106**, 121.

Multicentric osteolysis

Canun, S., Torres, P., Del Castillo, V. & Carnevale, A. (1976) Hereditary osteolysis with dominant transmission. *Excerpta medica: Fifth International Congress of Human Genetics*, p.63. Mexico.

Gorham, L. W. & Stout, A. P. (1955) Massive osteolysis. *Journal of Bone and Joint Surgery*, **37A**, 985.

Sage, M. R. & Allen, P. W. (1974) Massive osteolysis. *Journal of Bone and Joint Surgery*, **56B**, 130.
Torg, J. S., De George, A. M., Kirkpatrick, J. A. & Trujillo, M. M. (1969) Hereditary multicentric osteolysis with recessive transmission: a new syndrome. *Journal of Pediatrics*, **75**, 243.

Hypophosphatasia

Benzie, R., Doran, T. A., Escoffery, W., Gardner, H. A., Hoar, D. I. Hunter, A., Malone, R., Miskin, M. & Rudd, N. L. (1976) Prenatal diagnosis of hypophosphatasia. *Birth Defects: Original Article Series*, **12/6**, 271.
Danovitch, S. H., Baer, P. N. & Laster, L. (1968) Intestinal alkaline phosphatase activity in familial hypophosphatasia. *New England Journal of Medicine*, **278**, 1253.
Gorodischer, R., Davidson, R. G., Mosovich, L. L. & Yaffe, S. J. (1976) Hypophosphatasia: a developmental anomaly of alkaline phosphatase? *Pediatric Research*, **10/7**, 650.
Kozlowski, K., Sutcliffe, J., Barylak, A., Harrington, G., Kemperdick, H. & Nolte, K. (1976) Hypophosphatasia — a review of 24 cases. *Pediatric Radiology*, **5**, 103.
Macpherson, R. I., Krocker, M. & Houston, C. S. (1972) Hypophosphatasia. *Journal de l'association Canadïenne des Radiologistes*, **23**, 16.
Mehes, K., Klujiber, L., Lassu, G. & Kajtar, P. (1972) Hypophosphatasia: screening and family investigations in an endogamous Hungarian village. *Clinical Genetics*, **3**, 60.
Pimstone, B., Eisenberg, E. & Silverman, S. (1966) Hypophosphatasia: genetic and dental studies. *Annals of Internal Medicine*, **65**, 722.
Rathbun, J. C., MacDonald, J. W., Robinson, H. M. C. & Wanklin, J. M. (1961) Hypophosphatasia: a genetic study. *Archives of Diseases in Childhood*, **36**, 540.
Rubecz, I., Mehes, K., Klujber, L., Bozzay, L., Weisenbach, J. & Fenyvasi, J. (1974) Hypophosphatasia: screening and family investigation. *Clinical Genetics*, **6**, 155.
Silverman, J. L. (1962) Apparent dominant inheritance of hypophosphatasia. *Archives of Internal Medicine*, **110**, 191.
Svejcar, J. & Walther, A. (1975) The diagnosis of the early infantile form of hypophosphatasia tarda. *Humangenetika*, **28/1**, 49.

Vitamin-D resistant rickets

Albright, F., Butler, A. M. & Bloomberg, E. (1937) Rickets resistant to vitamin D therapy. *American Journal of Diseases of Children*, **54**, 529.
Arnaud, C. D., Maijer, R., Reade, T., Scriver, C. R. & Whelan, D. T. (1970). Vitamin D dependency: an inherited post-natal syndrome with secondary hyperparathyroidism. *Pediatrics*, **46**, 871.
Birtwell, W. M., Magsamen, B. F., Fenn, P. A., Torg, J. S., Tourtellotte, C. D. & Martin (1970) An unusual hereditary osteomalacic disease: pseudo-vitamin D deficiency. *Journal of Bone and Joint Surgery*, **52A**, 1222.
Burnett, C. H., Dent, C. E., Harper, C. & Warland, B. J. (1964) Vitamin D-resistant rickets. Analysis of twenty-four pedigrees with hereditary and sporadic cases. *American Journal of Medicine*, **36**, 222.
Cantu, J. M. (1974) Autosomal recessive nonhyperaminoaciduric vitamin D-dependent rickets. *Birth Defects: Original Article Series*, **10/4**, 294.
Dent, C. E., Friedman, M. & Watson, L. (1968) Hereditary pseudo-vitamin D deficiency rickets. *Journal of Bone and Joint Surgery*, **50B**, 708.
Graham, J. B., McFalls, V. W. & Winters, R. W. (1959) Familial hypophosphataemia with vitamin D-resistant rickets, II. Three additional families of the sex-linked dominant type with a genetic analysis of five such families. *American Journal of Human Genetics*, **11**, 311.
Harrison, H. E., Harrison, H. C., Lifshitz, F. & Johnson, A. D. (1966) Growth disturbance in hereditary hypophosphataemia. *American Journal of Diseases of Children*, **112**, 290.
McCance, R. A. (1947) Osteomalacia with Looser's nodes (Milkman's syndrome) due to raised resistance to vitamin D acquired about the age of 15 years. *Quarterly Journal of Medicine*, **16**, 33.
Pak, C. Y. C., Deluca, H. F., Bartter, F. C., Henneman, D. H., Frame, B., Simopoulos, A. & Delea, C. S. (1972) Treatment of vitamin D-resistant rickets with 25-hydroxycholecalciferol. *Archives of Internal Medicine*, **129**, 894.
Stamp, T. C. & Baker, L. R. I. (1976) Recessive hypophosphataemic rickets and possible aeteology of the Vitamin D-resistant syndrome. *Archives of Diseases of Childhood*, **51**, 360.

Pseudoparathyroidism

Albright, F., Burnett, C. H., Smith, P. H. & Parson, W. (1942) Pseudohypoparathyroidism: example of 'Seabright-bantam syndrome'; a report of three cases. *Endocrinology*, **30**, 922.
Brito Suarez, M., Herendez, C. & de la Rosa, J. (1975) Pseudohyperparathyroidism: three familial cases of Albright's hereditary osteodystrophy. *Revista española de reumatismo y enfermedades osteoarticulares*, **18/2**, 99.

Cederbaum, S. D. & Lippe, B. M. (1973) Probable autosomal recessive inheritance in a family with Albright's hereditary osteodystrophy and an evaluation of the genetics of the disorder. *American Journal of Human Genetics*, **25**, 638.

Drezner, M., Neelson, F. A. & Lebovitz, H. E. (1973) Pseudohypoparathyroidism type II: a possible defect in the reception of the cyclic AMP signal. *New England Journal of Medicine*, **289**, 1056.

Mann, J. B., Alterman, S. & Hill, A. G. (1962) Albright's hereditary osteodystrophy comprising pseudohypoparathyroidism and pseudo-pseudohypoparathyroidism, with a report of two cases representing the complete syndrome occurring in successive generations. *Annals of Internal Medicine*, **56**, 315.

Weinberg, A. G. & Stone, R. T. (1972) Autosomal dominant inheritance in Albright's hereditary osteodystrophy. *Journal of Pediatrics*, **79**, 996.

6. Osteoscleroses

The osteoscleroses share the feature of increased skeletal density, with little or no disturbance of the bony contours. The various forms of osteopetrosis are the commonest and most important osteoscleroses, but in view of the nature of the osseous changes, pycnodysostosis is conventionally included in this category.

1. Osteopetrosis — autosomal dominant, benign or tarda form
2. Osteopetrosis — autosomal recessive, malignant or congenita form
3. Pycnodysostosis.

The terms 'osteopetrosis' and 'Albers-Schönberg disease' are often used loosely and erroneously for any of the numerous osteoscleroses, craniotubular dysplasias and craniotubular hyperostoses. In the strict sense, osteopetrosis is applicable only to the specific autosomal dominant and autosomal recessive conditions which are described in this section. The nomenclature of the osteopetroses has been a source of confusion for many years. In the first half of this century, terminological problems were compounded by the use of the designations 'marble bones' (Schulze, 1921) and 'osteosclerosis fragilis generalisata' (Laurell and Wallgren, 1920). The situation was improved when Karshner (1926) proposed the term 'osteopetrosis' and clarified when McPeak (1936) recognised the existence of malignant and benign forms of the condition.

Genetic osteopetrosis occurs in laboratory animals, and experiments involving rodents are opening up promising lines of research into the pathogenesis and treatment of this group of disorders (Cotton et al., 1976; Loutit and Sansom, 1976; Marks, 1976). The histopathological changes in chondro-osseous tissue from patients with various forms of hyperostotic bone dysplasias have been reviewed by Kaitila and Rimoin (1976). The histological appearances are indicative of disparity in the basic defects in these conditions.

1. OSTEOPETROSIS — AUTOSOMAL DOMINANT, BENIGN OR TARDA FORM

Albers-Schönberg (1904) gave his name to posterity when he described a 26 year-old-male in whom an unusual increase in bone density had been detected radiologically. In subsequent reports of the same patient, Reiche (1915) used the term 'osteosclerosis' while Lorey and Reye (1923) employed the eponym 'Albers-Schönberg disease'. This individual was eventually lost to medical authors when he died at the age of 49. As the mother of Albers-Schönberg's original patient was affected, it is reasonable to reserve this eponym for the dominant form of the disorder. This form of osteopetrosis is comparatively common, with a wide ethnic and geographic distribution. More than 400 case reports have now appeared in the literature.

Clinical and radiographic features

Affected individuals may remain totally asymptomatic and the diagnosis is often reached by chance when radiographs are taken for some unrelated purpose. The facies, physique, mentality and lifespan are normal and general health is unimpaired. A mild anaemia is an infrequent complication. In a proportion of patients, the presenting feature is facial palsy or deafness, consequent upon cranial nerve compression by bony overgrowth. Pathological fractures due to bone fragility may occur, tooth extraction is sometimes difficult and osteomyelitis of the mandible occasionally develops.

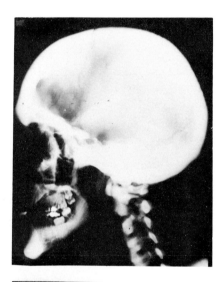

Fig. 6.1. Osteopetrosis — AD type; skull radiograph showing widening and increased density in the base and calvarium. (From Beighton, P., Horan, F. T. & Hamersma, H. (1977) *Postgraduate Medical Journal*, **53**, 507.)

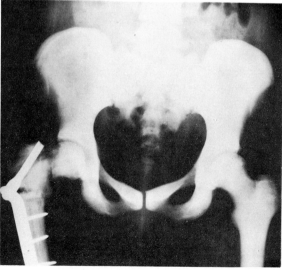

Fig. 6.2. Osteopetrosis — AD type; radiograph showing increased skeletal density without disturbance in bone contours. Fractures of the upper femoral region have been stabilised by fixation. In some patients the bones are fragile, while others have no problems of this type.

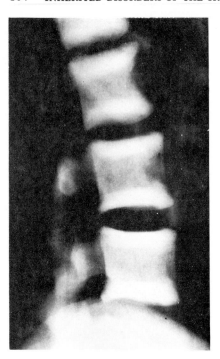

Fig. 6.3. Osteopetrosis —AD type; the 'rugger jersey' spine. Sclerosis of the end plates of the vertebral bodies produces a banded appearance. (From Beighton, P., Horan, F. T. & Hamersma, H. (1977) *Postgraduate Medical Journal*, **53**, 507.)

Radiologically, bone sclerosis becomes increasingly apparent as childhood progresses. Sclerotic foci, termed 'endobones' or 'bones within bone', are a striking feature. These changes usually disappear by the end of the second decade. Bone involvement is widespread but certain regions, particularly the extremities, are sometimes spared. The calvarium is dense and the sinuses may be obliterated. In the spine, thickening of the vertebral end plates gives rise to the characteristic 'rugger jersey' appearance. (The jersey worn by rugby players traditionally carries transverse bands as this configuration produces an illusion of increased body bulk, thereby disconcerting the opposition!)

Genetics

Johnston *et al*. (1968) reviewed 19 kindreds with 85 affected individuals and demonstrated that inheritance was autosomal dominant. There is considerable interfamilial variation and there is little doubt that the condition is heterogeneous. Although the manifestations are usually consistent within a particular kindred, there have been reports of anomalous situations, in which there has been disparity in the degree to which members of successive generations have been affected (Thomson 1949). The gene may occasionally be non-penetrant and skipped generations have been described. Johnston *et al*. (1968) concluded that dominant osteopetrosis was a single entity with variable manifestations. However, as case reports accumulate, the evidence lends support to the concept of heterogeneity. The dominant form of osteopetrosis has been recognised antenatally in a fetus during X-ray pelvimetry in late pregnancy. The diagnosis was confirmed in the mother at the same examination (Delahaye *et al*., 1976).

2. OSTEOPETROSIS — AUTOSOMAL RECESSIVE, MALIGNANT OR CONGENITA FORM

The autosomal recessive type of osteopetrosis is much less common than the autosomal dominant form. Although the radiographic changes are similar in these entities, their clinical features differ, and distinction is usually not difficult.

Clinical and radiographic features

The manifestations are evident during infancy. Bony overgrowth is associated with marrow dysfunction and presenting symptoms include failure to thrive, spontaneous bruising, abnormal bleeding and anaemia. The teeth become carious and hepatosplenomegaly develops. The bones are fragile and pathological fractures are a frequent complication. Palsies of the optic, oculomotor and facial nerves may occur in the later stages. Death from overwhelming infection or haemorrhage usually takes place in the first decade.

Generalised bone sclerosis is the predominant radiological feature. Penetrated films of the tubular bones reveal transverse bands in the metaphyseal regions and longitudinal striations in the shafts. The vertebrae show the classical 'rugger jersey' appearance and endobones are evident in the axial skeleton. As the condition progresses, the proximal humerus and distal femur develop a flask-shaped configuration. The skull becomes progressively thickened, with encroachment upon the foramina of the cranial nerves.

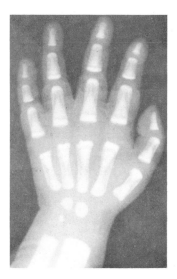 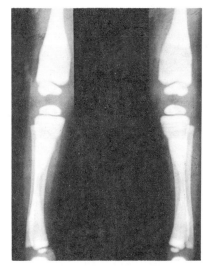

Fig. 6.4. (left) Osteopetrosis — AR type; endobones or 'bones within a bone' are evident in the metacarpals and phalanges.

Fig. 6.5. (right) Osteopetrosis — AR type; the lower ends of the femora have a club-like configuration. Transverse bands and longitudinal striations are seen in their shafts.

Genetics

There is little doubt that this form of osteopetrosis is inherited as an autosomal recessive. Affected sibs with normal parents have been observed in an inbred kindred (Enell and Pehrson, 1958), and parental consanguinity has been recorded (Tips and Lynch, 1962). The genes for one form of autosomal recessive osteopetrosis in the mouse have been assigned to chromosome 12 (Marks and Lane, 1976). As yet, there is no comparable human information.

A distinct autosomal recessive entity, in which osteosclerosis of moderate severity was associated with renal tubular acidosis, has been observed in three sibs (sly et al., 1972). This condition has also been encountered in two brothers with unaffected consanguineous parents (Guiband et al., 1972).

3. PYCNODYSOSTOSIS

Maroteaux and Lamy (1962) defined the characteristics of pycnodysostosis and established it as an entity in its own right. Previously, the predominant clinical feature of shortness of stature led to confusion with other types of dwarfism, while the generalised skeletal sclerosis and clavicular hypoplasia prompted some authors to regard the condition as a form of osteopetrosis or cleidocranial dyaplasia (Palmer and Thomas, 1958). However, the distinctive clinical and radiographic features permit accurate diagnosis. Recognition at a histological level is also possible, as ultrastructural studies of cartilage have revealed abnormal inclusions in the chondrocytes (Stanescu, Stanescu and Maroteaux, 1975). The kinetics of calcium metabolism in pycnodysostosis have been discussed by Cabrejas et al. (1976).

Clinical and radiographic features

Individuals with pycnodysostosis resemble each other. They have small faces with a hooked nose, receding chin and carious misplaced teeth. The cranium bulges and the anterior fontanelle remains patent. The terminal phalanges are short, with dysplasia of the fingernails. Bony fragility predisposes to spontaneous fracture (Roth, 1976). Other less consistent skeletal changes include narrowing of the thorax and spinal deformity. Adult height does not usually exceed 150 cm.

The impressionist painter Toulouse-Lautrec is thought to have had pycnodysostosis (Maroteaux and Lamy, 1965). Indeed, his appearance and medical history serve as a useful 'aide-memoire' to the manifestations of the disorder. It is well-known that he was of short stature, and the 'stove-pipe' hat which he habitually wore might have covered a patent fontanelle. Similarly, his beard may have been grown to conceal a receding chin. The stick which he carried is a reminder of bone fragility; he suffered two femoral fractures in childhood as a result of minor trauma. Finally, the fact that his aristocratic parents were first cousins is in keeping with the autosomal recessive inheritance of the condition.

Bone sclerosis becomes radiographically apparent in childhood and increases throughout the years of growth. Skeletal modelling and bony contours are undisturbed and neither striations nor endobones are seen. The calvarium is not particularly dense but patency of the fontanelles and the presence of multiple Wormian bones can usually be demonstrated. The facial bones and paranasal sinuses are hypoplastic and the angle of the mandible is obtuse. The terminal phalanges are

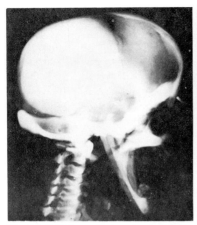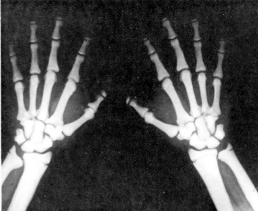

Fig. 6.6. (left) Pycnodysostosis; skull radiograph showing increased density of the base, wide fontanelles, hypoplasia of the sinuses and an obtuse angle to the mandibular ramus.

Fig. 6.7. (right) Pycnodysostosis; hand radiograph from an affected child. Although the bones are dense, their outlines are undisturbed. The terminal phalanges are shortened and irregular.

shortened, with distal irregularity, similar to that encountered in acro-osteolysis. The clavicles may be gracile, with underdevelopment of their lateral portions.

Genetics

Pycnodysostosis is inherited as an autosomal recessive. Following a review of the literature, Sedano, Gorlin and Anderson (1968) estimated that about 30 per cent of patients were the offspring of consanguineous unions. The majority of case descriptions have emanated from Europe and the USA. However, pycnodysostosis has been encountered in African Negroes (Palmer 1960; Wolpowitz and Matison, 1974), and in Japan (Siguira, Yamado and Koh, 1974). More than 50 kindreds have been reported (Diwan and Gogate, 1974).

An affected individual with a deletion of the short arm of a G group chromosome, probably chromosome 22, led Elmore *et al.* (1966) to speculate that the abnormal gene might be located at that particular chromosomal site. This finding has not been confirmed.

Danks *et al.* (1974) reported a six year old boy who had skeletal abnormalities which resembled those of pycnodysostosis, together with patches of dermal sclerosis and necrosis, and distinctive dental abnormalities. The authors considered that this condition was probably a distinct entity.

REFERENCES

Osteoscleroses – preamble

Cotton, W. R., Williams, G. A., Hargis, G. K. & Gaines, J. F. (1976) Parathyroid hormone as a possible causal factor in osteopetrosis of the TL rat. *Endocrinology*, **99**/3, 872.

Kaitila, I. & Rimoin, DL. L. (1976) Histologic heterogeneity in the hyperostotic bone dysplasias. *Birth Defects: Original Article Series*, **12**/6, 71.

Karshner, R. G. (1900) Osteopetrosis. *American Journal of Roentgenology, Radium Therapy and Nuclear Medicine*, **16**, 405.

Laurell, H. & Wallgren, A. (1900) Untersuchungen uber einen Fall einer eigenartigen Skeletterkrankung (osteosclerosis fragilis generalisata). *Upsala Läkareförenings Föhrandlingar*, **25**, 309.

Loutit, J. F. & Sansom, J. M. (1976) Osteopetrosis of micropthalmic mice — a defect of the hematopoietic stem cell? *Calcified Tissue Research*, **20/3**, 251.

Marks, S. C. (1976) Osteopetrosis in the IA rat cured by spleen cells from a normal littermate. *American Journal of Anatomy*, **146/3**, 331.

McPeak, C. N. (1936) Osteopetrosis. Report of eight cases occurring in three generations of one family. *American Journal of Roentgenology, Radium Therapy and Nuclear Medicine*, **36**, 816.

Schulze, F. (1921) Das Wesen des Krankheitsbildes der 'Marmorknochen' (Albers–Schönberg). *Archiv für klinische Chirugie*, **118**, 411.

Osteopetrosis, AD form

Albers–Schönberg, H. (1904) Röntgenbilder einer seltenen Knochenerkrankung. *Münchener medizinische Wochenschrift*, **51**, 365.

Johnston, C. C., Jr, Lavy, N., Lord, T., Vellios, F., Merritt, A. D. & Deiss, W. P., Jr. (1968) Osteopetrosis. A clinical, genetic, metabolic, and morphologic study of the dominantly inherited, benign form. *Medicine*, **47**, 149.

Lorey, A. & Reye, B. (1923) Über Marmorknochen (Albers–Schönbergsche krankheit). *Fortschritte auf dem Gebiete der Röntgenstrahlen und der Nukearmedizin*, **30**, 35.

Thomson, J. (1949) Osteopetrosis in successive generations. *Archives of Diseases in Childhood*, **24**, 143.

Osteopetrosis, AR form

Enell, H. & Pehrson, M. (1959) Studies on osteopetrosis. I. Clinical report of three cases with genetic considerations. *Acta paediatrica*, **47**, 279.

Guibaud, P., Labre, F., Freycon, M. T. & Genoud, J. (1972) Osteopetrose et acidose renale tubulaire. Deux cas de cette association dans une fratrie. *Archives Francaises de Pédiatrie*, **29**, 269.

Marks, S. C. & Lane, P. W. (1976) Osteopetrosis: a new recessive mutation on chromosome 12 of the mouse. *Journal of Heredity*, **67**, 11.

Sly, W. S., Lang, R., Avioli, L., Haddad, J., Lubowitz, H. & McAlister, W. (1972) Recessive osteopetrosis: a new clinical phenotype. *American Journal of Human Genetics*, **24**, 34.

Tips, R. L. & Lynch, H. T. (1962) Malignant congenital osteopetrosis resulting from a consanguineous marriage. *Acta paediatrica*, **47**, 279.

Pycnodysostosis

Cabrejas, M. L., Fromm, G. A., Roca, J. F. (1976) Pycnodysostosis. Some aspects concerning kinetics of calcium metabolism and bone pathology. *American Journal of Medical Science*, **271/2**, 215.

Danks, D. M., Mayne, V., Wettenhall, H. N. B. & Hall, R. K. (1974) Craniomandibular dermatodysostosis. *Birth Defects: Original Article Series*, **10/12**, 99.

Diwan, R. V. & Gogate, A. N. (1974) Pycnodysostosis (first report of a family from India). *Indian Journal of Radiology*, **28**, 268.

Elmore, S. M., Nance, W. E., McGee, B. J., Engel-de Montmollin, M. & Engel, E. (1966) Pycnodysostosis, with a familial chromosome anomaly. *American Journal of Medicine*, **40**, 273.

Maroteaux, P. & Lamy, M. (1965) The malady of Toulouse-Lautrec. *Journal of the American Medical Association*, **191**, 715.

Maroteaux, P. & Lamy, M. (1962) La pycnodysostose. *Presse Médicale*, **70**, 999.

Palmer, P. E. S. & Thomas, J. E. P. (1958) Osteopetrosis with unusual changes in the skull and digits. *British Journal of Radiology*, **31**, 705.

Palmer, P. E. S. (1960) Osteopetrosis with multiple epiphyseal dysplasia. *British Journal of Radiology*, **33**, 455.

Roth, V. G. (1976) Pycnodysostosis presenting with bilateral subtrochanteric fractures: case report. *Clinical Orthopaedics and Related Research*, **117**, 247.

Sedano, H. D., Gorlin, R. J. & Anderson, V. E. (1968) Pycnodysostosis. Clinical and genetic considerations. *American Journal of Diseases in Children*, **116**, 70.

Stanescu, R., Stanescu, V. & Maroteaux, P. (1975) Ultrastructural abnormalities of chondrocytes in pycnodysostosis. *Nouveautés medicales*, **4/37**, 247.

Sugiura, Y., Yamado, Y. & Koh (1974) Pycnodysostosis in Japan. Report of six cases and a review of the Japanese literature. *Birth Defects: Original Article Series*, **10/12**, 78.

Wolpowitz, A. & Matisson, A. (1974) A comparative study of pycnodysostosis, cleidocranial dysostosis, osteopetrosis and acro-osteolysis. *South African Medical Journal*, **48**, 1011.

7. Craniotubular dysplasias

The craniotubular dysplasias are a group of disorders in which abnormal modelling of the skeleton is the predominant feature. Increased radiological density of bone may be present and if the cranium is involved, complications include facial distortion and cranial nerve compression.

1. Metaphyseal dysplasia (Pyle)
2. Craniometaphyseal dysplasia
3. Craniodiaphyseal dysplasia
4. Frontometaphyseal dysplasia
5. Dysosteosclerosis
6. Tubular stenosis (Kenny–Caffey)
7. Osteodysplasty (Melnick–Needles).

1. METAPHYSEAL DYSPLASIA (PYLE)

Metaphyseal dysplasia or Pyle disease is a rare autosomal recessive disorder which is often the subject of semantic confusion with the craniometaphyseal dysplasias (*vide infra*). However, the very marked disturbance in bone modelling and the lack of cranial sclerosis serves to distinguish Pyle disease from these conditions.

Clinical and radiographic features
Valgus deformity of the knees may be the only obvious abnormality, but muscular weakness, scoliosis, and bone fragility are sometimes present. In contrast to the mild clinical stigmata, the radiographic changes are striking. The tubular bones of the legs show gross 'Erlenmeyer flask' flaring, particularly in the distal portions of the femora. The long bones of the arms are also undermodelled and the cortices are generally thin. The skull is virtually normal, apart from a supraorbital prominence. The bones of the pelvis and thoracic cage are expanded. The manifestations have been reviewed by Gorlin, Koszalka and Spranger (1970).

Genetics
Although the dramatic radiographic changes are unmistakeable, the clinical features may be mild, and Pyle disease is probably underdiagnosed. The evidence indicates that the disorder is inherited as an autosomal recessive. Affected sibs featured in the original report of Pyle (1931) and other sets were mentioned by Bakwin and Krida (1937), Hermal, Gershon-Cohen and Jones (1953) and Feld *et al.* (1955). Parental consanguinity was present in the kindred described by Daniel (1960). Temtamy *et al.* (1974) reported two sisters who had the clinical and radiological stigmata of Pyle

disease, together with dermal lesions and optic atrophy. As the unaffected parents were consanguineous, the authors speculated that this entity was inherited as an autosomal recessive.

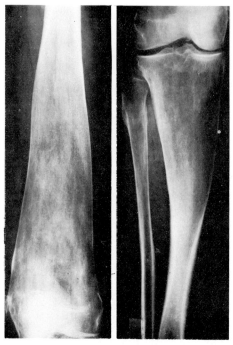

Fig. 7.1. (left) Pyle disease — the lower end of the femur is grossly expanded. In spite of the dramatic radiographic changes, there are few clinical manifestations.

Fig. 7.2. (right) Pyle disease; the tubular bones are undermodelled, and their cortices are thin.

2. CRANIOMETAPHYSEAL DYSPLASIA

The autosomal dominant form of craniometaphyseal dysplasia is relatively common in comparison with the other conditions in this group.

Clinical and radiographic features

Paranasal bossing develops during infancy and progressive expansion and thickening of the skull and mandible distort the jaw and face. These changes, which are very variable in degree, become static in the third decade. Paradoxically, the paranasal bossing diminishes with the passage of time. Bone encroachment leads to entrapment and dysfunction of the cranial nerves and some degree of facial palsy and deafness is usually present. Dental problems arise from malocclusion of the jaws and partial obliteration of the sinuses predisposes to recurrent nasorespiratory infection. The bones are not fragile and pathological fractures do not occur. Intelligence, height, general health and life span are normal.

The radiographic changes are age-related, usually becoming evident by the age of five. The main feature in the skull is sclerosis, which is maximal in the base, although

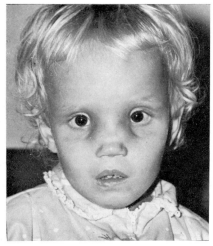

Fig. 7.3. (left) Craniometaphyseal dysplasia — AD type; an adult male with enlargement and asymmetry of the mandible. Deafness and facial palsy were additional problems.

Fig. 7.4. (right) Craniometaphyseal dysplasia — AD type; the daughter of the patient shown in Fig. 7.3. Paranasal bossing is a prominent feature. Curiously, this abnormality regresses in later life.

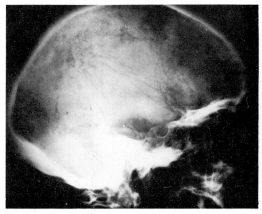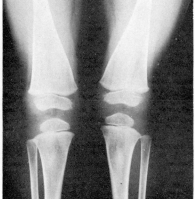

Fig. 7.5. (left) Craniometaphyseal dysplasia — AD type; a skull radiograph from the male shown in Fig. 7.3. The base is sclerotic.

Fig. 7.6. (right) Craniometaphyseal dysplasia — AD type; radiograph of the knees of the child shown in Fig. 7.4. The lower femoral metaphyses have a characteristic club-shaped configuration. (From Spiro, D. C., Hamersma, H. & Beighton, P. (1975) *South African Medical Journal*, **49**, 839.)

the cranium is always involved to some degree. The long bones have widened metaphyses and normal diaphyses, presenting a club-shaped configuration, particularly at the lower end of the femur. These changes are much less severe than those encountered in Pyle disease. Minor degrees of expansion and cortical thinning are evident in the ribs and clavicles, while the spine and pelvis are uninvolved.

5

Genetics

Semantic confusion with Pyle disease makes a literature survey difficult. Nonetheless, cases in successive generations can be recognised in the reports of Spranger, Paulsen and Lehmann (1965), Lejeune *et al*. (1966), Gladney and Monteleone (1970) and Stool and Caruso (1973). The author has personal knowledge of a kindred, with branches in England and South Africa, in which there all 11 patients in four generations. Male to male transmission has occurred in this family, and the pedigree is entirely consistent with autosomal dominant inheritance (Spiro, Hamersma and Beighton, 1975).

There have been a few reports of patients in which craniometaphyseal dysplasia has been inherited as an autosomal recessive. Lehman (1957) and Millard *el al*. (1967) described affected sibs born to normal parents, and Lievre and Fischgold (1956) mentioned parental consanguinity. The stigmata are of much greater severity and earlier onset than those of the dominant form. In particular, cranial nerve entrapment is universal and occurs at an early age. Facial distortion may be gross, even in childhood. All reports have concerned children or young adults and the long-term prognosis is as yet unknown.

3. CRANIODIAPHYSEAL DYSPLASIA

Craniodiaphyseal dysplasia was delineated by Joseph *et al*. (1958). As with the recessive form of craniometaphyseal dysplasia, earlier reports appeared in the literature under the designation 'leontiasis ossea' (Gemmell, 1935).

Clinical and radiographic features

Overgrowth of the skull results in grotesque deformation of the face. Bony encroachment leads to nasal obstruction and entrapment of the facial, auditory and optic nerves. Radiographic changes are maximal in the skull and mandible, where massive hyperostosis and sclerosis are evident. The ribs and clavicles are widened and the tubular bones are undermodelled. There is no metaphyseal flaring and the shape of the long bones has been likened to that of a policeman's truncheon.

Genetics

An affected child with consanguineous normal parents was studied by Halliday (1949). Isolated cases have been described by Joseph *et al*. (1958) and Stransky, Mabilangan and Lara (1962). It is reasonable to assume that inheritance is autosomal recessive. Macpherson (1974) reported three patients with widely disparate clinical and radiological features and emphasised the probable heterogeneity of craniodiaphyseal dysplasia.

4. FRONTOMETAPHYSEAL DYSPLASIA

Frontometaphyseal dysplasia (FMD) is yet another rare disorder which has recently been split off from the general category of craniotubular dysplasias.

Clinical and radiographic features

A prominent supraorbital ridge, which resembles a knight's visor, is the outstanding feature. In several published photographs the mandible appears to be hypoplastic, with anterior constriction. Dental anomalies are common, and deafness may develop in adulthood. Progressive contractures in the digits may simulate rheumatoid arthritis. General health is good and height is normal.

Radiographically, overgrowth of the supraorbital region is very marked. Sclerosis

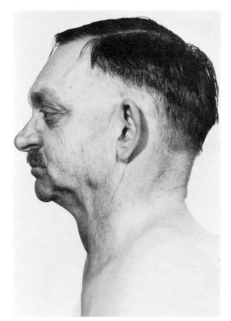

Fig. 7.7. Frontometaphyseal dysplasia; this patient presented with deafness at the age of 40. The supraorbital region is prominent and the mandible is constricted anteriorly.

 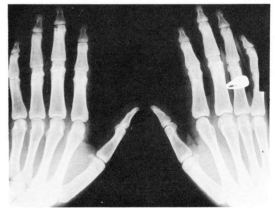

Fig. 7.8. (left) Frontometaphyseal dysplasia; skull radiograph showing massive thickening of the frontal region and patchy sclerosis throughout the skull.

Fig. 7.9. (right) Frontometaphyseal dysplasia; the tubular bones are undermodelled.

of the cranial vault is of mild degree, and may be patchy. The vertebral bodies are dysplastic but not sclerotic. The iliac crests are abruptly flared and the pelvic inlet is distorted. The femoral capital epiphyses are flattened with expansion of the femoral heads and a coxa valga deformity. Changes in the tubular bones are unremarkable. However, the bones of the fingers are undermodelled, and erosions and fusions may be present in the carpus.

Genetics

The majority of reports have concerned sporadic cases (Gorlin and Cohen, 1969; Holt, Thompson and Arenberg, 1972; Danks et al., 1972; Sauvegrain et al., 1975). There was doubt concerning the mode of transmission of the disorder until Weiss, Reynolds and Szymanowski (1975) described an affected mother and son and proposed that transmission was dominant. Kassner et al. (1977) produced further evidence to support this contention. However, Jarvis and Jenkins (1975) had encountered two mentally defective males who were born of different fathers to the same unaffected mother. This raises the question as to whether FMD is heterogeneous, existing in autosomal dominant and X-linked forms. As the majority of the previously reported sporadic cases have been males, the possibility of an X-linked type of frontometaphyseal dysplasia certainly merits serious consideration.

The detection of metachromasia in cultured fibroblasts from a patient suggests that a metabolic defect may be present (Danks et al., 1972). If similar changes are present in amniotic fluid cells, antenatal diagnosis may be possible.

5. DYSOSTEOSCLEROSIS

Dysosteosclerosis is a rare disorder which was delineated by Spranger et al. (1968). The manifestations have been reviewed by Liesti et al. (1975).

Clinical and radiographic features

Published reports have concerned children with short stature, fragile bones, enamel hypoplasia and dermal macular atrophy. Overgrowth of the skull may lead to cranial nerve compression. The clinical status of the affected adult is unknown. Sclerosis of the skull and axial skeleton, together with platyspondyly, are the major radiographic features. The long bones may be bowed, with metaphyseal expansion and epiphyseal sclerosis.

Genetics

Affected sibs with consanguineous parents are recognisable in the case reports of Ellis (1934) and Field (1939). Spranger et al. (1968) also recorded parental consanguinity. Inheritance is probably autosomal recessive.

6. TUBULAR STENOSIS

Kenny and Linnarelli (1966) reported a mother and son with proportionate dwarfism, low birth weight, delayed closure of the anterior fontanelle, narrowing of the tubular bones, hypocalcaemia and tetany. The radiographic changes in these patients were subsequently reviewed by Caffey (1967). The diameter of the shafts of the tubular bones was reduced, while the metaphyses were relatively flared. The external contours of these bones were irregular and the cortices were thickened, with narrowing of the medullary cavities.

Although only a single kindred has been described, tubular stenosis has achieved syndromic status. The mother to son transmission is consistent with dominant inheritance, either autosomal or X-linked.

7. OSTEODYSPLASTY

Less than 20 individuals with osteodysplasty have been reported. These include members of two kindreds investigated by Melnick and Needles (1966) and sporadic cases described by Coste, Maroteaux and Chouraki (1968), Wendler and Kellerer (1975) and Stoll *et al.* (1976). The manifestations of the condition were reviewed by Leiber *et al.* (1975). The designation 'osteodysplasty' pertains to the generalised nature of the skeletal dysplasia and it is intended to convey the meaning of 'badly formed'.

Clinical and radiographic features

The forehead is prominent and the mandible is small. Variable skeletal malformations include kyphoscoliosis, genu valgum and shortening of the distal phalanges. Deafness and degenerative osteoarthropathy may develop in adulthood. Radiographically, irregular ribbon-like constrictions of the ribs and tubular bones are a striking feature. The pelvis is distorted and coxa valga is present. The base of the skull may be thickened and patchy areas of sclerosis are seen in the cortices of the long bones.

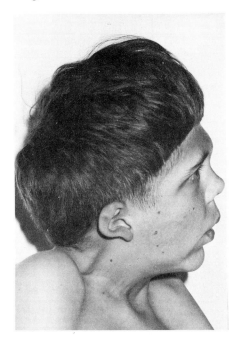

Fig. 7.10. Osteodysplasty; a 10-year-old boy with the characteristic prominent forehead and hypoplastic mandible.

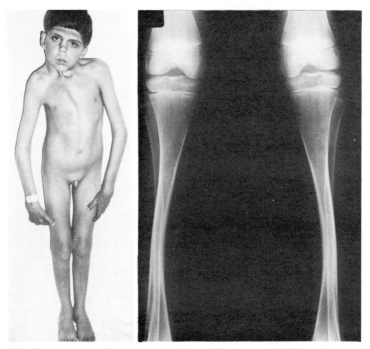

Fig. 7.11. (left) Osteodysplasty; short stature, spinal malalignment, torticollis and broadening of the thumbs.

Fig. 7.12. (right) Osteodysplasty; the skeleton is undermineralised and the bones of the shins have a wavy appearance.

Genetics

Transmission through four generations of one kindred and three generations of another was reported by Melnick and Needles (1966), and it seems likely that osteodysplasty is inherited as an autosomal dominant. In a personal conversation, Professor Jürgen Spranger proposed that osteodysplasty and at least one form of frontometaphyseal dysplasia might be the same entity.

A severe type of osteodysplasty, which is lethal in infancy, has been encountered in two Albanian sibs and a Polish girl. Danks, Mayne and Kozlowski (1974) designated this condition 'a precocious form of osteodysplasty', and suggested that inheritance was autosomal recessive.

Two brothers with osteodysplasty and mental retardation were reported by Ruvalcaba, Reichert and Smith (1971). The condition was partially expressed in two female maternal cousins and the authors raised the possibility of X-linked inheritance, with manifestations in the heterozygous female.

The genetic status and frequency of the craniotubular dysplasias

	Inheritance	Approximate number of reported cases
Metaphyseal dysplasia (Pyle)	AR	20
Craniometaphyseal	AD	50+
dysplasia	AR	10
Frontometaphyseal dysplasia	AD?	20
Dysosteosclerosis	AR	10
Tubular stenosis	AD	2
Osteodysplasty	AD	20
	AR form?	

REFERENCES

Metaphyseal dysplasia

Bakwin, H. & Krida, A. (1937) Familial metaphyseal dysplasia. *American Journal of Diseases of Children*, 53, 1521.

Daniel, A. (1960) Pyle's disease. *Indian Journal of Radiology*, 14, 126.

Feld, H., Switzer, R. A., Dexter, M. W. & Langer, E. W. (1955) Familial metaphyseal dysplasia. *Radiology*, 65, 206.

Gorlin, R. J., Koszalka, M. F. & Spranger, J. (1970) Pyle's disease (familial metaphyseal dysplasia). *Journal of Bone and Joint Surgery*, 52A, 347.

Hermel, M. B., Gershon-Cohen, J. & Jones, D. T. (1953) Familial metaphyseal dysplasia. *American Journal of Roentgenology, Radium Therapy and Nuclear Medicine*, 70, 413.

Pyle, E. (1931) Case of unusual bone development. *Journal of Bone and Joint Surgery*, 13, 874.

Temtamy, S. A., El-Meligy, M. R., Badrawy, H. A., Meguid, M. S. A. & Safwat, H. M. (1974) Metaphyseal dysplasia, anetoderma and optic atrophy: an autosomal recessive syndrome. *Birth Defects: Original Article Series*, 10/12, 61.

Craniometaphyseal dysplasia

Gladney, J. H. & Monteleone, P. L. (1970) Metaphyseal dysplasia. *Lancet*, 2, 44.

Lehmann, E. C. H. (1957) Familial osteodystrophy of the skull and face. *Journal of Bone and Joint Surgery*, 39B, 313.

Lejeune, E., Anjou, A., Bouvier, M., Robert, J., Vauzelle, J. L. & Jeanneret, J. (1966) Dysplasie cranio-metaphysaire. *Revue du Rhumatisme et des Maladies osteo-articulares*, 33, 714.

Lievre, J. A. & Fischgold, H. (1956) Leontiasis ossea chez l'enfant (osteopetrose partielle probable). *Presse médicale*, 64, 763.

Millard, D. R., Maisels, D. D., Batstone, J. H. F. & Yates, B. W. (1967) Craniofacial surgery in craniometaphyseal dysplasia. *American Journal of Surgery*, 113, 615.

Spiro, P. C., Hamersma, H. & Beighton, P. (1975) Radiology of the autosomal dominant form of craniometaphyseal dysplasia. *South African Medical Journal*, 49, 839.

Spranger, J., Paulsen, K. & Lehmann W. (1965) Die kraniometaphysaere Dysplasie (Pyle). *Zeitschrift für Kinderheilkunde*, 93, 64.

Stool, S. E. & Caruso, V. G. (1973) Cranial metaphyseal dysplasia. *Archives of Otolaryngology*, 97, 410.

Craniodiaphyseal dysplasia

Gemmell, J. H. (1935) Leontiasis ossea: a clinical and roentgenological entity. Report of a case. *Radiology*, 25, 723.

Halliday, J. (1949) Rare case of bone dystrophy. *British Journal of Surgery*, 37, 52.

Joseph, R., Lefebvre, J., Guy, E. & Job, J. C. (1958) Dysplasia cranio-diaphysaire progressive. Ses relations avec la dysplasie diaphysaire progressive de Camurati-Engelmann. *Annals of Radiology*, 1, 477.

Macpherson, R. I. (1974) Craniodiaphyseal dysplasia, a disease or group of diseases? *Journal of the Canadian Association of Radiologists*, 25, 22.

Stransky, E., Mabilangan, L. & Lara, R. T. (1962) On Paget's disease with leontiasis ossea and hypothyreosis, starting in early childhood. *Annals of Paediatrics*, 199, 399.

Frontometaphyseal dysplasia

Danks, D. M., Mayne, C., Hall, R. K. & McKinnon, M. C. (1972) Frontometaphyseal dysplasia. A progressive disease of bone and connective tissue. *American Journal of Diseases of Children*, 123, 254.

Gorlin, R. J. & Cohen, M. M. (1969) Frontometaphyseal dysplasia. A new syndrome. *American Journal of Diseases of Children*, 118, 487.

Holt, J. F., Thompson, G. R. & Arenberg, I. K. (1972) Fronto-metaphyseal dysplasia. *Radiology Clinics of North America*, 10, 225.

Jarvis, G. A. & Jenkins, E. C. (1975) In *Syndrome Identification*, Volume 3, Number 1, p.18. Ed. Bergsma. The National Foundation – March of Dimes, New York.

Kassner, E. G., Haller, J. O., Reddy, V. H., Mitarotundo, A. & Katz, I. (1977) Frontometaphyseal dysplasia: evidence for autosomal dominant inheritance. *American Journal of Roentgenology*, 127, 927.

Sauvegrain, J., Lombard, M., Garel, L. & Truscelli, D. (1975) Frontometaphyseal dysplasia. *Annals of Radiology*, 18/2, 155.

Weiss, L., Reynolds, W. A. & Szymanowski, R. T. (1975) Familial frontometaphyseal dysplasia: evidence for dominant inheritance. *Birth Defects: Original Article Series*, 11/5, 55.

Dysosteosclerosis

Ellis, R. W. B. (1934) Osteopetrosis (marble bones: Albers-Schönberg's distase: osteosclerosis fragilis generalisata: congenital osteosclerosis). *Proceedings of the Royal Society of Medicine*, 27, 1563.

Field, C. E. (1939) Albers-Schönberg disease. An atypical case. *Proceedings of the Royal Society of Medicine*, 32, 320.

Liesti, J., Kaitila, I., Lachman, R. S., Asch, M. J. & Rimoin, D. L. (1975) Dysosteosclerosis. *Birth Defects: Original Article Series*, 11/6, 349.

Roy, C., Maroteaux, P., Kremp, L., Courtrecuise, V. & Alagille, D. (1968) Un nouveau syndrome osseux avec anomalies cutanées et troubles neurologiques. *Archives Francaises de Pédiatrie*, 25, 985.

Spranger, J. W., Albrecht, C., Rohwedder, H. J. & Wiedemann, H. R. (1968) Die Hysosteosklerose: eine Sonderform der generalisierten Osteosklerose. *Fortschritte auf dem Gebiete der Röntgenstrahlen und der Nuklearmedizin*, 109, 504.

Tubular stenosis

Caffey, J. P. (1967) Congenital stenosis of medullary spaces in tubular bones and calvaria in two proportionate dwarfs, mother and son, coupled with transitory hypocalcaemic tetany. *American Journal of Roentgenology, Radium Therapy and Nuclear Medicine*, 100, 1.

Kenny, F. M. & Linarelli, L. (1966) Dwarfism and cortical thickening of tubular bones. Transient hypocalcaemia in a mother and son. *American Journal of Diseases of Children*, 111, 201.

Osteodysplasty

Coste, F., Maroteaux, P. & Chouraki, L. (1968) Osteodysplasty (Melnick and Needles' syndrome). Report of a case. *Annals of Rheumatic Diseases*, 27, 360.

Danks, D. M., Mayne, C. & Kozlowski, K. (1974) A precocious autosomal recessive type of osteodysplasty. In *The Clinical Delineation of Birth Defects*, Number 19. Baltimore: Williams and Wilkins.

Leiber, B., Olbrich, G., Moelter, N. & Walther, A. (1975) Melnick–Needles syndrome. *Monatsschrift für Kinderheilkunde*, 123/9, 178.

Melnick, J. C. & Needles, C. F. (1966) An undiagnosed bone dysplasia. A family study of four generations and three generations. *American Journal of Roentgenology, Radium Therapy and Nuclear Medicine*, 97, 39.

Ruvalcaba, R. H., Reichert, A. & Smith, D. W. (1971) A new familial syndrome with osseous dysplasia and mental deficiency. *Journal of Pediatrics*, 79, 450.

Stoll, C. L., Levy, J. M., Gardea, A. & Weil, J. (1976) L'osteodysplastie. *Pediatrie*, 31/2, 195.

Wendler, H. & Kellerer, K. (1975) Osteodysplastic syndrome (Melnick and Needles). *Fortschritte auf dem Gebiete der Röntgenstrahlen und der Nuklearmedizin*, 122/4, 309.

8. Craniotubular hyperostoses

Overgrowth of bone, which leads to alterations of contours and increase in radiological density of the skeleton, is the predominant feature of the craniotubular hyperostoses. In this group of conditions, hyperostosis is present in the skull and other regions in varying degrees and combinations. As with the craniotubular dysplasias, confusion with the osteopetroses is a recurring theme.

1. Endosteal hyperostosis (van Buchem)
2. Sclerosteosis
3. Diaphyseal dysplasia (Camurati–Engelmann)
4. Infantile cortical hyperostosis (Caffey)
5. Osteoectasia with hyperphosphatasia
6. Osteitis deformans (Paget).

1. ENDOSTEAL HYPEROSTOSIS (VAN BUCHEM)

Using the term 'hyperostosis corticalis generalisata familiaris', van Buchem, Hadders and Ubbens (1955) described two sibs in Holland with cranial sclerosis and widening of the diaphyses of the long bones. Further reports followed (van Buchem *et al.*, 1962; van Buchem, 1971) and the accumulated information was eventually published as a monograph (van Buchem, Prick and Jaspar, 1975). The eponymous designation is widely accepted.

Clinical and radiographic features
Overgrowth and distortion of the mandible and brow become evident during the latter part of the first decade. Subsequently, entrapment of the cranial nerves leads to facial palsy and deafness. Two sibs mentioned in the original case report were mentally defective, but other patients have been of normal intelligence. The disorder is progressive, and optic nerve involvement may be a late complication. However, the lifespan is not compromised, stature is normal and the bones are not fragile.

Widening and sclerosis of the calvarium, cranial base and mandible are the major radiographic features. Endosteal thickening is present in the diaphyses of the tubular bones. The external configuration of these bones is relatively undisturbed.

Genetics
The disorder described by van Buchem *et al.* (1955, 1962) is undoubtedly autosomal recessive. None of these Dutch patients had affected parents or offspring. Parental consanguinity was present in one kindred, and a pair of sibs and a set of dizygous twins were encountered in other families.

There have been reports of an autosomal dominant variety of endosteal hyperostosis (Worth and Wollin, 1966; Maroteaux *et al.*, 1971; Scott and Gautby, 1974). Owen (1976) studied the condition in six members of three generations of a British family and Beals (1977) investigated seven individuals in three generations of a Canadian kindred. Although the radiological features are very similar to those of the autosomal recessive type, the clinical course is milder, and cranial nerve involvement is not a problem. Nevertheless, it may not be easy to assign a sporadic case to a specific genetic category and counselling must be circumspect.

2. SCLEROSTEOSIS

Truswell (1958) recognised the existence of this disorder when he described six patients in an article entitled, 'Osteopetrosis with syndactyly; a morphological variant of Albers-Schönberg disease'. The designation 'sclerosteosis' is now in general use, although it does not accurately describe the manifestations of the condition. About 30 cases have been reported, the majority among the Afrikaner community of South Africa (Beighton, Durr and Hamersma, 1976). The only other known patients are a sibship in New York (Higinbotham and Alexander, 1941) and a young woman in Japan (Sugiura and Yasuhara, 1975).

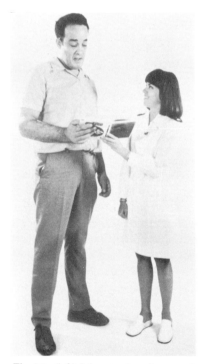 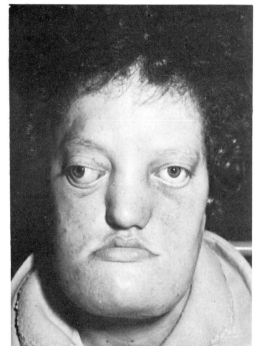

Fig. 8.1. (left) Sclerosteosis; tall stature is a notable feature. This young man is 214 cm in height.

Fig. 8.2. (right) Sclerosteosis; progressive bone thickening leads to distortion of the face and jaws, with proptosis, facial palsy and deafness. (From Beighton *et al.* (1977) *Clinical Genetics*, **11**, 1.)

Clinical and radiographic features

Progressive overgrowth and sclerosis of the skeleton, particularly the skull, develop in early childhood. Height and weight are often excessive. Indeed, a patient known to the author was a heavyweight boxing champion at the age of 14. The consistancy of his bones undoubtedly facilitated his pugilistic activities. Deafness and facial palsy due to cranial nerve entrapment may be a presenting feature. Distortion of the facies, which is apparent by the age of 10, eventually becomes very severe. In adulthood elevation of intracranial pressure may cause headache. Several adults have died suddenly from impaction of the brain stem in the foramen magnum. Cutaneous or bony syndactyly of the second and third fingers serves to distinguish sclerosteosis from the other disorders in this group. The terminal phalanges are deviated radially, with dystrophy of the finger nails. The bones are resistant to trauma and pathological fractures do not occur.

Gross widening and sclerosis of the skull is the predominant radiographic feature. Hypertrophy of the mandible and frontal regions leads to relative mid-facial hypoplasia. The vertebral bodies are spared although their pedicles are dense. The pelvic bones are sclerotic, but their contours are normal. The cortices of the long bones are sclerosed and hyperostotic. In distinction to endosteal hyperostosis, the tubular bones in sclerosteosis are markedly undermodelled with lack of the usual diaphyseal constriction (Beighton, Cremin and Hamersma, 1976).

Genetics

Sclerosteosis is inherited as an autosomal recessive. Consanguinity was present in five of 15 Afrikaner kindreds, into which a total of 25 affected individuals had been

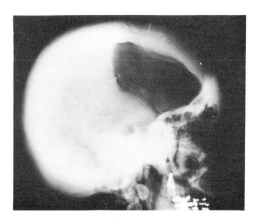

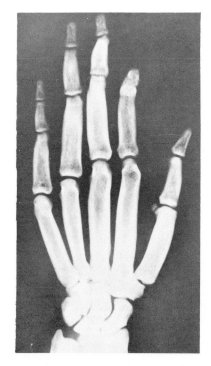

Fig. 8.3. (above) Sclerosteosis; the calvarium is sclerotic and hyperostotic. In this patient, craniotomy has been undertaken to relieve raised intracranial pressure. (From Beighton, P., Durr, L. & Hamersma, H. (1976) *Annals of Internal Medicine*, **84**, 393.)

Fig. 8.4. (right) Sclerosteosis; the tubular bones are undermodelled, with thickening of their cortices. (From Beighton, P., Cremin, B. J. & Hamersma, H. (1976) *British Journal of Radiology*, **49**, 934.)

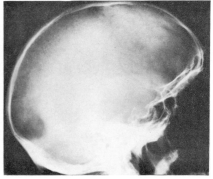

Fig. 8.5. (left) Sclerosteosis; the clinically normal heterozygote may have a minor degree of calvarial thickening.

Fig. 8.6. (right) Sclerosteosis; a normal skull, for comparison with Fig.8.5.

born (Beighton *et al.*, 1977). There were several sets of sibs in this series, but the parents were all normal. In this community, the minimum prevalance of sclerosteosis is 1:75,000, with a gene frequency of 0.0035. At least one in every 140 Afrikaners is a carrier of the gene and there are about 10,000 clinically normal heterozygotes in South Africa.

Calvarial thickening has been recognised in lateral skull radiographs from a number of obligatory heterozygotes. If this change proves to be a consistent feature, it could provide a basis for identification of the carrier of the gene. This finding may have important implications for the genetic counselling of the potentially heterozygous relatives of known patients.

3. DIAPHYSEAL DYSPLASIA (CAMURATI–ENGELMANN)

Diaphyseal dysplasia is comparatively well known and more than 100 cases have been reported. Although the eponym 'Camurati–Engelmann' disease is often used, McKusick (1975) contended that the disorder described by Camurati was a different condition and that the designation 'Engelmann disease' is more accurate. The clinical and genetic features have been reviewed by Sparkes and Graham (1972).

Clinical and radiographic features
Diaphyseal dysplasia presents in mid-childhood with muscular pain and weakness and wasting, typically in the legs. The condition is self-limiting and generally resolves by the age of thirty. Cranial nerve compression and raised intracranial pressure are occasional complications. The manifestations are variable and some patients are severely handicapped while others are virtually asymptomatic. There is no specific treatment, but Lindstrom (1974) reported that steroid therapy alleviated muscle pain and improved exercise tolerance in an affected 16 year old boy.

Marked thickening of the cortices of the leg bones is the predominant feature. The medullary canals are narrowed and the external bone contours are irregular. The changes are diaphyseal, and the metaphyses and epiphyses remain uninvolved. The long bones of the arms may be affected, but the extremities and the axial skeleton are usually spared. Infrequently, the skull is involved, with calvarial widening and basal sclerosis. As with the clinical features, the radiological changes are very variable.

Genetics
Autosomal dominant inheritance is well established and kindreds with cases in successive generations have been reported by Girdany (1959), Ramon and Buchner (1966), Allen *et al.* (1970) and Hundley and Wilson (1973). Thurmon and Jackson (1976) encountered dominantly transmitted Engelmann disease in the St Landry Mulattoes, a social isolate in Louisiana.

There is marked intrafamilial variation in severity. In some instances, phenotypic expression has been confined to minor radiological changes and complete non-penetrance of the gene has been recorded (Sparkes and Graham, 1972). The apparent excess of sporadic individuals, who might otherwise be assumed to represent new mutations, can be explained on a basis of this phenotypic inconsistency.

4. INFANTILE CORTICAL HYPEROSTOSIS

Infantile cortical hyperostosis, or Caffey disease, is an unusual but relatively well known disorder. Following the first report by Caffey and Silverman (1945), the features have been reviewed by Sherman and Hellyer (1950), Caffey (1957), Holman (1962) and Rademacher, Grossman and Wildner (1975).

Clinical and radiographic features
The condition presents with pain, swelling and inflammation of a localised area, often the mandible, shoulder girdle or limb, and the acute episode is accompanied by pyrexia, leucocytosis and a raised erythrocyte sedimentation rate. Spontaneous resolution takes place within a few weeks, but relapse is not unusual. Infantile cortical hyperostosis most frequently occurs before the age of six months, and it has been recognised radiologically in the fetus. There have been a few reports of presentation in later childhood, as in a 12 year old Moroccan girl studied by Gillet *et al.* (1974).

Radiographically, the cortices of affected bones are widened and sclerotic, with irregular contours. With remission of the illness the bones regain their normal appearance.

Genetics
The localisation and acute inflammatory nature of infantile cortical hyperostosis is most unlike a genetic disorder. Nevertheless, there have been several reports which strongly suggest autosomal dominant inheritance (Gerrard *et al.*, 1961; van Buskirk, Tampas and Peterson, 1961; Holman, 1962; Bull and Feingold, 1974). The sibs described by Clemett and Williams (1963) might be taken as evidence for autosomal recessive inheritance. However, non-penetrance of the gene in a parent could also explain this anomalous observation. These reports of familial aggregation are too numerous to be ignored, and there is little doubt that there is a strong genetic element in the pathogenesis of infantile cortical hyperostosis. However, deter-

mination of the actual mode of transmission may have to await the elucidation of the underlying disease process. In an account of 11 cases in two generations of a kindred, Fráňa and Sekanina (1976) aptly commented that the unfavourable genetic outlook was balanced by the benign character of the disease.

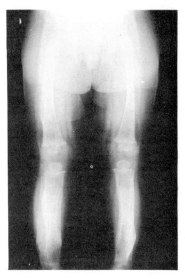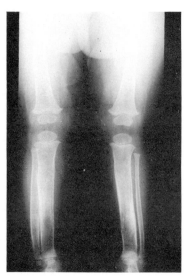

Fig. 8.7. (left) Infantile cortical hyperostosis; the cortices of the tibia and fibula are wide and irregular.

Fig. 8.8. (right) Infantile cortical hyperostosis; the patient depicted in Fig. 8.7, three months later. The condition is quiescent and the bone changes are resolving.

5. OSTEOECTASIA WITH HYPERPHOSPHATASIA

Virtually every case report concerning this disorder has appeared under a different designation, and there is therefore considerable semantic confusion. Amongst the titles which have been employed are endosteal hyperostosis, juvenile Paget disease, hyperostosis corticalis deformans, chronic congenital idiopathic hyper-phosphataemia, congenital hyperphosphatasia and osteoectasia with hyper-phosphatasia.

Caffey (1973) clarified the situation when he reviewed the literature and added a new case, bringing the total of reported patients to 14. He also mentioned seven other hitherto unreported children in two families in Puerto Rico.

Clinical and radiographic features

Osteoectasia has many features in common with the usual adult form of Paget disease, the main difference being the precocious onset and the severity. This serious progressive disorder usually begins in infancy, with swelling of the long bones and enlargement of the skull. The limbs become bowed, and the deformity may be compounded by repeated pathological fractures. Stature is diminished. Deafness and optic atrophy are late complications and death may be the consequence of vascular involvement. Angioid streaks in the retina and persistently elevated levels of serum alkaline phosphatase are reminiscent of Paget disease. Inconsistent biochemical changes include increased serum concentrations of uric acid and leucine amino peptidase.

Radiographically, the skeletal changes are asymmetrical, generalised demineralisation being the major feature. In the skull the calvarium is thickened with patchy areas of sclerosis, while the facial bones are spared. Platyspondyly and protrusio acetabulae may develop. Involvement of the tubular bones is severe and their width is greatly increased, with bowing and lack of modelling.

Genetics
Osteoectasia is inherited as an autosomal recessive. Affected sisters with consanguineous parents were reported by Bakwin, Golden and Fox (1964), while other sets of sibs were mentioned by Stemmermann (1966), and Eyring and Eisenberg (1968). Temtamy *et al*. (1974) studied an Egyptian girl with the condition who was the offspring of a consanguineous marriage.

Minor skeletal changes which were recognised on radiographic skeletal survey of the mother of the sibs described by Bakwin, Golden and Fox (1964) could represent phenotypic expression in the heterozygote. Caffey (1973) speculated that this woman might eventually develop the manifestations of the adult form of Paget disease.

There have been no reports of attempts at antenatal diagnosis of osteoectasia. However, as alkaline phosphatase levels are raised the condition might be recognisable in the fetus by determination of the activity of this enzyme in cultured amniotic fluid cells.

6. OSTEITIS DEFORMANS (PAGET)

Although more than a century has elapsed since Sir James Paget described osteitis deformans, the pathogenesis is still obscure. Indeed, in spite of the fact that Paget disease is comparatively common, there is still speculation as to whether or not the condition has a genetic basis.

Clinical and radiographic features
Individuals with the characteristic radiographic bone changes may remain totally asymptomatic. The prevalence increases with advancing age, and Paget disease is seldom encountered before middle life. Nager (1975) estimated that 13 per cent of the population over the age of 40 have the disorder. The abnormalities may be localised to one bone, or widely distributed throughout the skeleton. The skull, axial skeleton and proximal long bones are the sites of predilection. Bone pain is sometimes intractable and deafness, skeletal deformity, spontaneous fractures and osteosarcomatous changes are late complications. In severe cases, the vascular component acts as an arteriovenous shunt and cardiac failure may supervene. The serum alkaline phosphatase level is consistently raised. The management of Paget disease has been improved by the introduction of calcitonin therapy and in adequately treated patients, symptomatic remission can be induced.

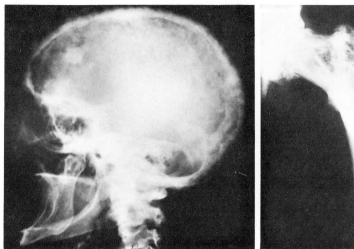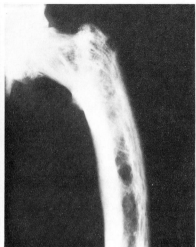

Fig. 8.9. (left) Osteitis deformans; skull radiograph of an elderly male, showing patchy sclerosis.

Fig. 8.10. (right) Osteitis deformans; radiograph of the upper end of the femur of an affected female. The bone is bowed and expanded, with areas of density and radiolucency.

Radiographically, bone involvement may be localised or widespread. Radiolucent areas appear in the early stages of the disease, followed by sclerosis, expansion and distortion of the bones. These changes sometimes mimic neoplastic metastases.

Genetics

The familial aggregation of Paget disease is well documented; nevertheless, the mode of genetic transmission has not yet been firmly established. The late onset often precludes the investigation of consecutive generations. The great variability of the manifestations and their potential for clinical silence makes any family study dependent upon radiology, with or without serum alkaline phosphatase determinations.

There are considerable geographic and ethnic discrepancies in the distribution of Paget Disease. In his large scale study in Australia, Barry (1969) observed that the condition was virtually confined to immigrants from Britain. Paget disease is rare in indigenous African populations, although isolated cases have been reported from Senegal, Nigeria and South Africa (van Meerdervoort and Richter, 1976). Simon *et al.* (1975) investigated the HLA antigenic status of 46 patients without finding any association with Paget disease.

Ashley-Montagu (1949) proposed that inheritance was X-linked, with clinical expression in a proportion of the female heterozygotes. However, this hypothesis is untenable, as male to male transmission was documented when Gutman and Kasabach (1936) analysed a series of 116 cases. McKusick (1972) reviewed the situation at length, published a collection of pedigrees of kindreds containing more than one affected member and concluded that Paget disease is inherited as an autosomal dominant, with varying clinical expression. Since Paget disease is predominantly a disorder of old age, it is unlikely that a patient would seek genetic counselling!

REFERENCES

Endosteal hyperostosis
Beals, R. K. (1976) Endosteal hyperostosis. *Journal of Bone and Joint Surgery*, **58**, 1172.
Buchem, F. S. P. van, Hadders, H. N. & Ubbens, R. (1955) Hyperostosis corticalis generalisata familiaris. *Acta radiologica*, **44**, 109.
Buchem, F. S. P. van, Hadders, H. N., Hansen, J. F. & Woldring, M. G. (1962) Hyperostosis corticalis generalisata. Report of seven cases. *American Journal of Medicine*, **33**, 387.
Buchem, F. S. P. van (1971) Hyperostosis corticalis generalisata. Eight new cases. *Acta medica scandinavica*, **189**, 257.
Buchem, F. S. P. van, Prick, J. J. G. & Jaspar, H. H. J. (1976) In *Hyperostosis Corticalis Generalisata Familiaris* (Van Buchem's Disease). Amsterdam, Oxford, New York: Excerpta Medica.
Maroteaux, P., Fontaine, G., Scharfman, W. & Farriaux, J. P. (1976) L'hyperostose corticale generalisée a transmission dominante. *Archives Francaises de Pédiatrie*, **28**, 685.
Owen, R. H. (1976) Van Buchem's disease (hyperostosis corticalis generalisata). *British Journal of Radiology*, **49**, 578.
Scott, W. C. & Gautby, T. H. T. (1974) Hyperostosis corticalis generalisata familiaris. *British Journal of Radiology*, **47**, 500.
Worth, H. M. & Wollin, D. G. (1966) Hyperostosis corticalis generalisata congenita. *Journal of the Canadian Association of Radiologists*, **17**, 67.

Sclerosteosis
Beighton, P., Cremin, B. & Hamersma, H. (1976) The radiology of sclerosteosis. *British Journal of Radiology*, **49**, 934.
Beighton, P., Durr, L. & Hamersma, H. (1976) The clinical features of sclerosteosis. A review of the manifestations in twenty-five affected individuals. *Annals of Internal Medicine*, **84/4**, 393.
Beighton, P., Davidson, J., Durr, L. & Hamersma, H. (1977) Sclerosteosis — an autosomal recessive disorder. *Clinical Genetics*, **11**, 1.
Higinbotham, N. L. & Alexander, S. F. (1941) Osteopetrosis, four cases in one family. *American Journal of Surgery*, **53**, 444.
Sugiura, Y. & Yasuhara, T. (1975) Sclerosteosis: a case report. *Journal of Bone and Joint Surgery*, **57/2**, 273.
Truswell, A. S. (1958) Osteopetrosis with syndactyly. A morphological variant of Albers-Schönberg disease. *Journal of Bone and Joint Surgery*, **40**, 208.

Diaphyseal dysplasia
Allen, D. T., Saunders, A. M., Northway, W. H., Williams, G. F. & Schafer, I. A. (1970) Corticosteroids in the treatment of Engelmann's disease: progressive diaphyseal dysplasia. *Pediatrics*, **46**, 523.
Girdany, B. R. (1959) Engelmann's disease (progressive diaphyseal dysplasia — a non-progressive familial form of muscular dystrophy with characteristic bone changes). *Clinical Orthopaedics and Related Research*, **14**, 102.
Hundley, J. D. & Wilson, F. C. (1973) Progressive diaphyseal dysplasia. Review of the literature and report of seven cases in one family. *Journal of Bone and Joint Surgery*, **55A**, 461.
Lindstrom, J. A. (1974) Diaphyseal dysplasia (Engelmann). Treated with corticosteroids. *Birth Defects: Original Article Series*, **10/12**, 504.
McKusick, V. A. (1975) Engelmann disease (progressive diaphyseal dysplasia) In *Mendelian Inheritance in Man*, 4th Edition, p.95. Baltimore and London: The Johns Hopkins University Press.
Ramon, Y. & Buchner, A. (1966) Camurati–Engelmann's disease affecting the jaws. *Oral Surgery*, **22**, 592.
Sparkes, R. S. & Graham, C. B. (1972) Camurati–Engelmann's disease. Genetics and clinical manifestations, with a review of the literature. *Journal of Medical Genetics*, **9**, 73.
Thurmon, T. H. & Jackson, J. (1976) Tumoral calcinosis and Engelmann disease. *Birth Defects: Original Article Series*, **12/5**, 321.

Infantile cortical hyperostosis
Bull, M. J. & Feingold, M. (1974) Autosomal dominant inheritance of Caffey disease. In *Skeletal Dysplasia*, Ed. Bergsma, D., p.139. Miami: The National Foundation; pub. Symposia Specialists.
Caffey, J. & Silverman, W. A. (1945) Infantile cortical hyperostosis: preliminary report on a new syndrome. *American Journal of Roentgenology, Radium Therapy and Nuclear Medicine*, **54**, 1.
Caffey, J. (1957) Infantile cortical hyperostosis: a review of the clinical and radiographic features. *Proceedings of the Royal Society of Medicine*, **50**, 347.

Clemett, A. R. & Williams, J. H. (1963) The familial occurrence of infantile cortical hyperostosis. *Radiology*, **80**, 409.

Fráňa, L. & Sekanina, M. (1976) Infantile cortical hyperostosis. *Archives of Diseases in Childhood*, **51**, 589.

Gerrard, J., Holman, G. H., Gorman, A. A. & Morrow, I. H. (1961) Infantile cortical hyperostosis: an inquiry into its familial aspects. *American Journal of Roentgenology, Radium Therapy and Nuclear Medicine*, **85**, 613.

Gillet, J., Imani, F., Benzakour, M. & Guignard, J. (1974) Infantile cortical hyperostosis (Caffey's disease); two cases. *Annals of Radiology*, **17/7**, 707.

Holman, G. H. (1962) Infantile cortical hyperostosis: a review. *Quarterly Review of Pediatrics*, **17**, 24.

Rademacher, K. H., Grossmann, I. & Wildner, G. P. (1975) Infantile cortical hyperostosis (IKH). Short survey and own observations. *Radiological Diagnosis*, **16/4**, 585.

Sherman, M. S. & Hellyer, D. T. (1950) Infantile cortical hyperostosis. Review of the literature and report of five cases. *American Journal of Roentgenology, Radium Therapy and Nuclear Medicine*, **63**, 212.

Van Buskirk, F. W., Tampas, J. P. & Peterson, O. S. (1961) Infantile cortical hyperostosis: an inquiry into its familial aspects. *American Journal of Roentgenology, Radium Therapy and Nuclear Medicine*, **85**, 613.

Osteoectasia with hyperphosphatasia

Bakwin, H., Golden, A. & Fox, S. (1964) Familial osteoectasia with macrocranium. *American Journal of Roentgenology, Radium Therapy and Nuclear Medicine*, **91**, 609.

Caffey, J. (1973) Familial hyperphosphatasemia with ateliosis and hypermetabolism of growing membranous bone: review of the clinical, radiographic and chemical features. In *Intrinsic Diseases of Bones*, Volume 4, p.438. Ed. Kaufmann, H. J. Basel: S. Karger.

Eyring, E. J. & Eisenberg, E. (1968) Congenital hyperphosphatasia. *Journal of Bone and Joint Surgery*, **50A**, 1099.

Stemmermann, G. N. (1966) An histologic and histochemical study of familial osteoectasia (chronic idiopathic hyperphosphatasia). *American Journal of Pathology*, **48**, 641.

Temtamy, S. A., El-Meligy, M., Salem, S. & Osman, N. (1974) Hyperphosphatasia in an Egyptian child. *Birth Defects: Original Article Series*, **10/12**, 196.

Osteitis deformans (Paget)

Ashley-Montagu, M. F. (1949) Paget's disease (osteitis deformans) and heredity. *American Journal of Human Genetics*, **1**, 94.

Barry, H. C. (1969) In *Paget's Disease of Bone*, p.14. London: E. & S. Livingstone.

Gutman, A. B. & Kasabach, H. (1963) Paget's disease (osteitis deformans). Analysis of 116 cases. *American Journal of Medical Science*, **191**, 361.

McKusick, V. A. (1972) In *Heritable Disorders of Connective Tissue*, 4th Edition, St. Louis: Mosby.

Nager, G. T. (1975) Paget's disease of the temporal bone. *Annals of Otolaryngology*, **84/4**, 22.

Simon, L., Blotman, F., Seignalet, J. & Claustre, J. (1975) The etiology of Paget's bone disease. *Revue du Rhumatisme et des Maladies Osteo-articulares*, **42/10**, 535.

Van Meerdervoort, H. F. P. & Richter, G. G. (1976) Paget's disease of bone in South African Blacks. *South African Medical Journal*, **50**, 1897.

9. Miscellaneous sclerosing and hyperostotic disorders

The condition in this group are uncommon disorders in which clinical silence contrasts with striking radiological manifestations.

1. Osteopathia striata
2. Osteopoikilosis
3. Melorheostosis
4. Pachydermoperiostosis.

1. OSTEOPATHIA STRIATA

The designation 'osteopathia striata' was used by Fairbank (1935) following the initial case description by Voorhoeve (1924). The term pertains to multiple lines of increased density which are radiologically apparent in the long bones and the pelvis. Striations of this type are a component of several conditions, including the osteopetroses, osteopoikilosis and focal dermal hypoplasia. In the precise sense, osteopathia striata is a specific bone disorder, which is probably inherited as an autosomal dominant.

 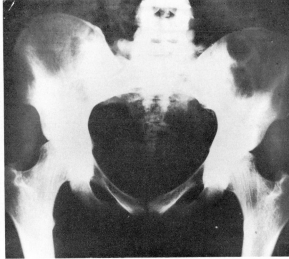

Fig. 9.1. (left) Osteopathia striata; radiograph of the lower end of the femur of a young woman, showing multiple parallel lines of sclerosis. The base of the skull was sclerotic and the patient was deaf, due to entrapment of the auditory nerves.

Fig. 9.2. (right) Osteopathia striata; striae are evident in the femoral necks and in the ilia. Inheritance was autosomal dominant as members of three generations of the kindred had similar changes.

Clinical and radiographic features

Individuals with osteopathia striata are usually aymptomatic. The radiographic abnormalities in Voorhoeve's two original patients were unchanged when Fermin (1962) re-examined them three decades later. Radiographically, multiple parallel lines of sclerosis run along the shafts of the tubular bones. In the ilia, the striations have a fan-shaped configuration. The contours of the bone are undisturbed.

Genetics

Osteopathia striata was observed by Voorhoeve (1924) in a father and his daughter. In the kindred reported by Rucker and Alfidi (1964) under the designation 'Fairbank disease', members of three generations were affected. Some patients with osteopathia striata have cranial sclerosis, mild facial distortion and cranial nerve palsies (Jones and Mulcahy, 1968; Walker, 1969). The author has personal knowledge of 4 kindreds with this combination of abnormalities. Father to son transmission has occurred and there is little doubt that inheritance is autosomal dominant, with inconsistent phenotypic expression.

2. OSTEOPOIKILOSIS

Osteopoikilosis or 'spotty bones' is an unusual but not uncommon condition. More than 300 cases have been reported (Szabo, 1971). The diagnosis is usually reached fortuitously, following radiographic examination for an unrelated purpose. Alternatively, the dermal lesions may draw attention to the disorder. (Hence the Lancashire proverb attributed to Granny Peard — 'what's bred in the bone comes out in the flesh'.)

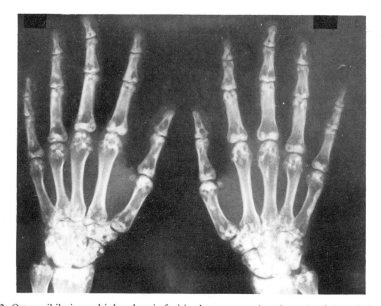

Fig. 9.3. Osteopoikilosis; multiple sclerotic foci in the carpus and at the ends of the tubular bones.

Clinical and radiographic features

Radiographically, small sclerotic foci are found throughout the skeleton. Many hundreds of these lesions may be present, and they tend to be congregated in the epiphyseal and metaphyseal regions of the tubular bones. Multiple sessile dermal naevi, known as 'dermatofibrosis lenticularis disseminata', are an associated feature. Osteopoikilosis is of no clinical importance.

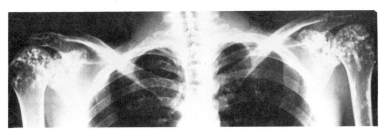

Fig. 9.4. Osteopoikilosis; in this patient, the diagnosis was made by chance, after routine chest X-ray.

Genetics

Reports of several large kindreds have been published and autosomal dominant inheritance is well established. Melnick (1959) described the disorder in 17 individuals in four generations, while father to son transmission was observed by Raque and Wood (1970). Clinical expression of the gene is variable and the dermal and osseous changes may occur together or separately in the same family (Berlin *et al.*, 1967). Schoenenberg (1975) observed the typical bone appearances in eight members of three generations of a kindred. Six also had the dermal manifestations.

3. MELORHEOSTOSIS

Melorheostosis is a rare disorder which is characterised by irregular cortical thickening, principally of the tubular bones. The appellation, which pertains to the 'streaming' appearance of the bones, is derived from the Greek. Campbell, Papademetriou and Bonfiglio (1968) described 14 cases and reviewed the literature. The condition has been recognised in a fibula from a prehistoric Alaskan Eskimo burial ground (Lester, 1967).

Clinical and radiographic features

The majority of individuals with melorheostosis are asymptomatic. However, vague pains in the bones and joint swelling may be the presenting feature. The skin over affected bones sometimes becomes indurated and thickened, and soft tissue contractures may develop. Onset usually occurs in early adulthood and progression is very slow (Murray, 1951; Patrick, 1969).

The radiographic manifestations are striking. The changes in the long bones have the appearance of melted wax flowing down the sides of a candle. The disorder is usually confined to a single bone or limb and limited to one side of the body. Ectopic bone occasionally forms in the adjacent soft tissues.

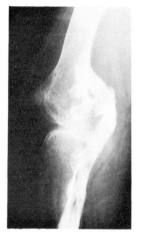 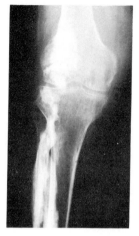 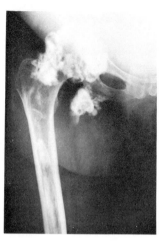

Fig. 9.5. (left) Melorheostosis; lateral radiograph of the knee of an affected woman. The overlying skin was indurated and sclerotic.

Fig. 9.6. (centre) Melorheostosis; the radiographic changes have the appearance of melted wax flowing down the side of a candle.

Fig. 9.7. (right) Melorheostosis; ectopic bone is present in the region of the ischium.

Genetics
There is no evidence that melorheostosis is a genetic disorder. As the condition is often clinically silent, family studies would have to be based upon skeletal surveys of relatives of known cases. So far, none have been reported.

4. PACHYDERMOPERIOSTOSIS

Pachydermoperiostosis is a relatively common disorder in which clubbing of the digits is associated with thickening and hyperhydrosis of the skin and extremities, and oiliness and seborrhoea of the scalp.

Clinical and radiographic features
The manifestations appear at puberty and are generally more severe in males. Radiographically, the cortices of the long bones are widened and sclerotic, with increased medullary trabeculation and periosteal thickening at the distal extremities. Clinical importance centres around the digital clubbing, which must be distinguished from the pulmonary osteoarthropathy of neoplastic or chronic cardiopulmonary disease. The individual with familial pachydermoperiostosis who was needlessly subjected to bronchoscopy and cardiac catheterisation would be in full agreement with this point!

Genetics
Clinical expression is very variable and individuals with the gene may manifest only certain components of the disorder. For this reason, family studies are difficult. Following a review of the literature, Rimoin (1965) concluded that pachy-

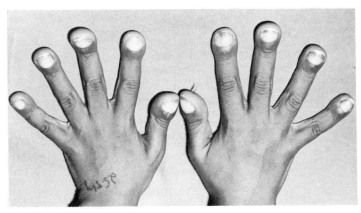

Fig. 9.8. Familial digital clubbing; several members of this boy's kindred also had clubbing of the fingers and toes. This familial condition must be distinguished from the clubbing which develops in chronic cardiopulmonary disease.

dermoperiostosis was inherited as an autosomal dominant. However, there have been several instances of affected sibs with normal parents and parental consanguinity has been reported (Findlay and Oosthuizen, 1951). It is therefore possible that there is an autosomal recessive variety of pachydermoperiostosis. McKusick (1975) has postulated that familial simple digital clubbing without skin changes might represent a separate genetic defect.

Digital clubbing, in association with widespread periosteal new bone formation in the shafts of the long bones and defects in cranial ossification, was encountered in two young Negro sisters by Currarino *et al*. (1961). The parents and two sibs were normal, but the development of the condition in a third sister was subsequently reported by Chamberlain, Whitaker and Silverman (1965). A further sporadic case was described by Cremin (1970). The cranial defects distinguish the disorder from the other familial digital clubbing syndromes. The mode of inheritance of this condition, which has been designated 'familial idiopathic osteoarthropathy of children', has not been elucidated.

REFERENCES

Osteopathia striata
Fairbank, H. A. T. (1935) Generalised disorders of the skeleton. *Proceedings of the Royal Society of Medicine*, **28**, 1611.
Fermin, H. E. A. (1962) Osteorhabdotose. Een voor het eerst door N. voorhoeve beschreven bijzondere vorm van osteopathia condensans disseminata. *Nederlansch Tijdschrift voor Geneeskunde*, **106**, 1188.
Jones, M. D. & Mulcahy, N. D. (1968) Osteopathia striata, osteopetrosis, and impaired hearing. A case report. *Archives of Otolaryngology*, **87**, 116.
Rucker, T. N. & Alfidi, R. J. (1964) A rare familial systematic affection of the skeleton: Fairbank's disease. *Radiology*, **82**, 63.
Voorhoeve, N. (1924) L'image radiologique non encore decrite d'une anomalie de squelette. *Acta radiologica*, **3**, 407.
Walker, B. A. (1969) Osteopathia striata with cataracts and deafness. In *Clinical Delineation of Birth Defects*, p.295. New York: National Foundation.

Osteopoikilosis
Berlin, R., Hedensio, B., Lilja, B. & Linder, L. (1967) Osteopoikilosis — a clinical and genetic study. *Acta medica scandinavica*, **181**, 305.

Melnick, J. C. (1959) Osteopathia condensans disseminata (osteopoikilosis). Study of a family of four generations. *American Journal of Roentgenology, Radium Therapy and Nuclear Medicine*, **82**, 229.
Raque, C. J. & Wood, M. G. (1970) Connective tissue naevus. Dermatofibrosis lenticularis disseminata with osteopoikilosis. *Archives of Dermatology*, **102**, 290.
Schoenenberg, H. (1975) Osteopoikilia with dermofibrosis lenticularis disseminata (Buschke Ollendorf syndrome). *Klinische Paediatrie*, **187/2**, 123.
Szabo, A. D. (1971) Osteopoikilosis in a twin. *Clinical Orthopaedics and Related Research*, **79**, 156.

Melorheostosis
Campbell, C. J., Papademetriou, T. & Bonfiglio, M. (1968) Melorheostosis. A report of the clinical, roentgenographic and pathological findings in fourteen cases. *Journal of Bone and Joint Surgery*, **50A**, 1281.
Lester, C. W. (1967) Melorheostosis in a prehistoric Alaskan skeleton. *Journal of Bone and Joint Surgery*, **49**, 142.
Murray, R. O. (1951) Melorheostosis associated with congenital arteriovenous aneurysms. *Proceedings of the Royal Society of Medicine*, **44**, 473.
Patrick, J. H. (1969) Melorheostosis associated with arteriovenous aneurysm of the left arm and trunk. *Journal of Bone and Joint Surgery*, **51B**, 126.

Pachydermoperiostosis
Chamberlain, D. S., Whitaker, J. & Silverman, F. N. (1965) Idiopathic osteoarthropathy and cranial defects in children. (Familial idiopathic osteoarthropathy). *American Journal of Roentgenology, Radium Therapy and Nuclear Medicine*, **93**, 408.
Cremin, B. J. (1970) Familial idiopathic osteoarthropathy of children: a case report and progress. *British Journal of Radiology*, **43**, 568.
Currarino, G., Tierney, R. C., Giesel, R. G. & Weihl, C. (1961) Familial idiopathic osteoarthropathy. *American Journal of Roentgenology, Radium Therapy and Nuclear Medicine*, **85**, 633.
Findlay, G. H. & Oosthuizen, W. J. (1951) Pachydermoperiostitis: syndrome of Touraine, Solente and Bole. *South African Medical Journal*, **25**, 747.
McKusick, V. A. (1975) In *Mendelian Inheritance in Man*, 4th Edition, p.246. Baltimore and London: The Johns Hopkins University Press.
Rimoin, D. L. (1965) Pachydermoperiostosis (idiopathic clubbing and periostosis). Genetic and physiologic considerations. *New England Journal of Medicine*, **272**, 923.

10. Mucopolysaccharidoses

The mucopolysaccharidoses (MPS) are a group of conditions in which defective enzymatic activity leads to storage of incompletely degraded glycosaminoglycans. A coarse facies, short stature and skeletal dysplasia are the major features. Other variable manifestations include progressive intellectual impairment, hepatosplenomagaly, corneal clouding and infiltration of the cardiac valves. In recent years the MPS have attracted considerable medical interest and effort. Indeed, there is some truth in the contention that there are more research workers currently investigating the MPS than there are patients suffering from these conditions!

Nosology
The MPS provide an object lesson in the way in which nosology is influenced by the development of knowledge. In the early part of this century, affected children were considered to resemble the grotesque gargoyles that traditionally embellished medieval buildings. The disorders were then grouped together under the unfortunate designation 'gargoylism'. Later, when it was appreciated that hepatosplenomegaly was often present in association with abnormalities of the skeleton, the more acceptable term 'lipochondrodystrophy' was employed. The recognition of consistent patterns of clinical stigmata led to the concept of heterogeneity, and eponyms such as Hunter, Hurler or Morquio syndrome were applied to the various subdivisions.

The generic term 'mucopolysaccharidosis' came into being when it became apparent that excess mucopolysaccharides were excreted in the urine. Subsequently, abnormal quantities of urinary constituents such as dermatan, heparan and keratan sulphate were identified in various forms of the disorder. It was then evident that the clinical categories of the MPS conformed to these biochemical subtypes and the individual conditions were therefore given numeric designations as alternatives to their eponyms.

In the past few years, the situation has become increasingly complex. Histochemical studies of cultured fibroblasts and the demonstration of cross-correction of the metabolic defect in mixed fibroblast cultures have had an enormous impact. New MPS have been delineated, numerical categories have been subdivided, allelism has been recognised, genetic compounds have been encountered and basic enzymatic defects have been identified. The present state of affairs may be summarised as follows:

Designation	Eponym	Enzymatic defect
MPS 1-H MPS 1-S MPS 1-H/S	Hurler syndrome Scheie syndrome Hurler–Scheie compound	} a-L-iduronidase
MPS 11 A MPS 11 B	Hunter syndrome, severe Hunter syndrome, mild	} L-iduronosulphate sulphatase
MPS III A MPS III B	Sanfilippo syndrome A Sanfilippo syndrome B	Heparan sulphate sulphatase N-acetyl-a-D-glucosaminidase
MPS IV	Morquio syndrome (allelic forms?)	N-acetylhexosaminidase- 6-SO_4 sulphatase
MPS V	(redesignated MPS 1-S)	
MPS VI A MPS VI B	Maroteaux–Lamy syndrome, severe form Maroteaux–Lamy syndrome, mild form	} Arylsulphatase B
MPS VII		β-glucuronidase

Radiographic features of the mucopolysaccharidoses

Spranger, Langer and Wiedemann (1974) pointed out that the fundamental radiographic changes were similar in each of the MPS and they proposed the non-specific term 'dysostosis multiplex' for these bone abnormalities. In dysostosis multiplex, the skeleton is osteoporotic. The skull shows calvarial thickening and a J-shaped pituitary fossa. The ribs are oar-shaped and the vertebrae are flat, with anterior beaking. The ilia are flared, and the acetabulae are dysplastic. The tubular bones are shortened, with defective diaphyseal modelling and irregularities of the metaphyses and epiphyses. As with many other skeletal dysplasias, the abnormalities in dysostosis multiplex are age-related. The individual MPS differ in the severity of skeletal involvement and recognition of characteristic features is often of diagnostic importance. The radiology of the MPS has been reviewed in detail by Grossman and Dorst (1973).

General considerations

With the exception of MPS II, the Hunter syndrome, which is X-linked, all forms of MPS are inherited as autosomal recessives. Antenatal diagnosis has been reported in MPS I-H, MPS II and MPS III B and the technology is available for the *in utero* recognition of the other types of the disorder.

In an attempt to replace the missing factor, treatment with infusions of plasma has been employed in MPS I-H and MPS II (Hussels *et al.*, 1974). Although some improvement was gained, it is now generally accepted that this approach does not have any practical value. In their review of the MPS, Pennock and Barnes (1976) pointed out that the discovery of the specific enzymatic defects may eventually provide the basis for therapy by enzyme replacement.

Rational genetic, orthopaedic and surgical management is strongly influenced by the long-term prognosis, which varies greatly in the different forms of the condition. In turn, prognostication is dependent upon diagnostic precision. Accuracy will become even more important in the future if enzyme replacement therapy becomes available. It is therefore imperative that every patient should be categorised at a biochemical level.

It must be emphasised that diagnosis is not always an easy matter. Apart from the problem of change in radiographic appearances with advancing years, the urinary excretion of mucopolysaccharides is also age-related. Excess mucopolysacchariduria may not be present in the first year of life, and in survivors, the urine may revert to normal at adolescence.

It is likely that further MPS remain to be delineated. Equally, a number of obscure skeletal dysplasias may well be the result of abnormalities of MPS metabolism in cartilage, in the absence of mucopolysacchariduria. The mucolipidoses (MLS), which resemble the MPS, have been recognised as distinct entities, and they are now classified separately (see Chapter 11).

1. MPS I-H (HURLER SYNDROME)

MPS I-H or the Hurler syndrome is probably the best known of the MPS. It has been estimated that the prevalence in British Columbia is of the order of 1 in 100,000 (Lowry and Renwick, 1971). The disorder has been encountered in many different ethnic groups and it is evident that the gene is widely distributed.

By the second year of life, infants with MPS I-H have a coarse facies, with a wide mouth, thick eyebrows and a protuberant tongue. The joints, particularly those of the hands, are stiff. The thorax is deformed and the aortic and mitral valves often become incompetent. Growth is retarded and intellectual function is impaired. Hepatosplenomegaly and corneal clouding develop in early childhood and death usually occurs by the end of the first decade. Radiographic changes of dysostosis multiplex of moderate severity are present in infancy. Urinary excretion of dermatan and heparan sulphate is excessive. The basic abnormality is defective activity of the enzyme a-L-iduronidase.

Genetics

In a comprehensive review of family data, Jervis (1950) showed that inheritance of the Hurler syndrome was autosomal recessive. Subsequently, Danes and Bearn (1966) observed metachromasia in cultured fibroblasts from six sets of obligatory heterozygote parents of affected children. Omura, Higami and Tada (1976) developed a method of homozygote and heterozygote detection based upon estimation of a-L-iduronidase activity in leucocytes.

The possibility that MPS I-H could be diagnosed antenatally was raised by Nadler (1968) and Fratantoni et al. (1969), when they demonstrated metachromasia in amniotic fluid cells from pregnancies in which affected infants were produced at term. An alternative method of antenatal diagnosis by the estimation of amniotic fluid glycosaminoglycan (GAG) content was developed by Matalon et al. (1970) and

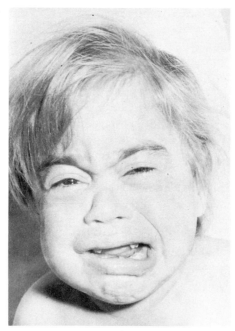

Fig. 10.1. MPS; a coarse facies is a feature of several of the mucopolysaccharidoses. This child has MPS I.

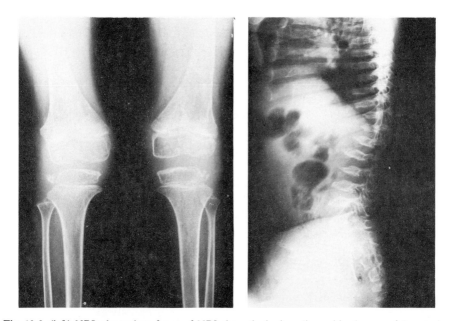

Fig. 10.2. (left) MPS; the various forms of MPS share the basic radiographic changes of dysostosis multiplex. The epiphyses and metaphyses of the tubular bones are irregular.

Fig. 10.3. (right) MPS; in dysostosis multiplex, the lumbar vertebrae are flattened, with anterior beaking.

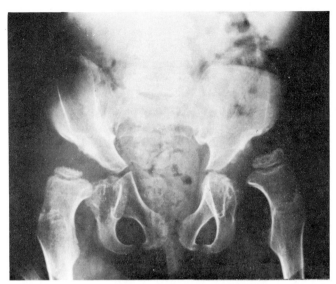

Fig. 10.4. MPS; the acetabulae are dysplastic, the femoral capital epiphyses are misshapen and the femoral necks have a valgus deformity.

Crawford *et al.* (1973). The results obtained by this technique were not entirely consistent, and although it has the advantage of saving time, it is not sufficiently accurate to be used alone. Currently, at risk pregnancies are monitored by measurements of the rate of uptake and clearance of isotopically labelled sulphate in cultured amniotic fluid cells, in conjunction with estimation of the ratios of various GAG in the amniotic fluid. Henderson and Nelson (1977) described the application of these techniques in the antenatal diagnosis of the Hurler syndrome in two successive pregnancies in the same patient. In each instance, the fetus was found to be affected and the pregnancy was terminated.

2. MPS I-S (SCHEIE SYNDROME)

MPS I-S, the Scheie syndrome, was initially designated 'MPS V'. However, when it was shown that activity of the enzyme α-L-iduronidase was defective in this condition, as well as in MPS I-H, the classification was adjusted and the two disorders were grouped together. MPS I-S is apparently rare, but in view of the relatively inconspicuous clinical features, it is likely that the condition not infrequently remains undiagnosed.

Individuals with Scheie syndrome have relatively normal height, intelligence and life span. Corneal clouding, rigidity of the digits, aortic incompetence and a tendency to develop the carpal tunnel syndrome are the major clinical problems. Radiographic changes are of minor degree. Urinary and enzymatic abnormalities are identical to those found in MPS I-H.

Genetics
Although there is doubt concerning the accuracy of the diagnosis in a number of early case descriptions, there are indisputable reports of affected sibs with normal

parents (Scheie, Hambrick and Barness, 1962; McKusick *et al.*, 1965). It is evident that MPS I-S is inherited as an autosomal recessive.

Activity of the enzyme α-L-iduronidase is defective in both MPS I-H and MPS I-S (Weismann and Neufeld, 1970). It is therefore likely that the determinant genes are allelic. McKusick *et al.* (1972) suggested that an individual with one MPS I-H gene and one MPS I-S gene would have stigmata which were intermediate between those of MPS I-H and MPS I-S. These authors described a group of seven patients with phenotypic features of this nature and postulated that they represented mixed heterozygotes or 'genetic compounds'. The results of fibroblast and enzymatic studies were consistent with this hypothesis and the condition was designated MPS I-H/S, the Hurler–Scheie compound. This condition was subsequently diagnosed in early adulthood in two Japanese brothers, who had dwarfism, Hurler-like facies and normal intelligence (Kajii *et al.*, 1974).

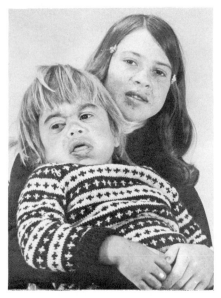

Fig. 10.5. MPS I; a two-year-old girl with the typical facies of MPS I. Her sister and the other members of the kindred are normal.

3. MPS II (HUNTER SYNDROME)

MPS II, the Hunter syndrome, is X-linked, and all patients are males. Two forms of the condition are recognised, the juvenile or severe type (MPS IIA) and the adult or mild type (MPS IIB). The Hunter syndrome is about as common as the Hurler syndrome.

Individuals with the severe type of MPS II usually die before adolescence, while those with the mild type survive into middle age. Stature is reduced, intelligence is impaired, the facies are coarse and the hands are clawed. Hepatosplenomegaly and involvement of the cardiac valves are frequent problems. Corneal clouding does not usually develop, but deafness is a common complication. Radiographic abnormalities are similar to those of MPS I-H. Excessive quantities of dermatan and heparan sulphate are excreted in the urine and the enzyme L-iduronosulphate sulphatase is defective in both forms of the condition. Liebaers and Neufeld (1976) have shown that estimation of the activity of this enzyme in serum, lymphocytes or fibroblasts is a relatively simple diagnostic procedure.

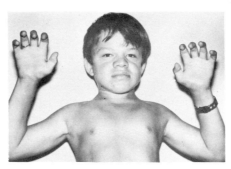

Fig. 10.6. MPS II; an 18-year-old male with the mild form of the Hunter syndrome. The fingers are fixed in flexion and stature is reduced.

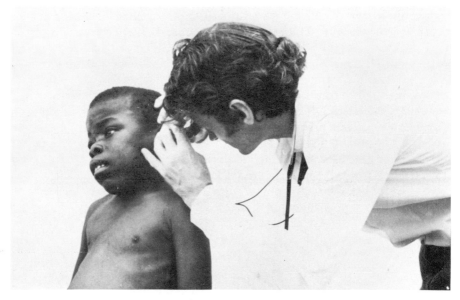

Fig. 10.7. MPS II; a 10-year-old male with the severe form of the Hunter syndrome. This boy is dwarfed and his mentality is impaired. He attends a special school for the deaf.

Genetics

The X-linked nature of MPS II was recognised by Noja (1946), when he reported 'a sex-linked type of gargoylism'. Subsequently several pedigrees showing X-linked transmission have been published. Those of Beebe and Formel (1954) and DiFerrante and Nichols (1972) are particularly impressive.

Danes and Bearn (1967) demonstrated that the clinically normal mothers of affected males have metachromasia in about 50 per cent of their cultured fibroblasts. Similarly, female heterozygotes can be detected by the recognition of two types of cells in tissue culture, with regard to the uptake of radioactive sulphate. These findings are consistent with the Lyon hypothesis of random X-chromosome inactivation. Linkage between the locus for the Hunter gene and the macroglobulin Xm locus, with an estimated recombination fraction of 0.09, was determined by Berg, Danes and Bearn (1968).

As in other X-linked conditions, fetal sexing is of importance in antenatal diagnosis. Niermeijer *et al.* (1976) monitored two pregnancies in which there was a likelihood that the unborn child would have the Hunter syndrome. In each instance, the fetal sex was shown to be female, and a normal outcome was succesfully predicted. In an at risk pregnancy, characterisation of amniotic fluid GAG ratios, the measurement of radioactive sulphate uptake and estimation of sulphatase activity in cultured amniotic fluid cells, following determination of fetal sex, would permit diagnostic confirmation. The substrate which has been prepared by Lim *et al.* (1974) may prove to be of value in this respect.

4. MPS III TYPES A AND B (SANFILIPPO SYNDROME)

The two forms of MPS III are clinically identical, and differ only in their basic enzymatic abnormalities. Profound mental retardation is the major feature and death occurs by early adulthood. Stature is relatively normal and there are few systemic ramifications. Some degree of joint rigidity and moderate hirsutism are usually present. Radiographic changes are very mild. Excess heparan sulphate is demonstrable in the urine. In MPS IIIA, heparan sulphate sulphamidase activity is deficient (Matalon and Dorfman, 1974), while in the MPS III B, the defect is in N-acetyl-α-D-glucosaminidase (O'Brien, 1972).

There are no firm figures for the prevalence of MPS III. However, more than 100 cases have been reported, including 68 which were reviewed by Spranger (1972). The disorder is probably underdiagnosed and institutionalised patients may remain unrecognised. The clinical variability of MPS III B was emphasised by Van de Kamp *et al.* (1976) following their studies of six definite and two probable patients in two related consanguineous Dutch sibships.

Genetics

A pair of affected sibs, with normal parents, was reported by Sanfilippo *et al.* (1963) and parental consanguinity was a feature of the complex sibship described by Maroteaux *et al.* (1966). MPS III A has been successfully diagnosed antenatally by the demonstration of defective enzyme activity in amniotic fluid cells and increased levels of GAG in the amniotic fluid (Harper *et al.*, 1974). The pregnancy in question was terminated at 23 weeks and the diagnosis was confirmed in the aborted fetus.

Obligatory heterozygotes for MPS III B have been shown to have a partial reduction in the activity of the N-acetyl-α-D-glucosaminidase (Figura *et al.*, 1973). Liem *et al.* (1976) undertook biochemical studies in 27 individuals from a kindred with MPS III B and recognised 6 homozygotes and 12 heterozygous sibs. These authors concluded that determination of N-acetyl-α-D-glucosaminidase activity in the plasma was a reliable method for the detection of the asymptomatic heterozygote.

Kress, von Figura and Bartsocas (1976) demonstrated metabolic cross-correction in co-cultivation studies with MPS III type A and B fibroblasts from two related Greek patients with the clinical stigmata of the Sanfilippo syndrome. On this evidence, the authors proposed that these individuals had a new disease entity, which they designated 'Sanfilippo syndrome type C'.

5. MPS IV (MORQUIO SYNDROME)

MPS IV or the Morquio syndrome was described independently in 1929 by Morquio and Brailsford. Although the conjoined eponym was popular for some years, the single designation is now preferred. The term 'Morquio syndrome' still leads to immense semantic confusion, as it is often used haphazardly and erroneously for any syndrome of dwarfism and spinal malalignment. MPS IV is much less common than MPS I or II.

Individuals with MPS IV are dwarfed, with a thoracolumbar gibbus, and a protuberant sternum. In distinction to MPS I and II, the joints of the hands are lax, the facies are relatively normal and the intellect is unimpaired. Corneal clouding and deafness develop in later childhood. Involvement of the cardiac valves contributes to cardiorespiratory embarrassment and death usually occurs in early adulthood. Genu valgum and spinal cord compression may necessitate operative intervention. The odontoid process may be hypoplastic, and atlantoaxial subluxation is a potentially lethal hazard during anasthaesia (Beighton and Craig, 1973).

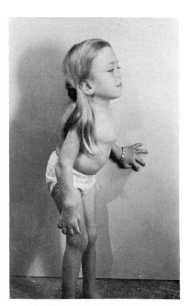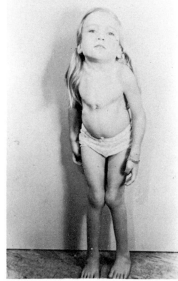

Fig. 10.8. (left) MPS IV; an eight-year-old girl with a crouching stance and thoracic asymmetry.

Fig. 10.9. (right) MPS IV; mild facial changes and genu valgum are present. Atlantoaxial dislocation occurred during general anaesthesia for a minor operation. (From Beighton, P. & Craig, J. (1973) *Journal of Bone and Joint Surgery*, 55, 478.)

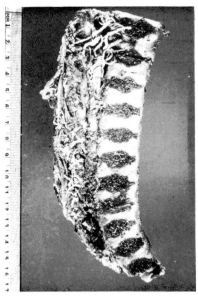

Fig. 10.10. MPS IV; a vertical section of the spine. Platyspondyly is clearly demonstrated.

The radiographic features are those of dysostosis multiplex in severe degree. Excess keratan sulphate is present in the urine and inclusion bodies may be demonstrated in leucocytes. Defective activity of the enzyme N-acetylhexosamine-6-sulphate sulphatase has been recognised in cultured fibroblasts (Matalon *et al.*,1974).

Genetics
The parents of the four affected sibs in the original family reported by Morquio (1929) were consanguineous. Gadbois, Moreau and Laberge (1973) investigated 48 cases in 27 kindreds in Quebec and found that the ratios of affected and unaffected sibs were consistent with autosomal recessive transmission. These authors claimed that they could recognise two phenotypes of the classical Morquio syndrome in this series of patients.

Skeletal dysplasias such as spondyloepiphyseal dysplasia, metatropic dwarfism and Dyggve–Melchior–Clausen syndrome have been split off from the general category of the Morquio syndrome. However, there may yet be a residual heterogeneity. A condition which closely resembles MPS IV, but in which the stigmata, particularly corneal clouding, are milder, and in which mucopolysacchariduria is absent, has been designated 'Morquio syndrome, non-keratosulphate excreting type.' This disorder is present in an inbred isolate living on the coast of Maryland, USA, and inheritance is evidently autosomal recessive. McKusick (1975) speculated that the genes for this condition and MPS IV may be allelic. If this supposition proves to be correct, genetic compounds as in MPS IH/S could occur.

6. MPS V

Vacant (formerly Scheie syndrome, which is now designated MPS I-S).

7. MPS VI (MAROTEAUX–LAMY SYNDROME)

MPS VI, the Maroteaux–Lamy syndrome, is one of the least common disorders in the MPS group. Following the original description by Maroteaux *et al.* (1963), fewer than 50 cases have been reported. The intellect is normal, but the clinical and radiographic stigmata are similar to those of the Hurler syndrome, although milder in degree. Excess dermatan sulphate is present in the urine. The leucocytes are packed with granules and cultured fibroblasts show metachromasia. Activity of the enzyme arylsulphatase-B is defective (Beratis *et al.*, 1975).

Genetics

There have been several reports of affected sibs and at least two instances of parental consanguinity (Spranger *et al.*, 1970). There is little doubt that inheritance is autosomal recessive. Antenatal diagnosis of MPS VI is theoretically possible. As there is disparity in radiographic changes in MPS VI, it has been suggested that there might be a mild variant (Spranger *et al.*, 1974). Studies of enzymatic activity in cultured fibroblasts have supported this contention (O'Brien, Cantz and Spranger, 1974).

8. MPS VII — β-GLUCURONIDASE DEFICIENCY

MPS VII or β-glucuronidase deficiency was recognised in a Negro infant with mild Hurler-like stigmata, who did not have mucopolysacchariduria. Abnormal accumulation of labelled sulphate and deficient activity of β-glucuronidase was demonstrated in cultured skin fibroblasts from this individual (Sly *et al.*, 1973). In keeping with autosomal recessive inheritance, the parents and several sibs were shown to have intermediate levels of enzyme activity. Additional patients with β-glucuronidase deficiency have been reported by Gehler *et al.* (1974) and Beaudet *et al.* (1975). There is some inconsistency in the clinical features, and it is possible that there are allelic forms of the abnormal gene.

REFERENCES

Mucopolysaccharidoses – preamble
Grossman, H. & Dorst, J. P. (1973) The mucopolysaccharidoses and mucolipidoses. *Progress in Pediatric Radiology*, **4**, 495.
Hussels, I. E., Eikman, E. A., Kenyon, K. R. & McKusick, V. A. (1974) Treatment of mucopolysaccharidoses. *Birth Defects: Original Article Series*, **10/12**, 212.
Pennock, C. A. & Barnes, I. C. (1976) The mucopolysaccharidoses. *Journal of Medical Genetics*, **13**, 169.
Spranger, J. W., Langer, L. O. & Wiedemann, H. R. (1974) In *Bone Dysplasias. An Atlas of Constitutional Disorders of Skeletal Development*, p.143. Stuttgart: Gustav Fisher Verlag.

MPS I-H (Hurler syndrome)
Crawford, M., Dean, M., Hunt, D., Johnson, D., MacDonald, R., Muir, H., Payling Wright, E. & Payling Wright, C. (1973) Early prenatal diagnosis of Hurler's syndrome with termination of pregnancy and confirmatory findings in the fetus. *Journal of Medical Genetics*, **10**, 144.
Danes, B. S. & Bearn, A. G. (1966) Hurler's syndrome: effect of retinol (vitamin A alcohol) on cellular mucopolysaccharides in cultured human skin fibroblasts. *Journal of Experimental Medicine*, **124**, 912.
Fratantoni, J. C., Neufeld, E. F., Uhlendorf, B. W. & Jacobson, C. B. (1970) Intrauterine diagnosis of the Hurler and Hunter syndromes. *New England Journal of Medicine*, **280**, 686.
Henderson, H. H. & Nelson, M. M. (1977) Antenatal diagnosis of Hurler's syndrome. *South African Medical Journal*, **51/8**, 241.

Jervis, G. A. (1950) Gargoylism: study of 10 cases with emphasis on the formes frustes. *Archives of Neurology and Psychiatry*, **63**, 681.

Lowry, R. B. & Renwick, S. H. G. (1971) The relative frequency of the Hurler and Hunter syndromes. *New England Journal of Medicine*, **284**, 221.

Matalon, R., Dorfman, A., Nadler, H. L. & Jacobson, C. B. (1970) A chemical method for the antenatal diagnosis of mucopolysaccharidoses. *Lancet*, **1**, 83.

Nadler, H. L. (1968) Antenatal detection of hereditary disorders. *Pediatrics*, **42**, 912.

Omura, K., Higami, S. & Tada, K. (1976) α-L-Iduronidase activity in leukocytes: diagnosis of homozygotes and heterozygotes of the Hurler syndrome. *European Journal of Pediatrics*, **112/2**, 103.

MPS I-S (Scheie syndrome)

Kajii, T., Matsuda, I., Ohsawa, T., Katsunama, H., Ichida, T. & Arashima, S. (1974) Hurler/Scheie genetic compound (mucopolysaccharidosis IH/IS) in Japanese brothers. *Clinical Genetics*, **6/5**, 394.

McKusick, V. A., Kaplan, D., Wise, D., Hanley, W. B., Suddarth, S. B., Sevick, M. E. & Maumenee, A. E. (1965) The genetic mucopolysaccharidoses. *Medicine*, **44**, 445.

McKusick, V. A., Howell, R. R., Hussels, I. E., Neuffeld, E. F. & Stevenson, R. (1972) Allelism, non-allelism and genetic compounds among the mucopolysaccharidoses: hypotheses. *Lancet*, **1**, 993.

Scheie, H. G., Hambrick, G. W. & Barness, L. A. (1962) A newly recognised forme fruste of Hurler's disease (gargoylism). *American Journal of Ophthalmology*, **53**, 753.

Wiesmann, U. & Neufeld, E. F. (1970) Scheie and Hurler syndromes: apparent identity of the biochemical defect. *Science*, **169**, 72.

MPS II

Beebe, R. T. & Formel, P. F. (1954) Gargoylism: sex-linked transmission in nine males. *Archives of Neurology and Psychiatry*, **63**, 681.

Berg, K., Danes, B. S. & Bearn, A. G. (1968) The linkage relation of the loci for the Xm serum system and the X-linked form of Hurler's syndrome (Hunter's syndrome). *American Journal of Human Genetics*, **20**, 398.

Danes, B. S. & Bearn, A. G. (1967) Hurler's syndrome: a genetic study of clones in cell culture, with particular references to the Lyon hypothesis. *Journal of Experimental Medicine*, **126**, 509.

DiFerrante, N. & Nichols, B. L. (1972) A case of the Hunter syndrome with progeny. *Johns Hopkins Medical Journal*, **130**, 325.

Liebaers, I. & Leufeld, E. F. (1976) Iduronate sulfatase activity in serum, lymphocytes, and fibroblasts: simplified diagnosis of the Hunter syndrome. *Pediatric Research*, **10/7**, 733.

Lim, T. W., Leder, I. G., Bach, G. & Neufeld, E. F. (1974) An assay for iduronate sulfatase (Hunter corrective factor). *Carbohydrate Research*, **37**, 103.

Niermeijer, M. F., Sachs, E. S., Jahodova, M., Tichelaar-Klepper, C., Kleijer, W. J. & Galjaard, H. (1976) Prenatal diagnosis of genetic disorders. *Journal of Medical Genetics*, **13**, 182.

Noja, A. (1946) A sex-linked type of gargoylism. *Acta pediatrica*, **33**, 267.

MPS III types A and B (Sanfilippo syndrome)

Figura, K. von, Logering, M., Mersmann, G. & Kresse, H. (1973) Sanfilippo B disease: serum assays for detection of homozygous and heterozygous individuals in three families. *Journal of Paediatrics*, **83**, 607.

Harper, P. S., Laurence, K. M., Parkes, A., Wusteman, F., Kresse, H., Figura, K. von, Ferguson-Smith, M., Duncan, D., Logan, R., Hall, F. & Whiteman, P, (1974) Sanfilippo A disease in the fetus. *Journal of Medical Genetics*, **11**, 123.

Kresse, H., Figura, K. von & Bartsocas, C. (1976) Clinical and biochemical findings in a family with Sanfilippo disease type C. *Clinical Genetics*, **10/6**, 364.

Liem, K. O., Giesberts, M. A. H., Van de Kamp, J. J. P., Van Pelt, J. F. & Hooghwinkel, G. J. M. (1976) Sanfilippo B disease in two related sibships. Biochemical studies in patients, parents and sibs. *Clinical Genetics*, **10/5**, 273.

Maroteaux, P., Frezal, J., Tahbaz-Zadeh & Lamy, M. (1966) Une observation familiale d'oligophrenie polydystrophique. *Journal de Génétique Humaine*, **15**, 93.

Matalon, R. & Dorfman, A. (1974) Sanfilippo A syndrome. Sulfamidase deficiency in cultured skin fibroblasts and liver. *Journal of Clinical Investigation*, **54**, 907.

O'Brien, J. S. (1972) Sanfilippo syndrome: profound deficiency of alpha-acetylglucosaminidase activity in organs and skin fibroblasts from type B patients. *Proceedings of the National Academy of Sciences of the United States of America*, **69**, 1720.

Sanfilippo, S. J., Podosin, R., Langer, L. O. & Good, R. A. (1963) Mental retardation associated with acid mucopolysacchariduria heparitin (sulfate type). *Journal of Paediatrics*, **63**, 837.

Spranger, J. (1972) The systemic mucopolysaccharidoses. *Ergebnisse der inneren Medizin und Kinderheilkunde*, 32, 165.
Van de Kamp, J. J. P., Van Pelt, J. F., Liem, K. O., Giesberts, M. A. H., Niepoth, L. T. M. & Staalman, C. R. (1976) Clinical variability in Sanfilippo B disease: a report on six patients in two related sibships. *Clinical Genetics*, 10/5, 279.

MPS IV (Morquio syndrome)
Beighton, P. & Craig, J. (1973) Atlanto-axial subluxation in the Morquio syndrome. *Journal of Bone and Joint Surgery*, 55B/3, 478.
Gadbois, P., Moreau, J. & Laberge, C. (1973) La maladie de Morquio dans la province de Quebec. *L'Union Medicale du Canada*, 102, 602.
Matalon, R., Arbogast, B., Justice, P., Brandt, I. & Dorfman, A. (1974) Morquio's syndrome: deficiency of a chondroitin sulfate N-acetyl hexosamine sulfate sulfatase. *Biochemical and Biophysical Research Communications*, 61, 759.
McKusick, V. A. (1975) Morquio Syndrome, non-keratosulfate excreting type. In *Mendelian Inheritance in Man*, 4th Edition, p.503. Baltimore, London: The Johns Hopkins University Press.
Morquio, L. (1929) Sur une forme de dystrophie osseouse familiale. *Bulletins de la Société de Pédiatrie de Paris*, 27, 145.

MPS VI (Maroteaux–Lamy syndrome)
Beratis, N. G., Turner, B. M., Weiss, R. & Hirschhorn, K. (1975) Arylsulfatase B deficiency in Maroteaux–Lamy syndrome. Cellular studies and carrier identification. *Pediatric Research*, 9, 475.
Maroteaux, R., Leveque, B., Marie, J. & Lamy, M. (1963) Une nouvelle dysostose avec elimination urinaire de chondroitinesulphate B. *Presse Medicale*, 71, 1849.
O'Brien, J. F., Cantz, M. & Spranger, J. (1974) Maroteaux–Lamy disease (mucopolysaccharidosis VI), subtype A; deficiency of N-acetyle galactosamine-4-sulfatase. *Biochemical and Biophysical Research Communications*, 60, 1170.
Spranger, J., Koch, F., McKusick, V. A., Natzschka, J., Wiedemann, H. R. & Zellweger, H. (1970) Mucopolysaccharidosis VI (Maroteaux–Lamy's disease). *Helvetica Paediatrica Acta*, 25, 337.
Spranger, J. W., Langer, L. O. & Wiedemann, H. R. (1974) Mucopolysaccharidosis VI. In *Bone Dysplasias*, p.166. Stuttgart: Gustav Fischer Verlag.

MPS VII — β-glucuronidase deficiency
Beaudet, A. L., DiFerrante, N. M., Ferry, G. D., Nichols, B. L. & Mullins, C. W. (1975) Variation in phenotypic expression of β-glucuronidase deficiency. *Journal of Paediatrics*, 86, 388.
Gehler, J., Cantz, M., Tolksdorf, M. & Spranger, J. (1974) Mucopolysaccharidosis VII: β-glucuronidase deficiency. *Humangenetika*, 23, 149.
Sly, W. S., Quinton, B. A., McAllister, W. H. & Rimoin, D. (1973) β-glucuronidase deficiency: report of clinical, radiologic and biochemical features of a new mucopolysaccharidosis. *Journal of Paediatrics*, 82, 249.

11. Mucolipidoses and sphingolipidoses

The mucolipidoses (MLS) are a group of rare disorders with phenotypic features which resemble the Hurler syndrome (Spranger and Wiedemann, 1970). The radiographic changes are those of dysostosis multiplex, in varying degree (Grossman and Dorst, 1973). The lymphocytes contain vacuoles and inclusions are present in cultured fibroblasts. However, with the exception of juvenile sulphatidosis, mucopolysacchariduria is absent. Accumulation of labelled sulphate has been demonstrated in cultured fibroblasts from several conditions in this group (Hieber *et al.*, 1975). Available evidence indicates that the MLS are all inherited as autosomal recessives. There has been some nosological confusion, but the following disorders are generally regarded as belonging to this category:

1. MLS I (lipomucopolysaccharidosis)
2. MLS II (I-cell disease)
3. MLS III (pseudo-Hurler polydystrophy)
4. Other mucolipidoses
    (a) Generalised gangliosidosis type I (pseudo-Hurler disease)
    (b) Generalised gangliosidosis type II (late infantile type)
    (c) Juvenile sulphatidosis (Austin type)
    (d) Fucosidosis
    (e) Mannosidosis
    (f) Miscellaneous MLS
5. Sphingolipidoses — Gaucher disease.

The sphingolipidoses are conventionally classified with the MPS and MLS as lysosomal storage disorders. Clinically important sphingolipidoses include Gaucher disease, Tay–Sachs disease, Fabry disease, Niemann–Pick disease and infantile metachromatic leucodystrophy. Significant involvement of the skeleton occurs only in Gaucher disease and discussion of the sphingolipidoses in this chapter is therefore confined to this condition.

1. MLS I

MLS I or lipomucopolysaccharidosis is a rare progressive disorder in which clinical and radiographic abnormalities resembling those of MPS I develop after infancy. Variability of life span might indicate that the condition is heterogeneous. The diagnosis may be suspected by recognition of vacuolated storage cells in the bone marrow and inclusions in cultured fibroblasts and diagnostic confirmation may be obtained by demonstration of characteristic electroencephalographic changes (Doose, Spranger and Warner, 1975).

There is considerable confusion in the literature, as a number of patients who were originally diagnosed as having various forms of MPS or MLS were subsequently thought to have MLS I. Conversely, an initial diagnosis of MLS I has been revised in several instances. In a recent review, Spranger (1975) commented that he knew of only three confirmed cases.

2. MLS II

The manifestations of MLS II or I-cell disease are reminiscent of a severe form of MPS. However, there is no mucopolysacchariduria. Irregular periosteal thickening and widening of the diaphyses of the long bones are important radiographic features. Death occurs in early childhood. Initially, the eponym 'Leroy' was applied to the disorder, following the early case descriptions by Leroy, Demars and Opitz (1969). The alternative term, I-cell disease, pertained to the inclusion bodies which are seen in cultured fibroblasts. As this histological abnormality is also present in the other MLS, the numerical designation MLS II is preferable.

MLS II is probably less rare than the other disorders in this group. Affected sibs and parental consanguinity have been recorded and abnormal inclusions have been demonstrated in cultured fibroblasts from obligatory heterozygotes. MLS II has been diagnosed antenatally in at risk fetuses by demonstration of increased activity of hydrolases in amniotic fluid and decreased activity of these enzymes in cultured amniotic fluid cells (Matsuda et al., 1975; Aula et al., 1975). In each instance, the diagnosis was subsequently confirmed by the recognition of typical inclusions in tissues and cultured fibroblasts from the aborted fetuses. The prenatal diagnosis of MLS II has been discussed by Gehler, Cantz and Spranger (1976). Rapola and Aula (1977) demonstrated that the characteristic ultrastructural changes are present in the placenta and suggested that histological study of this tissue might be of value in the investigation of recurrent unexplained spontaneous abortion.

3. MLS III

MLS III or pseudo-Hurler polydystrophy is compatible with survival into adulthood. Patients have a coarse facies, short stature, stiff joints and progressive mental deterioration (Maroteaux and Lamy, 1966). The radiographic features have been discussed by Nolte and Spranger (1976). With the accumulation of case reports, it has become obvious that the phenotype is very variable. As with other disorders in the general category of MLS, there has been some diagnostic confusion and overlap. Kelly et al. (1975) reviewed the features of 18 patients in whom the diagnosis was considered to be certain, and defined the evolution of the MLS phenotype. Several sets of affected sibs have been reported, including two adult brothers (Robinow, 1974) and two young sisters (Starreveld and Ashenhurst, 1975). Two affected sisters, in whom there was considerable disparity in the clinical stigmata, were reported by Gericke (1977). One of these patients had given birth to a normal daughter.

Cytoplasmic vacuolation is seen in fibroblasts cultured from conjunctival biopsy specimens, but obligatory heterozygotes are normal in this respect (Quigley and

Goldberg, 1971). Lysosomal enzyme activity is deficient in cultured fibroblasts and increased in the culture medium. Following electron microscopic studies of cultured fibroblasts from four patients with MLS III, Taylor *et al*. (1973) demonstrated that, at an ultrastructural level, this condition clearly differs from MLS II. McKusick (1975) suggested that there might be more than one genetic basis for the MLS III phenotype, and speculated that the MLS II and MLS III genes may be allelic.

4. OTHER MUCOLIPIDOSES

(a) Generalised gangliosidosis type I

In generalised gangliosidosis type I, also known as neurovisceral lipidosis and pseudo-Hurler disease, Hurler-like clinical and radiographic stigmata are present at birth. Widespread periosteal reaction in the long bones is an important diagnostic feature. Gangliosides accumulate in the liver and brain, and death occurs by the end of the second year. Activity of the enzyme β-galactosidase is deficient (Okada and O'Brien, 1968).

Metachromasia has been demonstrated in cultured skin fibroblasts from obligatory heterozygous parents of affected infants (Grossman and Danes, 1968). Antenatal diagnosis may be possible by microchemical assay of β-galactosidase activity in amniotic fluid cells (Niermeijer *et al*., 1976).

(b) Generalised gangliosidosis type II

In generalised gangliosidosis type II, manifestations do not become apparent until late infancy. Bone changes are of minor degree but accumulation of gangliosides in the brain leads to progressive deterioration and death by the end of the the first decade. Whereas all the isoenzymes of β-galactosidase are deficient in generalised gangliosidosis type I, only the B and C isoenzymes of β-galactosidase are deficient in type II (O'Brien *et al*., 1972).

(c) Juvenile sulphatidosis — Austin type

The facies become coarse in infancy and death from neurological infiltration occurs by adolescence. Stature is somewhat stunted but radiographic changes are mild. Excess mucopolysaccharides and sulphatides are present in the urine. The condition must be distinguished from metachromatic leucodystrophy, in which the enzyme arylsulphatase A is absent. In juvenile sulphatidosis, activity of arylsulphatides A, B and C is defective (Murphy *et al*., 1971). Less than 20 cases have been reported. Among these, there have been several sets of sibs, including those described by Mossakowski, Mathieson and Cummings (1961) and Austin (1965).

(d) Fucosidosis

Durand, Borrone and Della Cella (1969) reported two sibs who both died in early childhood from a progressive neurological disorder. These authors found excess fucose in the tissues, and termed the condition 'fucosidosis'. Deficient activity of the enzyme α-fucosidase was demonstrated in the livers of these infants. Patients with differing phenotypes, but with a similar deficiency of α-fucosidase in cultured skin fibroblasts were described by Patel, Watanabe and Zeman (1972). The condition is now conventionally subdivided into type I and II. However, Brill *et al.* (1975) concluded that it is not possible to differentiate between these two forms of fucosidosis on radiographic grounds alone.

A pregnancy in an Algerian family in which two sibs had proven fucosidosis was monitored by Poenaru *et al.* (1976). Normal levels of fucosidase were demonstrated in the amniotic fluid and amniotic fluid cells and the fetus was diagnosed as being unaffected. The validity of this assumption was confirmed after delivery.

(e) Mannosidosis

Ockerman (1967) described a boy in whom tissue α-mannosidase activity was defective. Subsequently, Ockerman, Autio and Norder (1973) mentioned two affected sisters in Hungary and three boys in Finland with the condition. Gehler *et al.* (1975) reported further cases, confirmed that the primary enzyme defect was detectable in cultured fibroblasts and suggested that prenatal diagnosis should be possible. Aylsworth *et al.* (1976) demonstrated reduced mannosidase activity in fibroblast cultures from obligatory heterozygotes. Booth, Chen and Nadler (1976) reported the clinical and biochemical manifestations in a kindred with affected adolescents and adults. The radiographic features of mannosidosis have been reviewed by Spranger, Gehler and Cantz (1976).

(f) Miscellaneous MLS

It is evident that there is further heterogeneity in the MLS group of disorders. For instance, Bundey, Ashenhurst and Dorst (1974) described two sisters who presented with the carpal tunnel syndrome in the middle of the first decade. They subsequently developed the radiographic features of dysostosis multiplex, together with cardiac and corneal changes. No mucopolysaccharides were detected in the urine. It seems possible that this condition could be an MLS. Similarly, Merlin *et al.* (1975) reviewed the features of four children with a disorder which was termed 'MLS IV'. Ocular problems and mental deficiency predominated and skeletal involvement was minimal.

5. SPHINGOLIPIDOSES (GAUCHER DISEASE)

Gaucher disease is the only member of the sphingolipidosis group of disorders in which bone changes are an important feature. In view of the fundamental relationship between the MLS and the sphingolipidoses, it is appropriate that Gaucher disease should be included in this section.

Three forms of Gaucher disease are recognised: the infantile cerebral type, the juvenile variety and the adult or chronic non-neuropathic type. They are all inherited as autosomal recessives and activity of the enzyme beta-glucosidase is defective in each. However, they differ in their clinical manifestations, and, at a phenotypic level, they are separate conditions.

The infantile and juvenile forms are lethal, due to accumulation of cerebrosides in the brain. Conversely, the adult form is compatible with a relatively normal life span. Splenomegaly is often the presenting feature of this type of Gaucher disease and skeletal complications include osteitis, pathological fractures and degenerative arthropathy. The radiographic features of 17 patients have been reviewed by Myers *et al.* (1975).

Genetics

The adult non-neuropathic form of Gaucher disease has a high prevalence in individuals of Ashkenazi Jewish stock. (The Ashkenazim are Jewish people of European derivation, while the Sephardic Jews had their origins in the Mediterranean region. This distinction is of clinical importance, as there are marked differences in the incidence of certain genetic disorders in these two populations.) Fried (1973) ascertained 100 patients with Gaucher's disease in Israel, estimated that the minimum prevalence was 1 in 100,000 and, after applying correction for various biases, calculated that the incidence at birth was 1 in 2500. The gene frequency of 0.02 for the Ashkenazim, which is based upon this figure, contrasts dramatically with the probable frequency of less than 0.005 in non-Ashkenazi Jewish populations (Fried, 1973). Gaucher disease is particularly common in the Jewish community of South Africa, in which there is a prevalence of at least 1 in 6000, with a carrier rate of between 1 in 20 and 1 in 30 (Beighton and Sacks, 1974).

The heterozygote can be detected by the recognition of diminished activity of β-glucosidase in leucocytes (Beutler and Kuhl, 1970). The results are not altogether consistent, and as the laboratory techniques are somewhat laborious, this method is not applicable to population screening. The enzymatic abnormality is expressed in cultured fibroblasts, and Gaucher disease should be detectable in the fetus by investigation of cultured amniotic fluid cells. However, to date there have been no reports of antenatal diagnosis by this means.

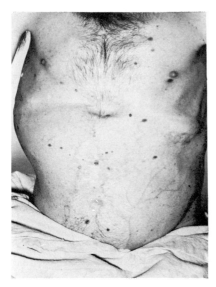 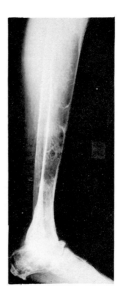

Fig. 11.1. (left) Gaucher disease; a middle-aged male with massive hepatomegaly, in the terminal stages of the disorder. Splenectomy had been undertaken several years previously.

Fig. 11.2. (right) Gaucher disease; diffuse rarifaction of the tibia and fibula give the characteristic 'soap bubble' appearance. (From Myers, H. S. *et al.* (1975) *British Journal of Radiology*, **48**, 465.)

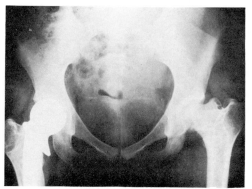

Fig. 11.3. (above) Gaucher disease; collapse of the femoral head is a frequent complication. Prosthetic replacement has proved to be of value. (Courtesy of Mr S. Sacks, FRCS, Johannesburg.)

Fig. 11.4. (right) Gaucher disease; autopsy specimen of the femora showing marrow infiltration and alteration of the normal bone contours.

REFERENCES

Mucolipidoses – preamble
Grossman, H. & Dorst, J. P. (1973) The mucopolysaccharidoses and mucolipidoses. *Progress in Paediatric Radiology*, **4**, 495.
Hieber, V., Distler, J., Jourdian, G. W. & Schmickel, R. (1975) Accumulation of 35 S-mucopolysaccharides in cultured mucolipidosis cells. *Birth Defects: Original Article Series*, **11/6**, 307.
Spranger, J. & Wiedemann, H. R. (1970) The genetic mucolipidoses: diagnosis and differential diagnosis. *Humangenitika*, **9**, 113.

MLS I
Doose, H., Spranger, J. & Warner, M. (1975) EEG in mucolipidosis I. *Neuropadiatrie*, **6/1**, 98.
Spranger, J. W. (1975) Mucolipidosis I. *Birth Defects: Original Article Series*, **11/6**, 279.

MLS II
Aula, P., Rapola, J., Autio, S., Raivio, K. & Karjalainen, O. (1975) Prenatal diagnosis and fetal pathology of I cell disease (mucolipidosis type II). *Journal of Paediatrics*, **87/2**, 221.
Gehler, J., Cantz, M., Stoeckenius, M. & Spranger, J. (1976) Prenatal diagnosis of mucolipidosis II (I cell disease). *European Journal of Paediatrics*, **122/3**, 201.
Leroy, J. G., Demars, R. I. & Optiz, J. M. (1969) Skeletal dysplasias. In *The Clinical Delineation of Birth Defects*, p. 174. New York: National Foundation.
Matsuda, I., Arashima, S., Mitsuyama, T., Oka, Y., Ikeuchi, T., Kaneko, Y. & Ishikawa, M. (1975) Prenatal diagnosis of I cell disease. *Humangenetika*, **30/1**, 69.
Rapola, J. & Aula, P. (1977) Morphology of the placenta in fetal I-cell disease. *Clinical Genetics*, **11**, 107.

MLS III
Gericke, G. S. (1977) Mucolipidosis III: two patients displaying genetic pleiotropism. *South African Medical Journal*, **51**, 140.
Kelly, T. E., Thomas, G. H., Taylor, H. A. & McKusick, V. A. (1975) Mucolipidosis III: clinical and laboratory findings. *Birth Defects: Original Article Series*, **11/6**, 295.

Leroy, J. G. & van Elsen, A. F. (1975) Natural history of a mucolipidosis. Twin girls discordant for MLS III. *Birth Defects: Original Article Series*, **11/6**, 325.
Maroteaux, P. & Lamy, M. (1966) La pseudo-polydystrophie de Hurler. *Presse Médicale*, **74**, 2889.
McKusick, V. A. (1975) Mucolipidosis III (pseudo-Hurler polydystrophy). In *Mendelian Inheritance in Man*, 4th Edition, p.504. Baltimore, London: The Johns Hopkins University Press.
Nolte, K. & Spranger, J. (1976) Early skeletal changes in mucolipidosis III. *Annals of Radiology*, **19/1**, 151.
Quigley, H. A. & Goldberg, M. F. (1971) Conjunctival ultrastructure in mucolipidosis III (pseudo-Hurler polydystrophy). *Investigative Ophthalmology*, **10**, 568.
Robinow, M. (1974) Mucolipidosis III. *Birth Defects: Original Article Series*, **10/10**, 267.
Starreveld, E. & Ashenhurst, E. M. (1975) Bilateral carpal tunnel syndrome in childhood: a report of two sisters with mucolipidosis III (pseudo-Hurler polydystrophy). *Neurology*, **25/3**, 234.
Taylor, H. A., Thomas, G. H., Miller, C. S., Kelly, T. E. & Siggers, D. (1973) Mucolipidosis III (pseudo-Hurler polydystrophy): cytological and ultrastructural observations of cultured fibroblast cells. *Clinical Genetics*, **4/5**, 388.

Generalised gangliosidosis type I

Grossman, H. & Danes, B. S. (1968) Neurovisceral storage disease: features and mode of inheritance. *Journal of Roentgenology, Radium Therapy and Nuclear Medicine*, **103**, 149.
Niermeijer, M. F., Sachs, E. S., Jahodova, M., Tichelaar-Klepper, C., Kleijer, W. J. & Galjaard, H. (1976) Prenatal diagnosis of genetic disorders. *Journal of Medical Genetics*, **13**, 182.
Okada, S. & O'Brien, J. S. (1968) Generalised gangliosidosis: beta-galactosidase deficiency. *Science*, **160**, 1002.

Generalised gangliosidosis type II

O'Brien, J. S., Ho, M. W., Veath, M. L., Wilson, J. F., Myers, G., Opitz, J. M., Zurhein, G. M., Spranger, J. W., Hartman, H. A., Haneberg, B. & Grosse, F. R. (1972) Juvenile Gm 1 gangliosidosis: clinical, pathological chemical and enzymatic studies. *Clinical Genetics*, **3**, 411.

Juvenile sulphatidosis

Austin, J. H. (1965) Metachromatic leukodystrophy. In *Medical Aspects of Mental Retardation*. Ed. Carter, C. C., p.768. Springfield Ill.: Charles C. Thomas.
Mossakowski, M., Mathieson, G. & Cummings, J. N. (1961) On the relationship of metachromatic leucodystrophy and amaurotic idiocy. *Brain*, **81**, 585.
Murphy, J. V., Wolfe, H. J., Balazs, E. A. & Moser, H. W. (1971) A patient with deficiency of arylsulfatases A, B, C, and steroid sulfatase, associated with storage of sulfatide, cholesterol sulfate and glycosaminoglycans. In *Lipid Storage Diseases: Enzymatic Defects and Clinical Implications*, p.67. Eds. Bernsohn, J. & Grossman, H. J. New York: Academic Press.

Fucosidosis

Brill, P. W., Beratis, N. G., Kousseff, B. G. & Hirschhorn, K. (1975) Roentgenographic findings in fucosidosis type II. *American Journal of Roentgenology, Radium Therapy and Nuclear Medicine*, **124/1**, 75.
Durand, P., Borrone, C. & Della Cella, G. (1969) Fucosidosis. *Journal of Paediatrics*, **75**, 665.
Patel, V., Watanabe, I. & Zeman, W. (1972) Deficiency of alpha-1-fucosidase. *Science*, **176**, 426.
Poenaru, L., Dreufus, J. C., Boue, J., Nicolesco, H., Ravise, N. & Bamberger, J. (1976) Prenatal diagnosis of fucosidosis. *Clinical Genetics*, **10/5**, 260.

Mannosidosis

Alsworth, A. S., Taylor, H. A., Stuat, C. M. & Thomas, G. H. (1976) Mannosidosis: phenotype of a severely affected child and characterization of a mannosidase activity in cultured fibroblasts from the patient and his parents. *Journal of Paediatrics*, **88/5**, 814.
Booth, C. W., Chen, K. K. & Nadler, H. L. (1976) Mannosidosis: clinical and biochemical studies in a family of affected adolescents and adults. *Journal of Paediatrics*, **88/5**, 821.
Gehler, J., Cantz, M., O'Brien, J. F., Tolksdorf, M. & Spranger, J. (1975) Mannosidosis: Clinical and biochemical findings. *Birth Defects: Original Article Series*, **11/6**, 269.
Ockerman, P. A. (1967) A generalized storage disorder resembling Hurler's syndrome. *Lancet*, **2**, 239.
Ockerman, P. A., Autio, S. & Norder, N. E. (1973) Diagnosis of mannosidosis. *Lancet*, **1**, 207.
Spranger, J., Gehler, J. & Cantz, M. (1976) The radiographic features of mannosidosis. *Radiology*, **119/2**, 401.

Miscellaneous MLS

Bundey, S. E., Ashenhurst, E. M. & Dorst, J. P. (1974) Mucolipidosis, probably a new variant with joint deformity and peripheral nerve dysfunction. *Birth Defects: Original Article Series*, **10/12**, 484.

Merlin, S., Livni, N., Berman, E. R. & Yatziv, S. (1975) Mucolipidosis IV: ocular, systemic and ultrastructural findings. *Investigative Ophthalmology*, **14/6**, 437.

Sphingolipidoses –Gaucher disease

Beutler, E. & Kuhl, W. (1970) The diagnosis of the adult type of Gaucher's disease and its carrier state by demonstration of deficiency of beta-glucosidase activity in peripheral blood leukocytes. *Journal of Laboratory and Clinical Medicine*, **76**, 747.

Beighton, P. & Sacks, S. (1974) Gaucher's disease in Southern Africa. *South African Medical Journal*, **48**, 1295.

Fried, K. (1973) Population study of chronic Gaucher's disease. *Israel Journal of Medical Science*, **9**, 1396.

Myers, H. S., Cremin, B. J., Beighton, P. & Sacks, S. (1975) Chronic Gaucher's disease: radiological findings in 17 South African cases. *British Journal of Radiology*, **48**, 465.

12. Craniofacial dysostoses

In this group of disorders, the major abnormalities are found in the bones of the skull and face. Categorisation is based upon the pattern of craniofacial changes and the extent of involvement of the extremities and other systems. Gorlin, Sedano and Boggs (1975) and Cohen (1975) have emphasised that the variations and combinations which may be present in these conditions preclude any simple classification.

1. Craniostenosis
2. Craniofacial dysostosis (Crouzon)
3. Acrocephalosyndactyly (Apert)
4. Acrocephalopolysyndactyly (Carpenter)
5. Mandibulofacial dysostosis (Treacher Collins)
6. Mandibular hypoplasia (Pierre Robin)
7. Oculomandibulofacial syndrome (Hallerman–Streiff).

Subcategories of several of these craniofacial dysostoses have been described, and there is no doubt that some are very heterogeneous. It must be emphasised that many conditions in which the skull and face are abnormal are not genetic, and these have been deliberately excluded from this account. Similarly, a number of genetic syndromes in which craniofacial dysostosis is a minor component have been considered in other chapters.

1. CRANIOSTENOSIS

Craniostenosis is the result of premature closure of the sutures of the skull. Cranial abnormalities of this type were recognised even in the distant past. For instance, Shou Lao, the Chinese God of Long Life was conventionally depicted with a tower-shaped skull, while Homer mentioned in the Iliad that the lame Thersites had a peaked head.

The abnormal configuration in craniostenosis is determined by the anatomical distribution of the sutures which are involved, the sequence in which the sutures close and by compensatory growth in non-affected regions. For this reason, in the familial forms of the disorder, there is often considerable variation in the manifestations between affected members of the same kindred. A number of descriptive designations have been applied to the various forms of skull configuration which result from craniostenosis.

Scaphocephaly   — a boat-shaped or long narrow skull
Acrocephaly or
Oxycephaly   — a sharp or pointed skull

Turricephaly	— a tower-shaped, high, broad and short head
Plagiocephaly	— asymmetry of the skull
Kleeblattschädel	— a trilobal or clover leaf skull (see thanatophoric dwarfism, Chapter 1).

Microcephaly and hydrocephaly are not included in this account as these abnormalities are secondary to alteration in the dimensions of the contents of the cranium and are not primary disturbances of growth of the bones of the skull. Craniostenosis may be associated with various skeletal and visceral abnormalities in a number of genetic syndromes. It can also result from intrauterine or perinatal trauma or infection. The pathogenesis of the craniostenoses has been reviewed in detail by Lepintre and Renier (1975).

The clinical importance of craniostenosis lies in the fact that growth and development of the brain may be impaired. Proptosis and visual disturbance are common complications and the facies may become grotesque. Radiographically, alteration in the shape of the cranium and obliteration of the sutures is apparent. Compensatory widening of non-affected sutures may occur, and 'digital' markings are often evident in the calvarium. Recent advances in surgical technology have had considerable impact upon the management of patients with conditions of this type (Archer *et al.*, 1974; Giuffre, Scarfo and Tomaccini, 1975).

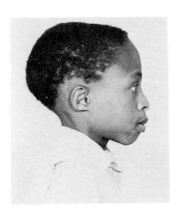 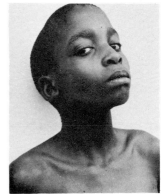

Fig. 12.1 (left) Craniostenosis; a boy with a long, narrow head (scaphocephaly).

Fig. 12.2. (right) Craniostenosis; a girl with a pointed skull (acrocephaly).

Genetics

The actual prevalence is unknown but isolated craniostenosis is probably relatively common. Gordon (1959) found six cases of craniostenosis when he undertook a survey of 600 African Negro children. Craniostenosis does not always have a genetic basis. Indeed, in a series of 519 patients in whom craniostenosis had been treated surgically, only about 5 per cent had affected kin (Shillito and Matson, 1968).

Generation to generation transmission of craniostenosis, consistent with dominant inheritance, has been reported by Murphy (1953), Gordon (1959), Bell, Clare and Wentworth (1961), and Nance and Engel (1967). Several of these reports were

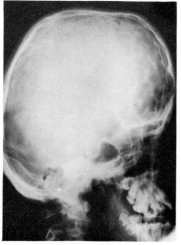 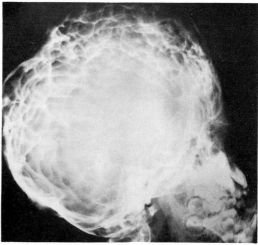

Fig. 12.3. (left) Craniostenosis; a short high skull (turricephaly).

Fig. 12.4. (right) Craniostenosis; the sutures are obliterated and digital markings are evident on the calvarium.

presented under the designation 'sarcocephaly'. However, there is often intrafamilial disparity in the anatomical abnormality, and the actual shape of the skull is probably not of great diagnostic importance. Three kindreds in which premature closure of the coronal suture was inherited as an autosomal dominant were reported by Kosnik, Gilbert and Sayers (1975). The existence of an autosomal recessive form of craniostenosis is indicated by reports of affected sibs with normal parents (Duguid, 1929; Gillot, Marchioni and Reibel, 1960; Gaudier et al., 1967; Armandares, 1970). Further evidence was provided by the recognition of many examples in an inbred religious isolate, the Amish of Ohio, USA (Cross and Opitz, 1969).

The parents of a child with uncomplicated craniostenosis may well seek counselling concerning the possibility of recurrence. In the absence of any genetic clues, such as affected kin, parental consanguinity or advanced paternal age, it is probably reasonable to give a relatively low risk estimate.

Craniostenosis — skeletal dysplasia syndromes
In addition to the uncomplicated inherited forms there are a number of atypical combinations of craniostenosis and abnormalities of the extremities. These conditions probably represent rare genetic entities.

(a) Gerold (1959) reported a brother and sister with turricephaly and radial aplasia.

(b) Lowry (1972) described two brothers with craniostenosis and congenital absence of the fibula. Their unaffected parents were consanguineous.

(c) Armendares et al. (1975) studied three brothers with craniostenosis, short stature and retinitis pigmentosa, and speculated that inheritance of this new entity was probably autosomal recessive.

(d) Sensenbrenner, Dorst and Owens (1975) described a brother and sister with premature closure of the sagittal suture, abnormal facies and hair and shortening of the fibula and phalanges.

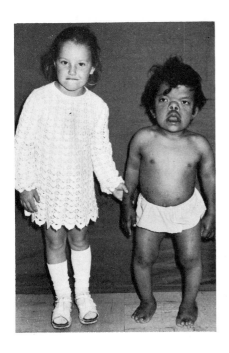
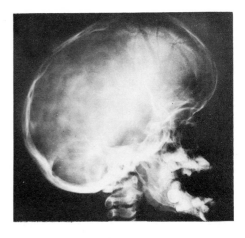

Fig. 12.5. (left) Craniostenosis; a 12-year-old girl with osteoglophonic dwarfism, depicted with a four-year-old child.

Fig. 12.6. (above) Craniostenosis; skull radiograph in osteoglophonic dwarfism, showing marked acrocephaly.

(e) Under the designation 'trigonocephaly' Hunter, Rudd and Hoffman (1976) reported a mother and son with craniostenosis, an unusual nose and deviation of the terminal phalanges of the fingers. The toes were broad, with reduplication of the terminal phalanges. Inheritance was apparently autosomal dominant.

(f) Keats, Smith and Sweet (1975) described a boy with acrocephaly, dwarfism and multiple fibrous metaphyseal defects. The author has encountered a 12 year old girl with the same condition. In view of the radiographic appearance of the bone lesions, the term 'osteoglophonic dwarfism', which has the Greek connotation of 'hollowed out', would be a suitable name for this disorder. The genetic basis has not been determined.

(g) Jackson *et al.* (1976) studied a large Amish kindred with craniostenosis, midfacial hypoplasia and foot abnormalities. This condition is apparently a private syndrome, which is inherited as an autosomal dominant. It seems to be a separate disorder from the autosomal recessive form of simple craniostenosis, which is also present in the Amish.

2. CRANIOFACIAL DYSOSTOSIS (CROUZON SYNDROME)

The main stigmata of the Crouzon syndrome or Crouzon craniofacial dysostosis are craniostenosis with midfacial hypoplasia, hypertelorism, proptosis and nasal beaking. Prominence of the eyes produces a 'frog face' appearance. Dental malocclusion and deafness are common concomitants. Although patients resemble each other, there is great diversity in the severity of the disorder, even in members of the same kindred. Radiographic changes include obliteration of sutures, digital impressions on the calvarium, shallowness of the orbits and small paranasal sinuses.

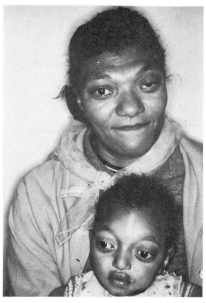

Fig. 12.7. Crouzon syndrome; a mother and child with proptosis, nasal beaking and hypertelorism. The extreme variability of the phenotype is of great importance in genetic counselling. (Courtesy of Dr S. Zieff, Cape Town.)

Genetics

The familial occurrence was mentioned in the initial case description of Crouzon (1912). Since that time, there have been many reports which are compatible with autosomal dominant inheritance. Kindreds with transmission through several generations were investigated by Shiller (1959) and Vulliamy and Normandale (1966). Appreciation of the potential variability of the stigmata is of importance in family studies and genetic counselling.

Franceschetti (1953), using the designation 'pseudo-Crouzon disease', described a disorder in which craniostenosis with prominent calvarial markings was associated with a normal facies. Dolivo and Gillieron (1955) reported a kindred in which members of four generations had the same condition. Inheritance is probably autosomal dominant.

3. ACROCEPHALOSYNDACTYLY (APERT SYNDROME)

Apert (1906), using the term 'acrocephalosyndactyly', described the association of craniostenosis and severe syndactyly. In a review of 54 cases, Blank (1960) distinguished 'typical' and 'atypical' forms of the disorder. The atypical forms were subsequently divided into a number of separate entities under the eponyms Vogt, Chotzen, Pfeiffer and Summit (*vide infra*). Spranger, Langer and Wiedemann (1974) and McKusick (1975) have also applied numerical designations to these conditions. However, there is overlap in these alternative classifications and some controversy concerning the assignment of certain patients to particular categories. Nevertheless,

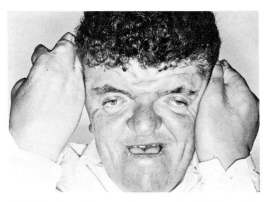

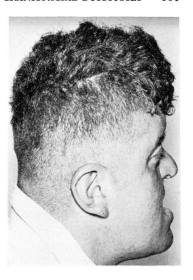

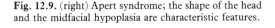

Fig. 12.8. (above) Apert syndrome; a male with the typical facies and syndactyly.

Fig. 12.9. (right) Apert syndrome; the shape of the head and the midfacial hypoplasia are characteristic features.

there is general agreement that the typical form or the 'Apert syndrome' should be designated acrocephalosyndactyly type I.

The major clinical features of the Apert syndrome are a high forehead, flat occiput, mid-facial hypoplasia and fusion of the second, third and fourth fingers and toes. Involvement of the first and fifth digits is variable. Other inconsistent abnormalities include mild shortening of the upper limbs, limitation of movement of large joints and cardiac, gastro-intestinal and renal anomalies. Ocular problems are not uncommon (Krueger and Ide, 1974). Many patients are mentally defective.

Genetics
More than 200 cases of the Apert syndrome have been reported and an incidence of 1 in 160,000 newborn has been estimated (Blank, 1960). The overwhelming majority

Fig. 12.10. Apert syndrome; an affected infant. In keeping with the concept of new dominant mutation, this child was the youngest of a large sibship and paternal age was increased.

of patients have been sporadic, with an equal sex frequency. Blank (1960) noted that the mean paternal age was increased. No sets of affected sibs have been encountered and there is no increased parental consanguinity. For these reasons, the evidence points to dominant inheritance, with the sporadic patients representing new mutations. Support for the dominant hypothesis is provided by descriptions of affected mothers and daughters (Hoover, Flatt and Weiss, 1970; Roberts and Hall, 1971 and Bergstrom, Neblett and Hemenway, 1972). In view of the mental deficiency and the severity of the malformations, it is not surprising that no male patient is known to have procreated.

Dodson et al. (1970) found an A-C translocation in an affected female and mentioned three other reports of abnormalities of A-group chromosomes. On the basis of these findings, it is possible that the abnormal gene in the Apert syndrome is located on a chromosome in this group.

Other forms of acrocephalosyndactyly
There is considerable variation in the stigmata of these uncommon disorders and the classification is neither exact nor complete. The problems involved in accurate diagnosis and categorisation have been discussed by Rochiccioli, Dutau and Marcou (1974). Attempts have been made to separate these conditions in terms of metabolic parameters based upon sarcosine loading (Minami, Olek and Wardenback, 1975).

(a) In the Vogt or Apert–Crouzon syndrome, the facies are said to resemble those of Crouzon syndrome, while involvement of the hands is of lesser degree than in the classical Apert form. All the reported patients have been sporadic. It is possible that these individuals had the Apert syndrome with unusually severe craniofacial manifestations, and that the Vogt syndrome does not exist as an entity in its own right.

(b) The Chotzen type of acrocephalosyndactyly is characterised by mild cranial changes and soft tissue syndactyly of the 2nd and 3rd fingers and 4th toes. There have been several reports of generation to generation transmission (Chotzen, 1932; Bartsocas, Weber and Crawford, 1970; Pruzansky et al., 1975). Pantke et al. (1975) reviewed the findings in six kindreds with 31 cases and concluded that inheritance was autosomal dominant.

(c) Pfeiffer (1964) reported a family with mild acrocephaly in association with cutaneous syndactyly of the second and third fingers and marked deviation of broadened thumbs and great toes. Other kindreds with an autosomal dominant pattern of inheritance have been described by Martsolf et al. (1971) and Saldino, Steinbach and Epstein (1972). This Pfeiffer type of acrocephalosyndactyly was probably present in six generations of a family described by Waardenburg, Franceschetti and Klein (1961). In a review, Navyeh and Friedman (1975) emphasised that the Pfeiffer syndrome was very rare.

(d) Two brothers with acrocephaly, obesity, cutaneous syndactyly and foot malformations were reported by Summitt (1969). Their unaffected parents were consanguineous. It is likely that this condition was inherited as an autosomal recessive.

4. ACROCEPHALOPOLYSYNDACTYLY (CARPENTER SYNDROME)

Acrocephalopolysyndactyly is distinguished from the acrocephalosyndactyly group of disorders by the presence of extra digits (Temtamy, 1966). Initially, two forms were recognised, type I, the Noack syndrome, and type II, the Carpenter syndrome. Subsequently, Pfeiffer (1969) restudied the kindred reported by Noack (1959) and suggested that the condition in question was acrocephalosyndactyly, Pfeiffer type. The same situation arose when Robinow and Sorauf (1975) investigated members of a large family in whom the Noack syndrome had been diagnosed and which conformed with the diagnostic criteria for the Pfeiffer syndrome. The numerical designations have now been discarded and the term 'Carpenter syndrome' has been accepted as being synonymous with 'acrocephalopolysyndactyly'.

Individuals with the Carpenter syndrome have craniostenosis, a high forehead, cutaneous syndactyly, and reduplication of the proximal phalanx of the thumb and great toe. Cardiac abnormalities, mental retardation, obesity and hypogonadism are additional features.

Genetics

Eaton *et al*. (1974) reported a pair of sibs and mentioned that 18 other cases could be identified in the literature. There has been no generation to generation transmission, but several other sets of affected sibs have been described (Carpenter, 1909; Temtamy, 1966). The author has examined an affected male infant whose sister died following cranial surgery in the neonatal period. The unaffected parents, who were of Afrikaner stock, were consanguineous. Inheritance of the Carpenter syndrome is autosomal recessive. This point is of practical significance, as the phenotype bears some resemblance to that of the Apert syndrome. This latter condition is transmitted as an autosomal dominant, and diagnostic accuracy is therefore vital when parents of an affected child request genetic counselling.

5. MANDIBULOFACIAL DYSOSTOSIS (TREACHER COLLINS SYNDROME)

Treacher Collins (1900) described a patient with an unusual configuration of the eyelids and malar bones. Subsequently, Franceschetti and Zwahlen (1944) and Franceschetti and Klein (1949) reported individuals with the condition, which they considered to be a new hereditary syndrome, under the title 'mandibulofacial dysostosis'. The features of 200 patients with the disorder were reviewed by Rogers (1964).

The main stigmata of the syndrome are deafness and a fish-like profile. The eyes have an anti-Mongoloid slant, the cheek bones are flattened and the mandible is poorly developed. Colobomata are present in the lower eyelids and the external ears are small and malformed. Roberts, Pruzansky and Aduss (1975) described the results of a radiocephalometric study in eight cases, and emphasised that patients with the condition tended to resemble each other, in spite of any diversity of their ethnic backgrounds. The pathogenesis of the Treacher Collins syndrome has been discussed from the embryological point of view by Poswillo (1975).

Genetics

Inheritance is autosomal dominant. Kindreds with generation to generation transmission have been reported by several authors, including Franceschetti and Klein (1949), Rovin *et al*. (1964), Rogers (1964) and Fazen, Elmore and Nadler (1967).

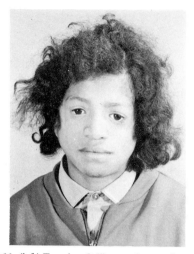 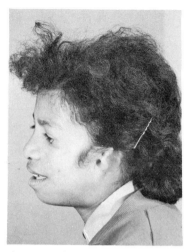

Fig. 12.11. (left) Treacher Collins syndrome; the eyes have an antimongoloid slant and a coloboma is present in the left lower eyelid. The cheek bones are flattened.

Fig. 12.12. (right) Treacher Collins syndrome; the external ear is malformed and the lower jaw is hypoplastic. Although this girl was profoundly deaf, a sibling with similar facial features had normal hearing.

Expression of the abnormal gene is notoriously variable. Indeed, in the author's experience, it is not unusual to reach the diagnosis in a mildly affected parent only after a severely affected infant has been produced. The Treacher Collins syndrome is not uncommon, and children with the disorder are encountered in the majority of special schools for the deaf. The fact that a severe hearing deficiency often compounds the handicap imposed by the abnormal facies is of importance when a parent with minimal manifestations seeks genetic guidance.

In an analysis of 15 cases, Herrmann *et al.* (1975) suggested that a 'Nager' form of acrofacial dysostosis could be differentiated by the presence of facial clefts and hypoplasia of thumb and radius. This disorder is inherited as an autosomal recessive and the implications for genetic counselling are therefore different to those in the Treacher Collins syndrome.

6. MANDIBULAR HYPOPLASIA (PIERRE ROBIN ANOMALAD)

Mandibular hypoplasia may occur in isolation or as a component of several well defined syndromes. The clinical importance of severe mandibular hypoplasia revolves around airway obstruction in the neonatal period and dental malocclusion and cosmetic appearance in later life.

In the strict sense, mandibular hypoplasia in association with a cleft palate and glossoptosis is known as the Pierre Robin anomalad. Hanson and Smith (1975) and Smith (1975) stressed that the term 'anomalad' is preferable to 'syndrome' in this context as the palatal cleft is secondary to the primary defect of mandibular development. For the sake of clarity, it must be emphasised that the designation 'Pierre Robin' or 'Robin' is often loosely applied to mandibular hypoplasia without palatal abnormalities.

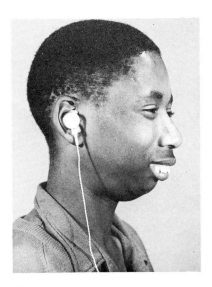

Fig.12.13. Mandibular hypoplasia; the lower jaw is very underdeveloped. No other members of this boy's large family had similar stigmata.

Genetics

Sibs with the Pierre Robin anomalad have been reported by Singh, Jaco and Voigna (1970), Shah, Pruzansky and Harris (1970) and Bixler and Christian (1971). Several had additional abnormalities, and it is possible that they were suffering from unrecognised syndromes. The majority of individuals with the isolated anomalad have been sporadic, and it must be assumed that the condition is usually non-genetic.

Other micrognathia syndromes

(a) In the cerebrocostomandibular syndrome, severe micrognathia is associated with rib defects and variable skeletal and cerebral abnormalities. As six of the seven reported cases have died in infancy, it is evident that the condition is potentially lethal. The disorder was delineated by Smith, Theiler and Schachenmann (1966) and subsequently encountered in a girl and her two brothers by McNicholl et al. (1970). Both groups of investigators found normal chromosomes in their patients. Other sporadic cases have been reported by Miller, Allen and Davis (1974) and Langer and Herrmann (1974). Inheritance is probably autosomal recessive.

(b) Micrognathia is a component of a number of dwarfism syndromes, including pycnodysostosis and the Langer form of mesomelic dwarfism. In the Weissenbacher–Zweymuller syndrome, micrognathia is associated with short limbs and dumbbell-shaped femora and humeri. Haller et al. (1975) reported two patients in whom the bone changes tended to regress with the passage of time. The genetic basis of this disorder has not been determined.

(c) Stanescu et al. (1963) described a kindred in which 11 members of three generations had brachycephaly and mandibular hypoplasia in association with shortening and massive thickening of the cortices of the long bones. The Stanescu craniofacial dysostosis syndrome, as it is now known, is evidently inherited as an autosomal dominant. The only other case report has emanated from Hall (1974), who reported a sporadic adult male. In this individual the typical cortical hyperostosis was lacking.

7. OCULOMANDIBULOFACIAL SYNDROME (HALLERMAN–STREIFF SYNDROME)

Reports concerning this disorder were published by Hallerman (1948) and Streiff (1950). The syndrome was finally delineated by Francois (1958), following a review of the literature and the eponym 'Hallerman–Streiff syndrome' is now in general use. An alternative descriptive designation which has found some favour is 'oculomandibulodyscephaly with hypotrichosis'.

Affected individuals have proportionate short stature, and facial features which include frontal bossing, micrognathia, microstomia, microphthalmia, congenital cataracts and a thin pointed nose. The hair is thin and sparse, and the skin is atrophic. Numerous inconsistent concomitants have been described. From the practical point of view, the ocular problems predominate.

Genetics

The genetic basis of the Hallerman–Streiff syndrome is uncertain. Although more than 60 cases have been reported, the majority have been sporadic. The sex distribution is equal. Dominant inheritance is suggested by the affected father and daughter studied by Guyard, Perdriel and Ceruti (1962). Nevertheless, as the father was married to a relative, the genetic situation in this kindred is still unclear. An atypical form of the syndrome in three generations of a kindred was reported by Koliopoulos and Palimeris (1975).

Fraser and Friedmann (1967) favoured the concept of dominant inheritance and suggested that the majority of patients represented new mutations. In view of the stigmata of the condition, fitness to reproduce is probably greatly diminished and if the Hallerman–Streiff syndrome is indeed dominant, the likelihood of transmission would be small. The chromosomes were normal in two cases studied by Forsius and de la Chapelle (1964).

Although the genetic background of the Hallerman–Streiff syndrome has not been elucidated, for the purposes of genetic counselling it is reasonable to assume that normal parents of an affected child have a low recurrence risk.

REFERENCES

Craniofacial dysostoses – preamble
Cohen, M. M. (1975) An etiologic and nosologic overview of craniosynostosis syndromes. *Birth Defects: Original Article Series*, **11/2**, 137.
Gorlin, R. J., Sedano, H. O. & Boggs, W. S. (1975) The face in the diagnosis of dysmorphogenesis. *Annales de Pédiatrie*, **4/3**, 10.

Craniostenosis
Archer, D. B., Gordon, D. S., Maguire, C. J. F. & Gleadhill, C. A. (1974) Ophthalmic aspects of craniosynostosis. *Transactions of the Ophthalmology Society of the United Kingdom*, **94/1**, 172.
Armendares, S. (1970) On the inheritance of craniostenosis. Study of thirteen families. *Journal de Génétique Humaine*, **18**, 121.
Armendares, S., Antillon, F., Del Castillo, V. & Jiminez, M. (1975) A newly recognised inherited syndrome of dwarfism, craniosynostosis, retinitis pigmentosa and multiple congenital malformations. *Birth Defects: Original Article Series*, **11/5**, 49.
Bell, H. S., Clare, F. B. & Wentworth, A. F. (1961) Case reports and technical notes on familial scaphocephaly. *Journal of Neurosurgery*, **18**, 239.
Cross, H. E. & Opitz, J. (1969) Craniostenosis in the Amish. *Journal of Paediatrics*, **75**, 1037.

Duguid, H. (1929) An instance of familial scaphocephaly. *Journal of Mental Science*, 75, 704.

Gaudier, B., Laine, E., Fontaine, G., Castier, C. & Farriaux, J. P. (1967) Les craniosynostoses (étude de vingt observations). *Archives Françaises de Pédiatrie*, 24, 775.

Gerold, M. (1959) Healing of a fracture in an unusual case of congenital anomaly of the upper extremities. *Zeitschrift für Chirurgie*, 34, 831.

Gillot, Marchioni, J. & Reibel, C. (1960) Craniostenose familiale. *Pédiatrie*, 15, 695.

Giuffre, R., Scarfo, C. B. & Tomaccini, D. (1975) Clinical and surgical notes on 78 cases of craniostenosis. *Minerva paediatrica*, 27/16, 949.

Gordon, H. (1959) Craniostenosis. *British Medical Journal*, 2, 792.

Hunter, A. G. W., Rudd, N. L. & Hoffmann, H. J. (1976) Trigonocephaly and associated minor anomalies in mother and son. *Journal of Medical Genetics*, 13/1, 77.

Jackson, C. E., Weiss, L., Renolds, W. A., Forman, T. F. & Petersen, J. A. (1976) Craniosynostosis, midfacial hypoplasia and foot abnormalities: an autosomal dominant phenotype in a large Amish Kindred. *Journal of Paediatrics*, 38, 963.

Keats, T. E., Smith, T. H. & Sweet, D. E. (1975) Craniofacial dysostosis with fibrous metaphyseal defects. *American Journal of Roentgenology, Radium Therapy and Nuclear Medicine*, 124, 271.

Kosnik, E. J., Gilbert, G. & Sayers, M. P. (1975) Familial inheritance of coronal craniosynostosis. *Developmental Medicine and Child Neurology*, 17/5, 630.

Lepintre, J. & Renier, D. (1975) Craniosynostosis. *Revue de Pédiatrie*, 11/8, 433.

Lowry, R. B. (1972) Congenital absence of the fibula and craniosynostosis in sibs. *Journal of Medical Genetics*, 9, 227.

Murphy, J. W. (1953) Familial scaphocephaly in father and son. *United States Armed Forces Medical Journal*, 4, 1496.

Nance, W. E. & Engel, E. (1967) Autosomal deletion mapping in man. *Science*, 155, 692.

Sensenbrenner, J. A., Dorst, J. P. & Owens, R. P. (1975) New syndrome of skeletal, dental and hair anomalies. *Birth Defects: Original Article Series*, 11/2, 372.

Shillito, J. & Matson, D. D. (1968) Craniostenosis: a review of 519 surgical patients. *Paediatrics*, 41, 829.

Crouzon syndrome

Crouzon, O. (1912) Dysostose cranio-facio hereditaire. *Bulletins et Mémoires de la Société Médicale des Hôpitaux de Paris*, 33, 545.

Dolivo, G. & Gillieron, J. D. (1955) Une famille de pseudo-Crouzon. *Confinia Neurologica*, 15, 114.

Franceschetti, A. (1953) Dysostose cranienne avec calotte cerebriforme (pseudo-Crouzon). *Confinia Neurologica*, 13, 161.

Shiller, J. G. (1959) Craniofacial dysostosis of Crouzon. A case report and pedigree with emphasis on hereditary. *Paediatrics*, 23, 107.

Vulliamy, D. G. & Normandale, P. A. (1966) Cranio-facial dysostosis in a Dorset family. *Archives of Diseases in Childhood*, 41, 375.

Acrocephalosyndactyly

Apert, M. E. (1906), De l'acrocephalosyndactylie. *Bulletins et Mémoires de la Société Médicale des Hôpitaux de Paris*, 23, 1310.

Bartsocas, C. S., Weber, A. L. & Crawford, J. D. (1970) Acrocephalosyndactyly type 3: Chotzen's syndrome. *Pediatrics*, 77, 267.

Bergstrom, L. V., Neblett, L. M. & Hemenway, W. G. (1972) Otologic manifestations of acrocephalosyndactyly. *Archives of Otolaryngology*, 96, 117.

Blank, C. E. (1960) Apert's syndrome (a type of acrocephalosyndactyly). Observations on a British series of thirty-nine cases. *Annals of Human Genetics*, 24, 151.

Chotzen, F. (1932) Eine eigenartige familiäre Entwicklungsstörung (akrocephalosyndaktylie, dysostosis craniofacialis und hypertelorismus). *Monatsschrift für Kinderheilkunde*, 55, 97.

Dodson, W. E., Museles, M., Kennedy, J. L. & Al-Aish, M. (1970) Acrocephalosyndactylia associated with a chromosomal translocation. *American Journal of Diseases of Children*, 120, 360.

Hoover, G. H., Flatt, A. E. & Weiss, M. W. (1970) The hand in Apert's syndrome. *Journal of Bone and Joint Surgery*, 52A, 878.

Krueger, J. L. & Ide, C. H. (1974) Acrocephalosyndactyly (Apert's syndrome). *Annals of Ophthalmology*, 6/8, 787.

Martsolf, J. T., Cracco, J. B., Carpenter, G. G. & O'Hara, A. E. (1971) Pfeiffer syndrome: an unusual type of acrocephalosyndactyly with broad thumbs and great toes. *American Journal of Diseases of Children*, 121, 257.

McKusick, V. A. (1975) Acrocephalosyndactyly type I (typical Apert syndrome). In *Mendelian Inheritance in Man*, 4th Edition, p.6. Baltimore, London: Johns Hopkins University Press.

Minami, R., Olek, K. & Wardenbach, P. (1975) Hypersarcosinemia with craniostenosis syndactylism syndrome. *Humangenetika*, **28**/2, 167.

Navyh, Y. & Friedman, A. (1976) Pfeiffer syndrome: report of a family and review of the literature. *Journal of Medical Genetics*, **13**/4, 277.

Pantke, O. A., Cohen, M. M. & Witkop, C. J. (1975) The Saethre Chotzen syndrome. *Birth Defects: Original Article Series*, **11**/2, 190.

Pfeiffer, R. A. (1964) Dominant erbliche Akrocephalosyndaktylie. *Zeitschrift für Kinderheilkunde*, **90**, 301.

Pruzansky, S., Pashayan, H., Kreiborg, S. & Miller, M. (1975) Roentgencephalometric studies of the premature craniofacial synosotoses: report of a family with the Saethre–Chotzen syndrome. *Birth Defects: Original Article Series*, **11**/2, 226.

Roberts, K. B. & Hall, J. B. (1971) Apert's acrocephalosyndactyly in mother and daughter. Cleft palate in the mother. *Birth Defects: Original Article Series*, **7**/7, 262.

Rochiccioli, P., Dutau, G. & Marcou, P. (1974) Acrocephalosyndactylia. Diagnostic problems with reference to three cases. *Journal de Génetique Humaine*, **22**/3, 269.

Saldino, R. M., Steinbach, H. L. & Epstein, C. J. (1972) Familial acrocephalosyndactyly (Pfeiffer syndrome). *American Journal of Roentgenology, Radium Therapy and Nuclear Medicine*, **116**, 609.

Spranger, J. W., Langer, L. O. & Wiedemann, H. R. (1974) In *Bone dysplasias*, p.261. Stuttgart: Gustav Fischer Verlag.

Summitt, R. L. (1969) Recessive acrocephalosyndactyly with normal intelligence. *Birth Defects: Original Article Series*, **5**/3, 35.

Waardenburg, P. J., Franschetti, A. & Klein, D. (1961) In *Genetics and Ophthalmology*, Volume 1, p.301. Springfield, Ill.: Charles C. Thomas.

Acrocephaly-polysyndactyly

Carpenter, G. (1909) Case of acrocephaly with other congenital malformations. *Proceedings of the Royal Society of Medicine*, **2**, 45.

Eaton, A. P., Sommer, A., Kontras, S. B. & Sayers, M. P. (1974) Carpenter syndrome — acrocephalopolysyndactyly type II. *Birth Defects: Original Article Series*, **10**/9, 249.

Noack, M. (1959) Ein Beitrag zum Krankheitsbild der Akrozephalosyndaktylie (Apert). *Archiv für Kinderheilkunde*, **160**, 168.

Pfeiffer, R. A. (1969) Associated deformities of the head and hands. *Birth Defects: Original Article Series*, **5**/3, 18.

Robinow, M. & Sorauf, T. J. (1975) Acrocephalopolysyndactyly type Noack in a large kindred. *Birth Defects: Original Article Series*, **11**/5, 99.

Temtamy, S. A. (1966) Carpenter's syndrome: acrocephalopolysyndactyly. An autosomal recessive syndrome. *Pediatrics*, **69**, 111.

Mandibulofacial dysostosis

Fazen, L. E., Elmore, J. & Nadler, H. L. (1967) Mandibulofacial dysostosis (Treacher Collins syndrome). *American Journal of Diseases of Children*, **113**, 405.

Franschetti, A. & Zwahlen, P. (1944) Un nouveau syndrome. La dysostose mandibulo-faciale. *Bulletin der Schweizerischen Akademie der Medizinischen Wissenschaften*, **1**, 60.

Franschetti, A. & Klein, D. (1949) Mandibulofacial dysostosis: new hereditary syndrome. *Acta ophthalmologica*, **27**, 143.

Hermann, J., Pallister, P. D., Kaveggia, E. G. & Opitz, J. M. (1975) Acrofacial dysostosis type Nager. *Birth Defects: Original Article Series*, **11**/5, 341.

Poswillo, D. (1975) The pathogenesis of the Treacher Collins syndrome (mandibulofacial dysostosis). *British Journal of Oral Surgery*, **13**/1, 1.

Roberts, F. G., Pruzansky, S. & Aduss, H. (1975) A radiocephalometric study of mandibulofacial dysostosis in man. *Archives of Oral Biology*, **10**/4, 265.

Rogers, B. O. (1964) Berry–Treacher Collins syndrome. A review of 200 cases. *British Journal of Plastic Surgery*, **17**, 109.

Rovin, S., Dachi, S. F., Borenstein, D. B. & Cotter, W. B. (1964) Mandibulofacial dysostosis, a familial study of five generations. *Pediatrics*, **65**, 215.

Treacher Collins, E. (1900) Cases with symmetrical congenital notches in the outer part of each lid and defective development of the malar bones. *Transactions of the Ophthalmology Society of the United Kingdom*, **20**, 190.

Mandibular hypoplasia

Bixler, D. & Christian, J. C. (1971) Pierre Robin syndrome occurring in two unrelated sibships. In *The Clinical Delineation of Birth Defects*, Volume 11, p.67. Baltimore: Williams and Wilkins.

Hall, J. G. (1974) Craniofacial dysostosis — either Stanesco dysostosis or a new entity. *Birth Defects: Original Article Series*, **10/12**, 521.

Haller, J. O., Berdon, W. E. & Robinow, M. (1975) The Weissenbacher Zweymuller syndrome of micrognathia and rhizomelic chondrodysplasia at birth with subsequent normal growth. *American Journal of Roentgenology, Radium Therapy and Nuclear Medicine*, **125/4**, 936.

Hanson, J. W. & Smith, D. W. (1975) U-shaped palatal defect in the Robin anomalad: developmental and clinical relevance. *Journal of Pediatrics*, **87/1**, 30.

Langer, L. O. & Herrmann, J. (1974) The cerebrocostomandibular syndrome. *Birth Defects: Original Article Series*, **10/7**, 167.

McNicholl, B., Egan-Mitchell, B. & Murray, J. P. (1970) cerebrocostomandibular syndrome. *Archives of Diseases in Childhood*, **45**, 421.

Miller, K. E., Allen, R. P. & Davis, W. S. (1972) Rib gap defects with micrognathia. *American Journal of Roentgenology, Radium Therapy and Nuclear Medicine*, **114**, 253.

Singh, R. P., Jaco, N. T. & Vigna, V. (1970) Pierre Robin syndrome in siblings. *American Journal of Diseases of Children*, **120**, 560.

Shah, C. V., Pruzansky, S. & Harris, W. S. (1970) Cardiac malformations with facial clefts. *American Journal of Diseases of Children*, **114**, 238.

Smith, D. W., Theiler, K. & Schachenmann, G. (1966) Rib-gap with micrognathia, malformed tracheal cartilages, and redundant skin: a new pattern of defective development. *Journal of Pediatrics*, **69**, 799.

Smith, D. W. (1975) Classification, nomenclature and naming of morphologic defects. *Journal of Pediatrics*, **87**, 162.

Stanesco, V., Maximilian, C., Poenaru, S., Florea, I., Stanesco, R., Ionesco, V. & Ioanitiu, D. (1963) Syndrome hereditaire dominant. *Revue Française d'Endocrinologie Clinique, Nutrition et Metabolisme*, **4**, 219.

Oculomandibulorfacial syndrome

Forsius, H. & De La Chapelle, A. (1964) Dyscephalia oculomandibulo-facialis. Two cases in which the chromosomes were studied. *Annals of Pediatrics*, **10**, 280.

François, J. (1958) A new syndrome: dyscephalia with bird face and dental anomalies, nanism, hypotrichosis, cutaneous atrophy, microphthalmia and congenital cataract. *Archives of Ophthalmology*, **60**, 842.

Fraser, G. R. & Friedmann, A. I. (1967) The causes of blindness in childhood. In *A Study of 776 Children with severe Visual Handicaps*, p.89. Baltimore: Johns Hopkins Press.

Guyard, M., Perdriel, G. & Ceruti, F. (1976) On two cases of cranial dysostosis with 'bird head'. *Bulletins et mémoires de la Société Française d'Ophthalmologie*, **62**, 443.

Hallerman, W. (1948) Vogelgesicht und Cataracta congenita. *Klinische Monatsblätter für Augenheilkunde*, **113**, 315.

Koliopoulos, J. & Palimerie, G. (1975) Atypical Hallerman–Streiff–Francois syndrome in three successive generations. *Journal of Pediatrics and Ophthalmology*, **12/4**, 235.

Streiff, E. B. (1950) Dysmorphie mandibulo-faciale (tete d'oiseau) et alteration oculaire. *Ophthalmologica*, **120**, 79.

13. Vertebral dysostoses

Involvement of the vertebral column is the predominant feature of the disorders in this group. The shoulder girdle and ribs are sometimes malformed and extra skeletal anomalies may be present.

1. Klippel–Feil anomalad
2. Cervicooculoacoustic syndrome (Wildervanck)
3. Oculoauriculovertebral syndrome (Goldenhar)
4. Spondylocostal dysostosis
5. Sprengel deformity.

Some of these conditions are relatively common. Nevertheless, they are not clearly defined and there is considerable heterogeneity and overlap among them. Although a few uncommon variants are inherited in a Mendelian fashion, the majority have a multifactorial background.

1. KLIPPEL–FEIL ANOMALAD

The primary defect in the Klippel–Feil anomalad is fusion of the cervical vertebrae. The anomalad may exist in isolation, or in association with a wide variety of concomitants. It is also a component of a number of specific disorders, such as the Wildervanck syndrome (*vide infra*).

The neck is short, with a low posterior hair line and a limited range of movement. Torticollis, webbing of the neck, scoliosis and asymmetrical elevation of the scapula are often present. It has been reported that up to 30 per cent of cases have some degree of deafness (Palant and Carter, 1972). Associated spinal defects include cervical meningomyelocele, syringomyelia and spinal dysraphism (Sherk, Shut and Chung, 1974). Mental deficiency, cleft palate and cardiac malformations are inconsistent features (Schey, 1976). Renal abnormalities were present in 30 per cent of a series of 50 patients studied by Hensinger, Lang and MacEwen (1974), and in 64 per cent of 39 patients investigated by Moore, Matthews and Rabinowitz (1975). These latter authors made the very reasonable suggestion that a routine intravenous pyelogram is indicated in all patients with the Klippel–Feil anomalad.

Genetics

The birth frequency is approximately 1 in 40,000, with a female preponderance of about 20 per cent (Gorlin, Pindborg and Cohen, 1976). This excess of affected girls might be the result of misdiagnosis of the Wildervanck syndrome (cervicooculoacoustic syndrome), which is virtually confined to females, and in which

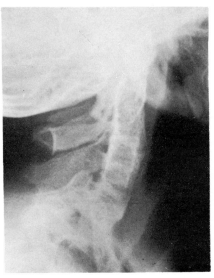

Fig. 13.1. (above) Klippel–Feil anomalad; the neck is short, and the posterior hair-line is low.

Fig. 13.2. (right) Klippel–Feil anomalad; fusion of the cervical vertebrae.

fusion of the cervical vertebrae is associated with perceptive deafness and abducens nerve palsy.

The Klippel–Feil anomalad is conventionally subclassified into three types, on a basis of the anatomical distribution of the vertebral abnormalities.

Type I   fusion of cervical and upper thoracic vertebrae.
Type II   localised fusion of cervical vertebrae.
Type III fusion of cervical and lower thoracic or lumbar vertebrae.

It is by no means proven that this subgrouping reflects true heterogeneity, although Gunderson *et al.* (1967) found a dominant pattern of inheritance in kindreds with the type II abnormality. However, the great majority of cases are sporadic. Dominant inheritance with incomplete expression has been advanced as an explanation of the presence of minor manifestations in relatives. Alternatively, the anomalad may be the result of the action of an unknown environmental agent upon a genetically predisposed fetus.

A few non-specific chromosomal abnormalities have been reported. For instance, a 4/14 chromosome translocation was detected in an obese boy with the Klippel–Feil anomalad (Berdel and Burmeister, 1974). The significance of these cytogenetic findings is uncertain.

Following an epidemiological study of 337 patients with congenital vertebral abnormalities, who had presented with scoliosis, Wynne-Davies (1975) suggested that multiple vertebral anomalies might be aetiologically related to spina bifida and anencephaly and estimated that subsequent sibs had a recurrence risk of about 10 per cent for any of these defects. This observation has important implications for pregnancy screening and antenatal diagnosis. However, it must be emphasised that the vertebral anomalies in question were generally more extensive than those of the Klippel–Feil anomalad.

2. CERVICOOCULOACOUSTIC SYNDROME (WILDERVANCK SYNDROME)

The cervicooculoacoustic or Wildervanck syndrome comprises the Klippel–Feil anomalad, congenital neural or conductive deafness, palsy of the sixth cranial nerve and retraction of the eyeball (Everberg, Ratjen and Sorenson, 1963; Wildervanck, Hoeksema and Penning, 1966). The ocular features, taken alone, are termed the Duane syndrome. Other inconsistent abnormalities include mental retardation, cleft palate, epibulbar dermoids and anomalies of the external ears. The condition is a well recognised cause of childhood deafness. There is considerable phenotypic overlap between the Wildervanck syndrome, the Klippel–Feil anomalad and the oculoauriculovertebral or Goldenhar syndrome (*vide infra*). Indeed, it is sometimes difficult to assign a patient to a specific category in this group of disorders.

Genetics
There is conjecture and controversy concerning the pathogenesis of the Wildervanck syndrome. As the vast majority of patients are females, it has been proposed that the condition might be an X-linked dominant, with lethality in the hemizygous male (McKusick, 1975). An alternative earlier explanation was dominant inheritance with non-penetrance and very variable expression (Kirkham, 1970). Konigsmark and Gorlin (1976) suggested that multifactorial inheritance would account for the great diversity of features and the familial tendency. However, neither of these latter hypotheses explains the female preponderance.

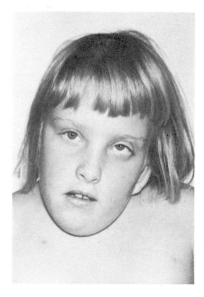

Fig. 13.3. Wildervanck syndrome; a girl with a short rigid neck, deafness and a palsy of the left sixth cranial nerve.

3. OCULOAURICULOVERTEBRAL DYSPLASIA (GOLDENHAR SYNDROME)

Oculoauriculovertebral dysplasia is also known as the Goldenhar syndrome. The components are facial asymmetry, mandibular hypoplasia, abnormal external ears, epibulbar dermoids, colobomata of the upper eyelids and vertebral abnormalities (Goldenhar, 1952). The condition is clinically similar to the Treacher Collins syndrome and the Wildervanck or cervicooculoacoustic syndrome. The similarity of the descriptive designations in this group of disorders is a source of confusion, and for this reason, there is obvious advantage in the retention of the eponyms.

Genetics
The Goldenhar syndrome was inherited as an autosomal dominant in the large kindred studied by Summit (1969). Conversely, autosomal recessive inheritance was possible in the affected sibs with normal parents reported by Saraux, Grignon and Dhermy (1963) and Krause (1970). This anomalous situation could be the result of heterogeneity or diagnostic confusion. Many patients have been sporadic, including those mentioned by Labrune and Choulot (1974). Papp, Gardó and Walawska

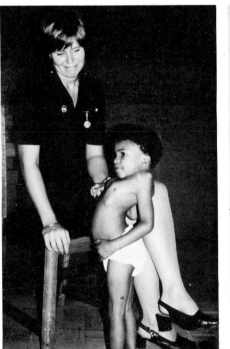 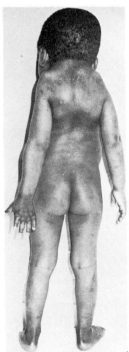

Fig. 13.4. (left) Spondylocostal dysplasia; an eight-year-old girl with gross abnormalities of the vertebrae and ribs.

Fig. 13.5. (right) Spondylocostal dysplasia; the neck is short and rigid. An elder sister is similarly affected, but the parents and her other sibs are normal. Inheritance in this kindred is autosomal recessive.

(1974) described monozygotic twins who were discordant for the condition. A deletion of the short arm of chromosome 18 was reported by Buffoni *et al.* (1976). Konigsmark and Gorlin (1976) suggested that inheritance of the Goldenhar syndrome was probably multifactorial.

Monoud, Klein and Weber (1975) described a young boy with the Goldenhar syndrome whose mother had accidently ingested a large quantity of vitamin A during the second month of pregnancy. These authors discussed the possible teratogenic effects of vitamin A and extensively reviewed the other aetiological factors which had previously been reported in the syndrome.

Four patients with a unilateral blepharooculocranial dysplasia, which bore some resemblance to the Goldenhar syndrome, were reported by Lund (1974). There was no evidence to indicate that this particular disorder had a genetic basis.

4. SPONDYLOCOSTAL DYSOSTOSES

The spondylocostal dysostoses or dysplasias are a heterogeneous group of disorders in which abnormalities of vertebral segmentation are associated with distortion, fusion or absence of the ribs. Clinically, the trunk is shortened, with thoracic asymmetry and spinal deformity. Spinal cord compression is a potentially dangerous complication. However, prognostication is not easy, as the characteristics of the individual entities within this category have not been clearly defined.

Genetics

A kindred with autosomal dominant inheritance of a spondylocostal dysplasia, which presented as short trunked dwarfism, was reported by Rimoin, Fletcher and McKusick (1968). Transmission was probably autosomal recessive in a consanguineous family in which four severely affected sibs died in infancy (Lamy, Palmer and Merrit, 1967). Subsequently, Langer and Moe (1975) described an Iranian brother and sister with abnormalities of the thoracic vertebrae and ribs, in association with anomalies in the carpus. The parents were first cousins and the condition, which seems to be yet another distinct entity, was apparently inherited as an autosomal recessive.

Sibs with multiple hemivertebrae, with or without rib abnormalities, have been described by Caffey (1967) and Bartsocas *et al.* (1974). Norum (1969) reported affected sibships in an inbred community. The relationship of these disorders and the other spondylocostal dysplasias is uncertain. However, inheritance was evidently autosomal recessive.

Hemivertebrae and rib anomalies were present in an infant born to a woman who had taken lysergic acid diethylamide (LSD) in early pregnancy (Eller and Morton, 1970), and in a mother and daughter with 14-15 chromosomal translocations (DeGrouchy *et al.*, 1963). It is not known if the spondylocostal malformations in these individuals were causally related to these exogenous or endogenous factors.

Other forms of spondylocostal dysplasia

(a) Under the designation 'cerebrofaciothoracic dysplasia', Pascual Castroviejo *et al.* (1975) described two girls and a boy with mental retardation, an unusual facies, vertebral anomalies and deformity of the ribs. The authors considered that this disorder represented a new syndrome, and suggested that inheritance might be autosomal recessive.

(b) Perez Comas and Garcia Castro (1974) reviewed the features of 'occipito-facial-cervico-thoracico-abdomino-digital dysplasia'. This anatomical litany is somewhat cumbersome and the original eponymous designation 'Jarcho–Levin syndrome' has considerable merit. In this rare condition, which is lethal in infancy, the ribs are often decreased in number and the thorax has a radiographic crab-like configuration.

(c) Abnormalities of the ribs and vertebrae, together with short metacarpals and mandibular cysts, are a feature of the naevoid basal cell carcinoma syndrome (Koutnik et al., 1975). The skeletal abnormalities are overshadowed by the dermal naevi, which are prone to malignant degeneration. There have been over 200 reports of the condition, which is inherited as an autosomal dominant (Gorlin and Goltz, 1960; Ferrier and Hinrichs, 1967; Lile, Rogers and Gerald, 1968).

5. SPRENGEL DEFORMITY

Sprengel's shoulder, or congenital elevation of the scapula, may occur as an isolated anomaly. However, it is often accompanied by abnormalities of the vertebrae or ribs and it is a frequent concomitant of the Klippel–Feil anomalad. The deformity may be unilateral or bilateral. The literature concerning congenital abnormalities of the scapula, including the Sprengel deformity, has been reviewed by McClure and Raney (1975).

Genetics

There have been reports of members of several generations of a kindred having uncomplicated Sprengel deformity (Engel, 1943), and it is evident that the condition can be transmitted as a dominant trait. Nevertheless, the majority of cases are sporadic and the disorder does not usually have a genetic basis.

REFERENCES

Klippel–Feil anomalad
Berdel, D. & Burmeister, W. (1974) Pickwickian and Klippel–Feil syndromes in a boy aged 12 years. Klinische Paediatrie, 106/6, 467.
Gorlin, R. J., Pindborg, J. J. & Cohen, M. M. (1976) In Syndromes of the Head and Neck, 2nd Edition, p.408. New York: McGraw-Hill Book Company.
Gunderson, C. G., Greenspan, R. H., Glaser, G. H. & Lubs, H. A. (1967) The Klippel–Feil syndrome: genetic and clinical re-evaluation of cervical fusion. Medicine, 46, 491.
Hensinger, R. N., Lang, J. E. & Macewan, G. D. (1974) Klippel–Feil syndrome. A constellation of associated anomalies. Journal of Bone and Joint Surgery, 56/6, 1246.
Moore, W. B., Matthews, T. J. & Rabinowitz, R. (1975) Genitourinary anomalies associated with Klippel–Feil syndrome. Journal of Bone and Joint Surgery, 57/3, 355.
Palant, D. J. & Carter, B. L. (1972) Klippel–Feil syndrome and deafness. American Journal of Diseases of Children, 123, 218.
Sherk, H. H., Shut, L. & Chung, S. (1974) Iniencephalic deformity of the cervical spine with Klippel–Feil anomalies and congenital elevation of the scapula. Report of three cases. Journal of Bone and Joint Surgery, 56/6, 1254.

Schey, W. (1976) Vertebral malformations and associated somaticovisceral abnormalities. *Clinical Radiology*, **27**, 341.
Wynne-Davies, R. (1975) Congenital vertebral anomalies: aetiology and relationship to spina bifida cystica. *Journal of Medical Genetics*, **12/3**, 280.

Cervicooculoacoustic syndrome
Everberg, G., Ratjen, E. & Sorensen, H. (1936) Klippel–Feil's syndrome associated with deafness and retraction of the eyeball. *British Journal of Radiology*, **36**, 562.
Kirkham, T. H. (1970) Inheritance of Duane's syndrome. *British Journal of Ophthalmology*, **54**, 323.
Konigsmark, B. W. & Gorlin, R. J. (1976) Klippel–Feil anomolad and abducens paralysis with retracted bulb and sensorineural or conduction deafness. In *Genetic and Metabolic Deafness*, p.188. Philadelphia: W. B. Saunders Company.
McKusick, V. A. (1975) In *Mendelian Inheritance in Man*, p.663. 4th Edition. Baltimore and London: Johns Hopkins University Press.
Wildervanck, L. S., Hoeksema, P. E. & Penning, L. (1966) Radiological examination of the inner ear of deaf-mutes presenting the cervicooculoacusticus syndrome. *Acta otolaryngology*, **61**, 445.

Oculoauriculovertebral dysplasia
Buffoni, L., Tarateta, A. & Aicardi, G. (1976) Hypophyseal nanism and Goldenhar multiple deformities in a subject with deletion of the short arm of chromosome 18. *Minerva paediatrica*, **28/12**, 716.
Goldenhar, M. (1952) Associations malformatives de l'oeil et de l'orielle. En particulier, le syndrome: dermoide epibulbaire-appendices auriculaires-fistula auris congenita et ses relations avec la dysostose mandibulo-faciale. *Journal de Génetique Humaine*, **1**, 243.
Konigsmark, B. W. & Gorlin, R. J. (1976) Klippel–Feil anomalad and abducens paralysis with retracted bulb and sensorineural or conduction deafness. In *Genetic and Metabolic Deafness*, p.189. Philadelphia: W. B. Saunders Company.
Krause, U. (1970) The syndrome of Goldenhar affecting two siblings. *Acta ophthalmology*, **48**, 494.
Labrune, B. & Choulot, J. J. (1974) Goldenhar's oculoauriculovertebral syndrome. *Semaine des Hôpitaux de Paris*, **50/39**, 713.
Lund, O. E. (1974) Blepharooculocranial dysplasia: a hitherto unknown syndrome. *Medizinische Klinik*, **69/42**, 1715.
Mounoud, R. L., Klein, D. & Weber, F. (1975) A case of Goldenhar's syndrome: acute maternal vitamin A poisoning during pregnancy. *Journal de Génetique Humaine*, **23/2**, 135.
Papp, Z., Gardo, S. & Walawska, J. (1974) Probable monozygotic twins with discordance for Goldenhar syndrome. *Clinical Genetics*, **5/2**, 86.
Saraux, H., Grignon, J. L. & Dhermy, P. (1963) À propos d'une observation familiale de syndrome de Franceschetti-Goldenhar. *Bulletins et Mémoires de la Société Française d'Ophthalmologie*, **63**, 705.
Summitt, R. L. (1969) Familial Goldenhar syndrome. In *The Clinical Delineation of Birth Defects*. 5/2 p.106. New York: National Foundation.

Spondylocostal dysostoses
Bartsocas, C. S., Kiossoglou, K. A., Papas, C. V., Xanthou-Tsingoglou, M., Anagnostakis, D. E. & Daskalopoulou, H. D. (1974) Costovertebral dysplasia. *Birth Defects: Original Article Series*, **10/9**, 221.
Caffey, J. P. (1967) Normal vertebral column. In *Pediatric X-ray diagnosis*, 5th Edition, p.1101. Chicago: Year Book Medical Publishers.
De Grouchy, J., Mlynarski, J. C., Maroteaux, P., Lamy, M., Deshaies, G., Benichou, C. & Salmon, C. (1963) Syndrome polydysspondylique par translocation 14-15 et dyschondrosteose chez un meme sujet. Segregation familiale. *Comptes Rendus Hebadomadaires des Séances de l'Académie*, **256**, 1614.
Eller, J. L. & Morton, J. M. (1970) Bizarre deformities in offspring of user of lysergic acid diethylamide. *New England Journal of Medicine*, **283**, 395.
Ferrier, P. E. & Hinrichs, W. L. (1967) Basal-cell carcinoma syndrome. *American Journal of Diseases of Children*, **113**, 538.
Gorlin, R. J. & Goltz, R. W. (1960) Multiple nevoid basal-cell epithelioma, jaw cysts and bifid rib: a syndrome. *New England Journal of Medicine*, **262**, 908.
Koutnik, A. W., Kolodny, S. C., Hooker, S. P. & Roche, W. C. (1975) Multiple nevoid basal cell epithelioma, cysts of the jaw and bifid rib syndrome: report of a case. *Journal of Oral Surgery*, **33/9**, 686.
Lamy, N. W., Palmer, C. G. & Merrit, A. D. (1966) A syndrome of bizarre vertebral anomalies. *Journal of Pediatrics*, **69**, 1121.
Langer, L. O. & Moe, J. H. (1975) A recessive form of congenital scoliosis different from spondylothoracic dysplasia. *Birth Defects: Original Article Series*, **11/6**, 83.

Lile, H. A., Rogers, J. F. & Gerald, B. (1968) The basal cell nevus syndrome. *American Journal of Roentgenology, Radium Therapy and Nuclear Medicine*, 103, 214.

Norum, R. A. (1969) Costovertebral anomalies with apparent recessive inheritance. In *The Clinical Delineation of Birth Defects*, Volume 4, p.326. New York: National Foundation.

Pascual Castroviejo, I., Santolaya, J. M. & Lopez Martin, V. (1975) Cerebro-facio-thoracic dysplasia: report of three cases. *Development Medicine and Child Neurology*, 17/3, 343.

Perez Comas, A. & Garcia Castro, J. M. (1974) Occipitofacial-cervico-thoracic-abdomino-digital dysplasia: Jarcho-Levin syndrome of vertebral anomalies. *Journal of Pediatrics*, 85/3, 388.

Rimoin, D., Fletcher, B. D. & McKusick, V. A. (1968) Spondylocostal dysplasia: a dominantly inherited form of short-trunked dwarfism. *American Journal of Medicine*, 45, 948.

Sprengel deformity

Engel, D. (1943) The etiology of the undescended scapula and related syndromes. *Journal of Bone and Joint Surgery*, 25, 613.

McClure, J. G. & Raney, B. R. (1975) Anomalies of the scapula. *Clinical Orthopaedics and Related Research*, 110, 22.

14. Stiff joint syndromes

Articular rigidity is a component of many well defined genetic entities, such as diastrophic dwarfism, the Freeman–Sheldon syndrome and the Schwartz syndrome. These conditions fall into a variety of nosological categories and several have been considered elsewhere. The following subgroups form the subject of this section:

1. Arthrogryposis multiplex congenita
2. Rigidity syndromes
   (a) Cerebro-oculofacio-skeletal syndrome
   (b) Goodman camptodactyly syndrome
   (c) Emery–Nelson syndrome
   (d) Liebenberg syndrome
   (e) Mietens syndrome
   (f) Ankylosis–pulmonary hypoplasia syndrome
   (g) Kuskokwim disease
3. Stiff hand–foot syndromes
   (a) Camptodactyly–cleft palate–clubfoot syndrome
   (b) Digitotalar dysmorphism
   (c) Digital-ulnar drift syndrome
   (d) Digital deviation–clubfoot syndrome
   (e) Tel-Hashomer camptodactyly syndrome
   (f) Dominant metatarsus varus
4. Flexed digit syndromes
5. Adducted thumb syndromes.

Restricted joint movement may result from neurological damage, as in meningomyelocele and cerebral palsy, and from rheumatic disease during childhood. Diminution in articular mobility may also be the consequence of dislocation or subluxation, which are common complications in a number of the syndromes in which joint laxity predominates. These 'secondary' forms of joint stiffness enter into the differential diagnosis of any patient with articular rigidity.

1. ARTHROGRYPOSIS MULTIPLEX CONGENITA

The term 'arthrogryposis multiplex congenita' (AMC) is often used loosely for any disorder in which restricted articular mobility is present at birth. However, this abnormality has many causes and there is value in reserving the designation 'AMC' for a specific syndrome, after known genetic entities and localised predisposing factors have been excluded.

'Primary AMC', in the narrow sense of the definition, is a condition in which infants are born with a limited range of movement or with fixed deformities of multiple joints. Muscular hypoplasia is often a prominent feature, and soft tissue webbing may be present. Due to the rigidity, fractures of the long bones may be sustained during delivery. Congenital dislocation of the hips and talipes equinovarus are common concomitants. AMC is not progressive, but affected individuals are severely disabled. Radiographically, the skeleton is undermineralised and gracile, and the hip joints may be dysplastic.

There is disagreement as to whether the basic defect lies in the spinal cord, the nerves or the muscles. It is possible that there are neuropathic and myopathic forms of the disorder. However, the phenotypic features of these entities have not been clearly defined.

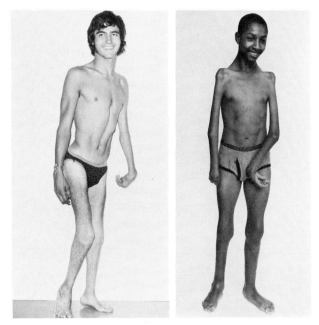

Fig. 14.1. (left) AMC; this young man has the primary or classical form of the condition. Widespread articular rigidity causes considerable disability.

Fig. 14.2. (right) AMC; the muscles may be hypoplastic. However, there are no systemic ramifications and general health is good. (From Davidson, J. & Beighton, P. (1976) *Journal of Bone and Joint Surgery,* 58, 492.)

Genetics

AMC is not uncommon, and large series of patients have been reviewed by Fried-lander, Westin and Wood (1968), Lloyd-Roberts and Lettin (1970), and Hall *et al.* (1976). The overwhelming majority of cases have been sporadic and it is probable that the condition is non-genetic. A maternal age effect was noted by Wynne-Davies and Lloyd-Roberts (1976), in their study of 66 cases. Drachman (1971) suggested that the clinical manifestations might be the consequence of inadequate develop-ment of articular and periarticular tissues, due to impairment of foetal movements.

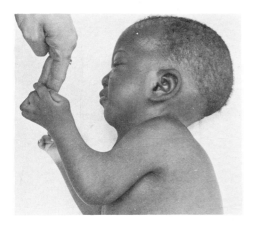

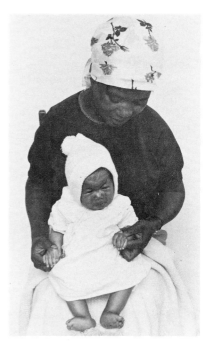

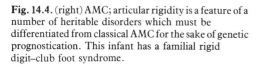

Fig. 14.3. (above) AMC; soft tissue webbing may be present across the flexor surfaces of rigid joints.

Fig. 14.4. (right) AMC; articular rigidity is a feature of a number of heritable disorders which must be differentiated from classical AMC for the sake of genetic prognostication. This infant has a familial rigid digit–club foot syndrome.

Evidence for this attractive hypothesis was provided by experiments in which chick embryos, previously paralysed with curare, were found to have an arthrogrypotic-like condition when they hatched (Drachman and Coulombre, 1962). Further support came from the observations of Jago (1970), who described an infant with AMC born to a mother who had contracted tetanus during early pregnancy, and had been treated with curare. If an environmental determinant does exist, it must be ubiquitous, as cases of AMC have been reported from all parts of the world. In this context, demographic data obtained during an investigation of 26 patients suggests that the condition is increasing in prevalence in South Africa (Davidson and Beighton, 1976).

There have been a few reports which are consistent with autosomal recessive inheritance. Affected sibs were described by Pena et al. (1968) while parental consanguinity was mentioned by Bargeton et al. (1961). It is likely that these patients had a distinct genetic form of AMC. Similarly, a phenotypic appearance which resembles that of primary AMC may be present in autosomal recessive congenital muscular dystrophy (Emery and Walton, 1967) and type I spinal muscular atrophy (Emery, 1971).

A variety of AMC has been recognised in 23 members of an inbred Arab community living near Tel Aviv in Israel (Lebenthal et al., 1970). The knees and elbows were mainly involved and several patients had cardiac lesions. There was little muscle wasting and their clinical features were different from those of classical primary AMC. Histological and electromyographical studies indicated that the defect lay in the muscles rather than the nerves. This condition is certainly a specific autosomal recessive entity.

Other reports of the familial occurrence of arthrogryposis in association with additional abnormalities probably concern separate genetic disorders. For instance, Srivastava (1968) described two brothers with multiple vertebral anomalies and arthrogryposis, while Daentl *et al.* (1974) mentioned an affected father and daughter in a report entitled 'a new familial arthrogryposis without weakness'.

Diagnostic precision is crucial for effective counselling. When familial AMC is encountered, it is virtually certain that the condition represents a distinct genetic entity, which may or may not have been previously delineated. Conversely, sporadic individuals with the classical phenotypic features of primary AMC do not have a heritable disorder and the recurrence risk is therefore very low.

2. RIGIDITY SYNDROMES

Articular rigidity is present in numerous heritable conditions. In some, non-articular manifestations predominate, while in others, the changes are virtually confined to the joints. Flexion contracture of the digits (camptodactyly) is a feature of many of these disorders.

(a) Cerebro-oculo-facio-skeletal syndrome

The cerebro-oculo-facio-skeletal syndrome (COFS) was delineated by Pena and Skokeir (1974) following their studies of 10 affected infants. The major manifestations were restriction on movements of the knees and elbows, coxa valga, acetabular dysplasia, 'rocker bottom' feet due to vertical talus, clenched hands and camptodactyly. Microcephaly, microphthalmia, micrognathia and large ears were additional features. These infants failed to thrive, and death from respiratory infection took place before the age of three.

Seven of the children were members of a consanguineous American Indian kindred living in Manitoba, Canada. Pena and Schokeir (1974) suggested that the abnormal gene might have been introduced by two French sisters who married into this community in the late nineteenth century. In further reports Preus and Fraser (1974) described an affected infant with consanguineous Italian parents and Lurie *et al.* (1976) mentioned a girl born into an inbred Gipsy kindred.

There is ample evidence that the COFS syndrome is inherited as an autosomal recessive. A similar disorder was encountered by Lowry *et al.* (1971) in a pair of sibs. The hand and foot deformities were less severe in these two infants, and this condition might be a separate entity.

(b) Goodman camptodactyly-fibrous tissue hyperplasia-skeletal dysplasia syndrome

Goodman, Bat-Miriam Katznelson and Manor (1972) described a young man and his two sisters in whom digital contractures had become evident at the end of the first decade. A Marfanoid habitus, mild thoracic scoliosis, knuckle pads and low intelligence were additional features. The parents, who were of Iranian–Jewish stock, were first cousins, and it is therefore probable that inheritance was autosomal recessive.

(c) Emery–Nelson syndrome

Emery and Nelson (1970) reported a Scottish mother and daughter with clawing of the fingers and toes, extension deformities of the thumbs, mild pes cavus, short

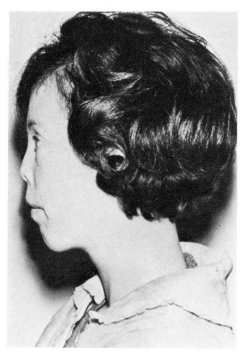

Fig. 14.5 Emery–Nelson syndrome; the face is flat and the forehead is high. (From Emery, A. E. H. & Nelson, M. M. (1970) *Journal of Medical Genetics*, **3**, 379.)

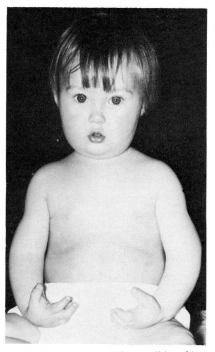

Fig. 14.6. Liebenberg elbow–wrist–hand syndrome; abnormalities of bone structure produce flexion deformities in the arms and hands. The skeleton is otherwise normal.

stature, a high forehead, depressed nasal bridge, and flattened malar regions. Using the designation 'Emery–Nelson syndrome', Gorlin, Pindborg and Cohen (1976) mentioned that they had also encountered a mother and two sons with the disorder. The mode of inheritance is dominant.

(d) Liebenberg elbow–wrist–hand syndrome
Liebenberg (1973) investigated a South African kindred, of Afrikaner stock, in whom flexion deformities of the elbow, wrist and digits were present in five generations. The restricted joint movements were the consequence of structural abnormalities of the bones of the forearms and hands, and the skeleton was otherwise normal. This disorder is inherited as an autosomal dominant.

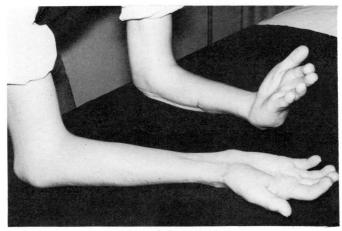

Fig. 14.7. Liebenberg elbow–wrist–hand syndrome; the father of the infant depicted in Fig.14.6. This autosomal dominant 'private' syndrome is present in five generations of an Afrikaner kindred.

(e) Mietens syndrome
Mietens and Weber (1966) studied four sibs with flexion contractures of the elbows, in association with low intelligence, corneal opacity and nasal narrowing. The tubular bones of the forearms were short and the radial heads were dislocated. Thoracic asymmetry, varus feet and abnormalities in the hip and knee joints were present in some of these children. The parents and two other sibs were normal. There was a suggestion of consanguinity in the kindred, and it is likely that inheritance was autosomal recessive.

(f) Ankylosis–facial anomaly–pulmonary hypoplasia syndrome
Pena and Shokeir (1974) described two sisters with camptodactyly, clubfeet, ankylosis of the hips and knees, an unusual facies and pulmonary hypoplasia. The affected infants died in the perinatal period from respiratory problems. Subsequently, Pena and Shokier (1976) encountered a further three cases, including a brother and sister who were the offspring of an incestuous relationship between uncle and niece. The total of reported cases rose to seven when a pair of affected brothers were described by Mease et al. (1976). The evidence points to autosomal recessive transmission.

(g) Kuskokwin disease

Kuskokwim disease is an arthrogrypotic-like disorder which is present in a consanguineous Eskimo community in Alaska. Fixed deformities of the large weight bearing joints cause severe disability. The condition is inherited as an autosomal recessive (Petajan *et al.*, 1969; Wright and Aase, 1969).

3. STIFF HAND–FOOT SYNDROMES

Deformity of the hands and feet, due to rigidity of the joints, is the predominant feature of a group of genetic disorders. Several of these conditions have been reported in only single kindreds and they are therefore regarded as 'private' syndromes. There is little doubt that many others await delineation.

(a) Camptodactyly–cleft palate–clubfoot syndrome

Gordon, Davies and Berman (1969) reported a large Cape Town family of mixed ancestry, in whom flexion deformities of the fingers and talipes equinovarus were variably associated with cleft palate. Since the original report, which concerned members of three generations, further affected individuals have been born into this kindred. The mode of transmission is autosomal dominant.

Fig. 14.8. Camptodactyly–cleft palate–clubfoot syndrome; an affected father and daughter.

(b) Digitotalar dysmorphism

Digitotalar dysmorphism was recognised in nine members of five generations of a South African kindred of English stock (Sallis and Beighton, 1972). These individuals had flexion deformities of the fingers with narrowing of the middle phalanges. The feet had a 'rocker bottom' configuration, due to vertical talus. The author has subsequently encountered digitotalar dysmorphism in a large African Negro family and in a father and son in England. Inheritance is clearly autosomal

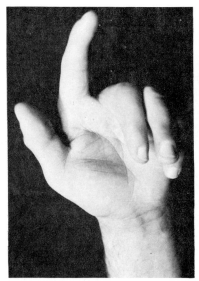 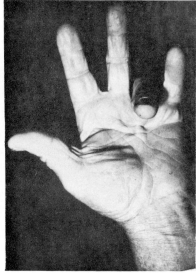

Fig. 14.9. (left) Digitotalar dysmorphism; flexion deformity of the digits. Vertical talus is the other component of the syndrome. (From Sallis, J. & Beighton, P. (1972) *Journal of Bone and Joint Surgery*, **54**, 509.)

Fig. 14.10. (right) Dupuytren contracture; this abnormality can mimic the various flexed digit syndromes.

dominant. A similar condition has been studied in 15 members of four generations of a kindred and reported under the designation 'thumb-clutched hand syndrome' by Canale *et al.* (1976).

(c) Digital-ulnar drift syndrome

Stevenson, Scott and Epstein (1975) reported an American kindred with hand abnormalities which resembled those of digitotalar dysmorphism. However, the feet were normal, and the authors concluded that the condition was a distinct autosomal dominant disorder, which they designated 'congenital ulnar drift with webbing and flexion contractures of the fingers.' Powers and Ledbetter (1976) and Fried and Mundel (1976) encountered kindreds with a similar syndrome.

(d) Digital deviation–clubfoot syndrome

Fisk, House and Bradford (1974) described a family with dominantly inherited deformities of the extremities. The term 'congenital ulnar deviation of the fingers, with clubfoot deformities' was used in the report, in which it was stated that maldevelopment of the digital extensor tendons was probably an underlying pathogenic factor.

(e) Tel-Hashomer camptodactyly syndrome

Goodman, Bat-Miriam Katznelson and Manor (1972) studied a brother and sister of Jewish Moroccan stock, in whom flexion contractures of the fingers were associated with digital spindling, clubbing of the feet, short stature, thoracic scoliosis, muscular hypoplasia and an unusual facies. Subsequently, Goodman *et al.* (1976) reported two affected sibs from a different family using the designation 'Tel-

Hashomer camptodactyly syndrome', and confirmed that inheritance was autosomal recessive.

(f) Dominantly inherited metatarsus varus

Metatarsus varus is a form of clubfoot, in which the anterior portion of the extremity is angulated medially. There are various types of this anomaly, which are defined on a basis of their clinical manifestations and response to treatment. As with talipes equinovarus, inheritance is generally multi-factorial. Juberg and Touchstone (1974) reported a family in which nine members of four generations had metatarsus varus, together with variable minor anomalies, including fixed extension of digits and other joints. The authors recognised that their pedigree data were consistent with auto-somal dominant inheritance. They suggested that metatarsus varus was yet another congenital deformity which displayed heterogeneity, and that it exists in common polygenic and uncommon monogenic forms.

4. FLEXED DIGIT SYNDROMES

Flexion deformities of single or multiple digits may occur in isolation or as a component of a large number of malformation syndromes. Gordon, Davies and Berman (1969) listed the conditions in which camptodactyly may be present.

The term 'camptodactyly' implies flexion contraction, usually at the proximal interphalangeal joints, while 'clinodactyly' denotes radial curvature, typically of the fifth finger. In practice, it is not always easy to draw a precise distinction between these abnormalities and the designations are often used synonymously. The alternative term 'streblodactyly' pertains to a digit which is twisted or bent, while 'streblomicrodactyly' denotes a shortened, crooked digit. Temtamy and McKusick (1969), in their modification of the classification of dominant uncomplicated brachydactyly syndromes of Bell (1951), listed clinodactyly as 'type A-3' (*vide* Chapter 17).

(a) Isolated camptodactyly or clinodactyly of the fifth finger is often sporadic. However, there have been several reports of large kindreds with dominant inheritance of the deformity (Hefner, 1941; Spear, 1946; Welch and Temtamy, 1966). Emphasising the anatomical disparity and the semantic imprecision concerning flexion deformities of the fifth finger, Katz (1970) reported an American kindred with 17 affected individuals in five generations. Of these, seven had incurving of the distal interphalangeal joint, while 10 had streblomicrodactyly. Alluding to the report of Kirner (1972), which concerned flexion and radial bowing of the terminal phalanges of the fifth finger, David and Burwood (1972) described nine English kindreds with 18 affected members. The authors discussed the radiological features of the bone changes in the phalanges in 'Kirner's deformity' and concluded that inheritance was autosomal dominant with variable penetrance.

(b) Familial streblodactyly and aminoaciduria was reported by Parish, Horn and Thompson (1963). Ten females in three generations were affected, while nine males were normal. It was suggested that the condition might be inherited as a sex-linked dominant.

(c) A kindred with autosomal dominant inheritance of congenital flexion contractures of shortened fingers was described by Edwards and Gale (1972) under the

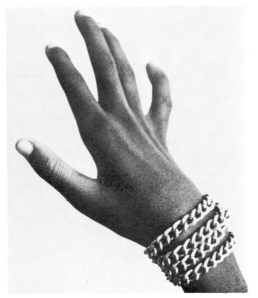 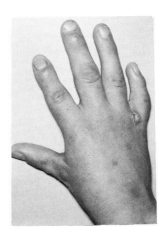

Fig. 14.11 (left) Clinodactyly; flexion contracture of the fifth fingers.

Fig. 14.12. (right) Streblomicrodactyly; the hands of a child with Down's syndrome. The fifth fingers are short and crooked.

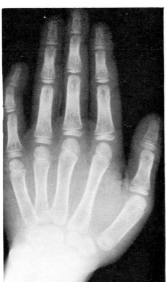 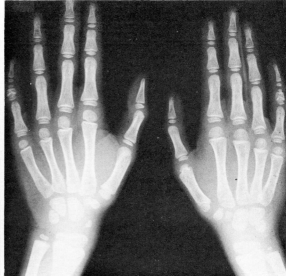

Fig. 14.13. (left) Clinodactyly; radiograph showing flexion and radial deviation of the fifth finger.

Fig. 14.14. (right) Clinodactyly; the term 'Kirner deformity' denotes flexion of the fifth finger due to malformation of the middle phalynx. This abnormality is also designated 'brachydactyly type A3'.

designation 'camptobrachydactyly'. Syndactyly, polydactyly and genitourinary malformations were present in some individuals. It is possible that two severely affected children within this family might have been homozygous for the abnormal gene.

(d) Murphy (1926) described a large family in the USA, in whom camptodactyly was associated with instability of the knee joints. The condition was recognised in five generations and inheritance was evidently autosomal dominant.

(e) Pronounced lateral deviation of the fifth toes was observed in members of two kindreds during the course of an epidemiological survey of bone and joint disorders in a tribal Xhosa population in Transkei, Southern Africa (Schwartz *et al.*, 1974). The condition which was designated 'Transkei foot', was present in successive generations, and it is possible that inheritance was dominant.

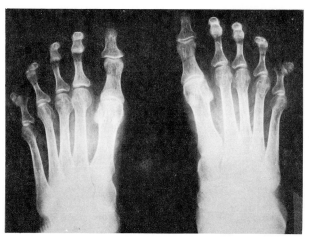

Fig. 14.15. Transkei foot; lateral deviation of the fifth toe, which is apparently inherited as an autosomal dominant in two Xhosa kindreds in Transkei, Southern Africa. (From Schwartz, P. A. *et al.* (1974) *South African Medical Journal*, **48**, 961.)

5. ADDUCTED THUMB SYNDROMES

Adduction deformity of the thumbs is a prominent feature of a number of genetic syndromes. These have been reviewed by Fitch and Levy (1975). Autosomal dominant entities in this group include the 'clasped thumb' syndrome described by Weckesser, Reed and Heiple (1968) and an adducted thumb-mental deficiency syndrome (MASA) reported by Bianchine and Lewis (1974). Adduction contracture of the thumb due to congenital hypoplasia of the muscles of the thenar eminence was observed in eight individuals, including three members of a kindred, by Strauch and Spinner (1976). The authors suggested that inheritance was probably autosomal dominant. Christian *et al.* (1971) described an autosomal recessive adducted thumb syndrome which was present in an inbred Amish community. Joint contractures, craniostenosis and cleft palate were additional features.

A clenched hand, with adduction of the thumb across the palm, may be the presenting feature of trisomy 18 in the newborn. A phenocopy of the disorder, in which the chromosomes are normal, has been reported (Hongre *et al.*, 1972). In addition, an adducted thumb is a useful diagnostic feature of X-linked hydrocephalus (Edwards, 1961).

REFERENCES

Arthrogryposis

Bargeton, E., Nezelop, C., Guran, P. & Job, J. C. (1961) Étude anatomique d'un case d'arthrogrypose multiple congenitale et familiale. *Revista Neurologia,* **104,** 479.

Daentl, D. L., Berg, B. O., Layzer, R. B. & Epstein, C. J. (1974) A new familial arthrogryposis without weakness. *Neurology,* **24,** 55.

Davidson, J. & Beighton, P. (1976) Whence the arthrogrypotics? *Journal of Bone and Joint Surgery,* **58B,** 492.

Drachman, D. B. & Coulombre, A. J. (1962) Experimental clubfoot and arthrogryposis multiplex congenita. *Lancet,* **ii,** 523.

Drachman, D. B. (1971) The syndrome of arthrogryposis multiplex congenita. *Birth Defects: Original Article Series,* 7/2, 90.

Emery, A. E. H. & Walton, N. (1967) The genetics of muscular dystrophy. *Progress in Medical Genetics,* 5, 116.

Emery, A. E. H. (1971) The nosology of the spinal muscular atrophies. *Journal of Medical Genetics,* **8,** 481.

Friedlander, H. L., Westin, G. W. & Wood, W. L. (1968) Arthrogryposis multiplex congenita — review of 45 cases. *Journal of Bone and Joint Surgery,* **50A,** 89.

Hall, J. G., Greene, G., Shinkoskey, S. & McIlvaine, R. (1976) Arthrogryposis — clinical and genetic heterogeneity. In *Fifth International Congress of Human Genetics,* p.78. Mexico.

Jago, R. H. (1970) Arthrogryposis following treatment of maternal tetanus with muscle relaxants. Case report. *Archives of Diseases in Childhood,* **45,** 277.

Lebenthal, E., Shochet, S. B., Adam, Seelenfreund, M., Fried, A., Majenson, T., Sandbank, U. & Matoth, Y. (1970) Arthrogryposis multiplex congenita — 23 cases in an Arab kindred. *Pediatrics,* **46,** 891.

Lloyd-Roberts, G. C. & Lettin, A. W. F. (1970) Arthrogryposis multiplex congenita. *Journal of Bone and Joint Surgery,* **52B,** 494.

Pena, C. E., Miller, F., Budzilovich, G. N. & Feign, I. (1968) Arthrogryposis multiplex congenita: report of two cases of a radicular type with familial incidence. *Neurology,* **18,** 926.

Srivastava, R. N. (1968) Arthrogryposis multiplex congenita. Case report of two siblings. *Clinical Genetics,* 7, 691.

Wynne-Davies, R. & Lloyd-Roberts, G. C. (1976) Arthrogryposis multiplex congenita. *Archives of Diseases in Childhood,* **51,** 618.

Rigidity syndromes

(a) COFS syndrome

Lowry, R. B., Maclean, R., McLean, D. M. & Tischler, B. (1971) Cataracts, microcephaly, kyphosis and limited joint movement in two siblings: a new syndrome. *Journal of Pediatrics,* **79,** 282.

Lurie, I. W., Cherstvoy, D. E., Lazjuk, G. I., Nedzued, M. K. & Usoev, S. S. (1971) Further evidence for the autosomal recessive inheritance of the COFS syndrome. *Clinical Genetics,* **10,** 343.

Pena, S. D. J. & Shokeir, M. H. K. (1974) Autosomal recessive cerebro-oculo-facio-skeletal (COFS) syndrome. *Clinical Genetics,* **5,** 285.

Preus, M. & Fraser, F. C. (1974) The cerebro-oculo-facio-skeletal syndrome. *Clinical Genetics,* **5,** 294.

(b) Goodman camptodactyly syndrome

Goodman, R. M., Bat-Miriam Katznelson, M. & Manor, E. (1972) Camptodactyly: occurrence in two new genetic syndromes and its relationship to other syndromes. *Journal of Medical Genetics,* **9,** 203.

(c) Emery–Nelson syndrome

Emery, A. E. H. & Nelson, M. M. (1970) A familial syndrome of short stature, deformities of the hands and feet, and an unusual facies. *Journal of Medical Genetics,* **7,** 379.

Gorlin, R. J., Pindborg, J. J. & Cohen, M. M. (1976) In *Syndromes of the Head and Neck*, 2nd Edition, p.738. New York: McGraw-Hill Book Company.

(d) Liebenberg elbow–wrist–hand syndrome

Liebenberg, F. (1973) A pedigree with unusual anomalies of the elbows, wrist and hands in five generations. *South African Medical Journal,* **47,** 745.

(e) Mietens syndrome
Mietens, C. & Weber, H. (1966) A syndrome characterised by corneal opacity, nystagmus, flexion contracture of the elbows, growth failure, and mental retardation. *Journal of Pediatrics*, **69**, 624.

(f) Ankylosis–facial anomaly–pulmonary hypoplasia syndrome
Mease, A. D., Yeatman, G. W., Pettett, G. & Merenstein, G. B. (1976) A syndrome of ankylosis, facial anomalies and pulmonary hypoplasia secondary to fetal neuromuscular dysfunction. *Birth Defects: Original Article Series*, **12/5**, 193.
Pena, S. D. J. & Shokeir, M. H. K. (1974) Syndrome of camptodactyly, multiple ankyloses, facial anomalies, and pulmonary hypoplasia: a lethal condition. *Journal of Pediatrics*, **85**, 373.
Pena, S. D. J. & Shokeir, M. H. K. (1976) Syndrome of camptodactyly, multiple ankyloses, facial anomalies and pulmonary hypoplasia — further delineation and evidence for autosomal recessive inheritance. *Birth Defects: Original Article Series*, **13/5**, 201.

(g) Kuskokwim disease
Petajan, J. H., Momberger, G. L., Aase, J. M. & Wright, D. G. (1969) Arthrogryposis syndrome (Kuskokwim disease) in the Eskimo. *Journal of the American Medical Association*, **209**, 1481.
Wright, D. G. & Aase, J. (1969) The Kuskokwim syndrome: an inherited form of arthrogryposis in the Alaskan Eskimo. In *The Clinical Delineation of Birth Defects*, Volume 3, p.91. New York: National Foundation.

Stiff hand–foot syndromes

(a) Camptodactyly–cleft palate–clubfoot syndrome
Gordon, H., Davies, D. & Berman, M. (1969) Camptodactyly, cleft palate and club foot. Syndrome showing the autosomal dominant pattern of inheritance. *Journal of Medical Genetics*, **6**, 266.

(b) Digitotalar dysmorphism
Canale, S. T., Ingram, A. J., Tipton, R. E. & Wilroy, R. S. (1976) The thumb-clutched hand syndrome. In *Fifth International Congress of Human Genetics*, Mexico.
Sallis, J. G. & Beighton, P. (1972) Dominantly inherited digito-talar dysmorphism. *Journal of Bone and Joint Surgery*, **54B**, 509.

(c) Dominantly inherited ulnar drift
Fried, K. & Mundel, G. (1976) Absence of distal interphalangeal creases of fingers with flexion limitation. *Journal of Medical Genetics*, **13/2**, 127.
Powers, R. C. & Ledbetter, R. H. (1976) Congenital flexion and ulnar deviation of the metacarpo-phalangeal joints of the hand: a case report. *Clinical Orthopaedics and Related Research*, **116**, 173.
Stevenson, R. E., Scott, C. I. & Epstein, M. (1975) Dominantly inherited ulnar drift. *Birth Defects: Original Article Series*, **11/5**, 75.

(d) Digital deviation–clubfoot syndrome
Fisk, J. R., House, J. H. & Bradford, D. S. (1974) Congenital ulnar deviation of the fingers with clubfoot deformities. *Clinical Orthopaedics and Related Research*, **104**, 200.

(e) Tel-Hashomer camptodactyly
Goodman, R. M., Bat-Miriam Katznelson, M. & Manor, E. (1972) Camptodactyly: occurrence in two new genetic syndromes and its relationship to other syndromes. *Journal of Medical Genetics*, **9**, 203.
Goodman, R. M., Bat-Miriam Katznelson, M., Hertz, M. & Katznelson, A. (1976) The Tel-Hashomer camptodactyly syndrome: a new heritable disorder of connective tissue. In *Fifth International Congress of Human Genetics*, Mexico.

(f) Dominantly inherited metatarsus varus
Juberg, R. C. & Touchstone, W. J. (1974) Congenital metatarsus varus in four generations. *Clinical Genetics*, **5/2**, 127.

Flexed digit syndromes

Bell, J. (1951) On brachydactyly and symphalangism. In *Treasury of Human Inheritance*, Volume 5, p.1. London: Cambridge University Press.
David, T. J. & Burwood, R. L. (1972) The nature and inheritance of Kirner's deformity. *Journal of Medical Genetics*, **9/4**, 430.

Edwards, J. A. & Gale, R. P. (1972) Camptobrachydactyly: a new autosomal dominant trait with two probable homzygotes. *American Journal of Human Genetics*, **24**, 464.

Gordon, H., Davies, D. & Berman, M. (1969) Camptodactyly, cleft palate and club foot. Syndrome showing the autosomal dominant pattern of inheritance. *Journal of Medical Genetics*, **6**, 266.

Hefner, R. A. (1941) Crooked little finger (minor streblodactyly). *Journal of Heredity*, **32**, 37.

Katz, G. (1970) A pedigree with anomalies of the little finger in five generations and 17 individuals. *Journal of Bone and Joint Surgery*, **52A/4**, 717.

Kirner, J. (1927) Doppelseitige Verkrummungen des Kleinfingerendgliedes als selbstandiges Krankheitsbild. *Fortschritte auf dem Gebiete der Roentgenstrahlen und Nuklearmedizin*, **36**, 804.

Murphy, D. P. (1926) Familial finger contracture and associated familial knee-joint subluxation. *Journal of the American Medical Association*, **86**, 395.

Parish, J. G., Horn, D. B. & Thompson, M. (1963) Familial streblodactyly with amino-aciduria. *British Medical Journal*, **2**, 1247.

Schwartz, P. A., Shlugman, D., Daynes, G. & Beighton, P. (1974) Transkei foot. *South African Medical Journal*, **48**, 961.

Spear, G. S. (1946) The inheritance of fixed fingers. *Journal of Heredity*, **37**, 189.

Temtamy, S. & McKusick, V. A. (1969) Synopsis of hand malformations with particular emphasis on genetic factors. *Birth Defects: Original Article Series*, **5/3**, 125.

Welch, J. P. & Temtamy, S. A. (1966) Hereditary contracture of the finger (camptodactyly). *Journal of Medical Genetics*, **3**, 104.

Adducted thumb syndrome

Bianchine, J. & Lewis, R. (1974) The MASA syndrome: a new heritable mental retardation syndrome. *Clinical Genetics*, **5**, 298.

Christian, J. C., Andrews, P. A., Conneally, P. M. & Muller, J. (1971) The adducted thumbs syndrome. *Clinical Genetics*, **2**, 95.

Edwards, J. H. (1961) The syndrome of sex-linked hydrocephalus. *Archives of Diseases in Childhood*, **36**, 486.

Fitch, N. & Levy, E. P. (1975) Adducted thumb syndromes. *Clinical Genetics*, **8**, 190.

Hongre, J. F., Toursei, M. F., Staquet, J. P., Farriaux, J. P. & Walbraum, R. (1972) Phenocopie de la trisomie 18. *Annals de Pediatrie*, **19**, 830.

Strauch, B. & Spinner, M. (1976) Congenital anomaly of the thumb: absent intrinsics and flexor pollicis longus. *Journal of Bone and Joint Surgery*, **58/1**, 115.

Weckesser, E., Reed, J. & Heiple, K. (1968) Congenital clasped thumb (congenital flexion-adduction deformity of the thumb). *Journal of Bone and Joint Surgery*, **50A**, 1417.

15. Digital abnormalities

The nosology and genetics of digital anomalies have been reviewed at length by Temtamy and McKusick (1969). The following groups of conditions are discussed in this section:

1. Polydactyly
2. Syndactyly
3. Symphalangism
4. Brachydactyly
5. Ectrodactyly.

These terms pertain to anatomical abnormalities rather than to specific disorders. Each of these anomalies may represent a heterogeneous heritable condition or a sporadic non-genetic malformation. Mixed malformations, in which extra, fused or dysplastic digits occur in conjunction with each other, are not uncommon. Dysplastic digits are also a component of numerous distinct syndromes:

6. Digital dysplasia syndromes
    (a) Holt–Oram syndrome
    (b) Orofacial–digital syndrome
    (c) Oculodental–osseous dysplasia
    (d) Aase syndrome
    (e) Aglossia–adactylia syndrome
    (f) Tetramelic monodactyly
    (g) Goltz syndrome (focal dermal hypoplasia).

1. POLYDACTYLY

The presence of an extra digit is termed 'polydactyly'. Polydactyly is conventionally classified into preaxial and postaxial types. In the former, the thumb and great toe are duplicated, while in the latter, the additional digit is adjacent to the fifth finger and toe. If digital fusion is also present, the term 'polysyndactyly' is employed. In a survey of 10,000 consecutive newborn infants in Turkey, Tuncbilek and Say (1976) found the incidence of all forms of polydactyly to be 2.6 per thousand. Following a study of 13 human embryos with preaxial polydactyly, Yasuda (1975) concluded that the pathogenesis was disordered interaction between the limb ectoderm and mesoderm.

A unilateral extra thumb or hallux is usually non-genetic, but there are several autosomal dominant forms of uncomplicated, bilateral, preaxial digital duplication

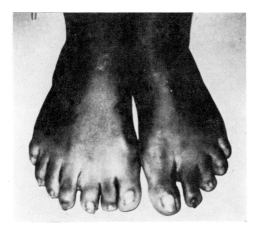 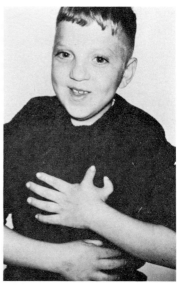

Fig. 15.1. (left) Polydactyly; the right foot has seven toes. This abnormality is rare.

Fig. 15.2. (right) Preaxial polydactyly; an extra digit lateral to the thumb. (Preaxial implies that the additional digit is anterior to the anatomical axis of the limb.) This isolated abnormality is non-genetic.

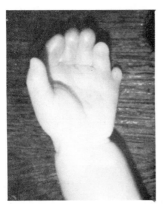 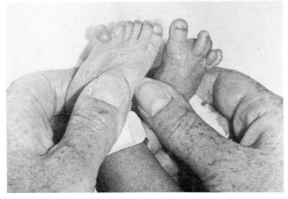

Fig. 15.3. (left) 'Postaxial polydactyly'; an extra digit on the medial side of the hand. This common abnormality is probably inherited as an autosomal dominant with variable phenotypic expression. (Postaxial implies that the additional digit is posterior to the anatomical axis of the limb.)

Fig. 15.4. (right) Postaxial polydactyly; an extra digit on the lateral side of the foot.

(Hefner, 1940; Swanson and Brown, 1962; James and Lamb, 1963). Atasu (1976) observed an extra index finger in members of four generations of a Turkish family. Preaxial polydactyly is a feature of a rare dominantly inherited condition in which syndactyly and bilateral aplasia of the tibia are variable components. Affected kindreds have been reported by Eaton and McKusick (1969) and Pashayan *et al.* (1971).

A sporadic female with polydactyly and triphalangeal thumbs in association with upper limb and pectoral dysplasia was described by Temtamy and Dorst (1975).

Although no other members of the family had the abnormality, the authors commented that triphalangeal thumbs with or without preaxial polydactyly could be inherited as an autosomal dominant. Say *et al.* (1976) described a mother and three daughters with preaxial polydactyly of the hands and feet, triphalangeal thumbs, brachydactyly, camptodactyly, congenital dislocation of the patella, short stature and low intelligence. The authors listed the conditions in which triphalangeal thumbs were a feature and commented that inheritance of their patient's syndrome was either autosomal dominant or X-linked.

Postaxial polydactyly is more common than the preaxial type. A minute sixth digit is often present in individuals of African Negro stock. Indeed, Okeahialam (1974) commented that this was the commonest anomaly encountered during a 12 month survey of congenital malformations in Dar-es-Salaam, East Africa. The reason for this high prevalence is unknown, but it has been suggested that inheritance is autosomal dominant, with variable expression (Johnstone and Davis, 1953). Large kindreds with autosomal dominant inheritance of extra digits of greatly varying size and distribution have been reported by several authors, including Woolf and Woolf (1970). Autosomal recessive forms of postaxial polydactyly may also exist. Cantu *et al.* (1975) recognised this anomaly in four out of six sibs in a consanguineous Mexican kindred, while Temtamy and Rogers (1975) reported an affected Turkish boy with consanguineous parents. This child had multiple defects, including brachydactyly, craniofacial asymmetry, ear malformations and cardiac anomalies.

Postaxial polydactyly is a component of various potentially lethal conditions such as trisomy 13, asphyxiating thoracic dystrophy and the Saldino–Noonan syndrome. It is also a feature of chondroectodermal dysplasia (Ellis–van Creveld syndrome).

Postaxial polysyndactyly is found in the Bardet–Biedl syndrome, where it is associated with mental retardation, retinitis pigmentosa and hypogenitalsim. Ammann (1970) distinguished this autosomal recessive disorder from the Laurence–Moon syndrome, in which the digits are normal. Holt (1975) described two English families with autosomal dominant inheritance of postaxial polydactyly in association with variable shortening of the metacarpals and metatarsals.

Laurence *et al.* (1975) reported a kindred in which two male infants had been born with ulnar polydactyly and broad great toes, together with Hirschsprung's disease and cardiac malformations. A subsequent pregnancy was monitored by fetoscopy at the eighteenth week. The observation that the fetus had normal digits was confirmed when an unaffected boy was delivered. Although leakage of amniotic fluid was troublesome in later pregnancy, there is no doubt that fetoscopy was of real value in this particular case.

2. SYNDACTYLY

The term 'syndactyly' pertains to bone or soft tissue union of two or more digits. The extent of the anomaly is variable. In severe forms bone fusion of adjacent digits may be complete, while in mild types minimal skin webbing at the digital roots is the only abnormality. Syndactyly is a component of several genetic syndromes. It may also exist in isolation, either as a genetic or non-genetic entity. Temtamy and McKusick (1969) listed five types of autosomal dominant syndactyly. In general, the genetic forms are bilateral and fairly symmetrical, while the non-genetic varieties are unilateral.

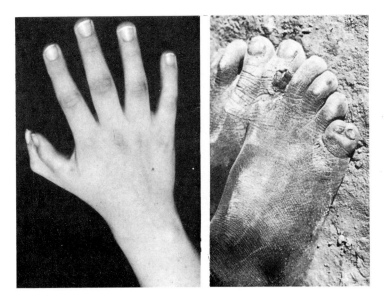

Fig. 15.5. (left) Preaxial polysyndactyly; the thumb is duplicated and fused.

Fig. 15.6. (right) Postaxial polysyndactyly; the fifth toe is duplicated and fused.

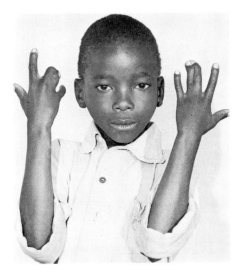

Fig. 15.7. Syndactyly; the soft tissues of the third and fourth fingers are fused.

Fig. 15.8. Syndactyly; asymmetrical soft tissue fusion of the digits.

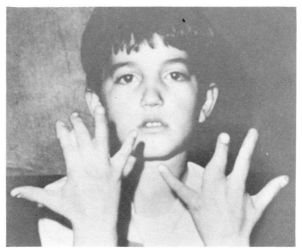

Fig. 15.9. Syndactyly; soft tissue fusion in the right hand and bony union of the third and fourth digits in the left hand.

The association of unilateral cutaneous synbrachydactyly with ipsilateral absence of the pectoralis minor muscle and the sternal portion of the pectoralis major is known as the Poland anomalad. Rib defects, together with absence of part or all of the hand, are occasional features. The Poland anomalad is relatively common and most cases are sporadic. However, the condition was recognised in successive generations by Trier (1965). In an investigation of possible environmental determinants David (1972) noted that several mothers of affected children had made unsuccessful attempts to abort their pregnancies. Bouvet, Maroteaux and Briard Guillemot (1976) reviewed the clinical and genetic features of the anomalad and concluded that inheritance was probably multifactorial.

Fitch, Jequier and Papageorgiou (1976) reported a pair of half brothers with syndactyly and various cranial, facial, oral and skeletal abnormalities. As the mother had minor deviation deformities of her digits, the authors postulated that inheritance could be either autosomal dominant with variable penetrance or X-linked, with manifestations in the female heterozygote.

Goldberg and Pashayan (1976) described a kindred in which 10 individuals in three generations had syndactyly of the hallux and adjacent toes, a small sixth digit on the ulnar side of the hands and an abnormal configuration of the earlobes. This condition seems to be a 'private syndrome' in which inheritance is autosomal dominant, with variable phenotypic expression of the gene.

3. SYMPHALANGISM

Symphalangism is far less common than syndactyly. In symphalangism, inter-phalangeal joints are absent, so that movements of the digits are restricted. Unlike syndactyly, there is no fusion between neighbouring digits. One form of symphalangism is inherited as an autosomal dominant. Drawing upon historical sources, Elkington and Huntsman (1967) described how this abnormality had been present in several generations of the aristocratic Talbot family in Britain.

A syndrome of symphalangism and coalition of the small bones of the wrist and ankle is transmitted as an autosomal dominant. Affected kindreds have been reported by Austin (1951) and Strasburger et al. (1965). Symphalangism with metacarpophalangeal fusions and elbow abnormalities was described by Kassner, Katz and Qazi (1976). This condition was present in three generations of a kindred, and inheritance was evidently autosomal dominant. Walbaum, Hazard and Cordier (1976) reported the association of brachydactyly and symphalangism, which was probably inherited as an autosomal recessive.

4. BRACHYDACTYLY

The term 'brachydactyly' pertains to shortening of a single or multiple digits, due to bone maldevelopment. There are many types of brachydactyly, some of which are components of syndromes or which exist in conjunction with other defects in the extremities. The various forms of dominantly inherited isolated brachydactyly were analysed and listed by Bell (1951). Temtamy and McKusick (1969) subsequently modified and extended this classification. A number of these conditions are 'private' syndromes, while others are relatively common. An outline is given below:

Type A-1 (Farabee)
The main features are digital shortening, which is maximal in the middle phalanges, dysplasia of the femoral heads and small stature. Haws and McKusick (1963) restudied the kindred first reported by Farabee at the turn of the century, and confirmed that the condition was inherited as an autosomal dominant.

Type A-2 (Mohr–Wreidt)
In this disorder the middle phalanges of the second finger and toe are shortened, with radial deviation of the terminal phalanx. Mohr and Wreidt (1919) reported a kindred in which two affected parents had produced a daughter with total adactylia. It is possible that this child was homozygous for the abnormal gene.

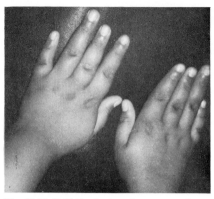 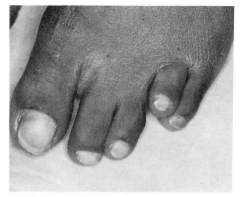

Fig. 15.10. (left) Brachydactyly; apparent shortening of a finger is often the result of reduction in length of the corresponding metacarpal. In this patient, the third metacarpals are short, while the phalanges are normal.

Fig. 15.11. (right) Brachydactyly; the fourth metatarsal is short. This relatively common abnormality may occur in isolation or as a non-specific component of a number of genetic syndromes.

Type A-3

A shortened and distorted middle phalanx leads to flexion and incurving of the fifth finger. This is clinodactyly or the Kirner anomaly, as discussed in Chapter 14.

Type A-4

In this type of brachdactyly the middle phalanges of the second and fifth fingers are shortened. In addition, the middle phalanges of the lateral four toes are sometimes absent.

Type A-5

Bass (1968) described 13 members of four generations of a large kindred who had absence of the middle phalanges, dysplasia of the nails and duplication of the terminal phalanx of the thumb.

Type B

The major characteristics are shortening of the middle phalanges, with hypoplasia of the terminal phalanges and dystrophy of the nails of the second to fifth digits, broadening or partial duplication of the terminal phalanx of the thumb and minor cutaneous syndactyly. MacArthur and McCullough (1932) reported this abnormality in three generations of a kindred, under the designation 'apical dystrophy'. Zavala, Hernandez Ortiz and Lisker (1975) described generation to generation transmission of brachydactyly type B, together with symphalangism, in a Mexican family.

Type C

In this form of brachydactyly, shortening and deformity of the proximal and middle phalanges of the second and third fingers predominates. Haws (1963) identified the abnormality in 86 members of a Mormon kindred. Rimoin, Hollister and Lachman (1975) encountered type C brachydactyly together with limited flexion of the distal

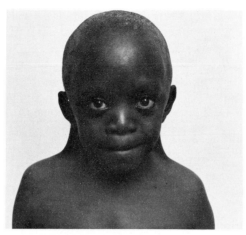

Fig. 15.12. Turner's syndrome; webbing of the neck and shortening of the fourth metacarpal are often present in this chromosomal disorder.

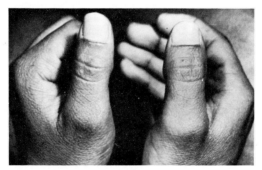

Fig. 15.13. Brachydactyly type D, 'murderer's thumbs'. The distal phalynx is short and broad.

interphalangeal joints in 10 members of five generations of an American Negro family.

Type D (stub thumb)

Shortening and broadening of the terminal phalanges of the first digit is a common abnormality which is sometimes alluded to as 'murderer's thumb'. By coincidence, and perhaps not inappropriately, J. W. Jailer was a co-author of a paper in which a family with this anomaly was described (Sayles and Jailer, 1934). Stub thumb has been studied in kindreds from different ethnic groups by Goodman, Adam and Sheba (1965) and the familial nature of the abnormality has been discussed by Davies (1975). Broad thumbs are also a component of several syndromes, including those bearing the eponyms Larsen, Apert, Carpenter and Rubinstein-Taybi. Poznanski, Garn and Holt (1971) have published an extensive review of the thumb in congenital malformation syndromes.

Type E

The commonest brachydactyly syndrome involves shortening of the metacarpals and metatarsals. There is intra and interfamilial variation in the distribution and severity of this abnormality. The condition is heterogeneous, and kindreds with recognisably different forms of the disorder have been described by several authors, including McKusick and Milch (1964), Cuevas-Sosa and Carcia-Segur (1971) and Gnamey *et al.* (1975). Newcombe and Keats (1969) reported two large kindreds with autosomal dominant inheritance of peripheral dysostosis and emphasised that although this condition resembles type E brachydactyly, the disorders are actually different entities. According to these authors, peripheral dysostosis can be distinguished by the involvement of the phalanges and the radiographic appearance of cone-shaped epiphyses in these bones.

Shortening of one or more metacarpals and metatarsals, usually the fourth and fifth, is a fairly common sporadic anomaly. A short fourth metacarpal is also encountered in females with an XO chromosome constitution (Turner's syndrome). All the metacarpals may be shortened in pseudohypoparathyroidism.

Other brachydactyly syndromes

There have been a number of reports of other brachydactyly syndromes. For instance, dominant inheritance of short digits in association with hypertension has been described (Bilginturan *et al.*, 1973). Similarly, a kindred with short adducted thumbs and great toes in four generations was studied by Christian *et al.* (1972). A mother and five offsprings with brachydactyly, which would now be classified as type B, together with pigmented colobomata of the macula, were investigated by Sorsby (1935). Gorlin and Sedano (1971) described a kindred with type E brachydactyly and multiple impacted teeth, using the term 'cryptodontic brachymetacarpalia'. Villaverde and Silva (1975) found distal brachyphalangy of the thumbs in more than 3 per cent of 852 mentally retarded individuals. Several had additional features such as abnormalities of the feet, cranium and sella turcica.

5. ECTRODACTYLY

Maldevelopment of the central rays of the limb buds may produce longitudinal splitting of the extremities. In some instances, the split hand or foot has a 'lobster claw' configuration. The term 'ectrodactyly', which has the Greek connotation of 'aborted finger', is used loosely to include abnormalities of this type.

Split hand and foot may be sporadic or inherited as an autosomal dominant. Expression is very variable and skipped generations are not uncommon. Large families with dominant transmission have been reported by MacKenzie and Penrose (1951) and Stevenson and Jennings (1960). The latter authors commented that in several kindreds, affected males had a tendency to produce an excess of affected sons. There is no obvious explanation for this anomalous situation.

It is likely that there is an uncommon autosomal recessive form of ectrodactyly. Mosavy and Vakhsuri (1975) described a pair of Iranian sisters with splitting of the hands and feet. Nine sibs and the consanguineous parents were normal. Similarly, Verma *et al.* (1976) reported the deformity in two related consanguineous Indian kindreds and Gemme *et al.* (1976) observed two affected sibs with consanguineous parents. There do not seem to be any phenotypical features which distinguish these dominant and recessive forms of ectrodactyly.

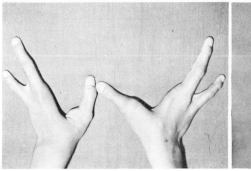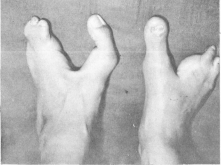

Fig. 15.14. (left) Ectrodactyly; the split-hand deformity in a 14-year-old girl.

Fig. 15.15. (right) Ectrodactyly; the feet of the patient depicted in Fig.15.14.

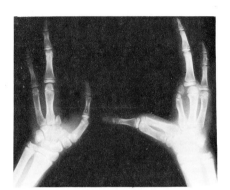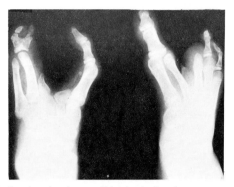

Fig. 15.16. (left) Ectrodactyly; radiograph showing the abnormalities in the hands.

Fig. 15.17. (right) Ectrodactyly; radiograph showing the abnormalities in the feet.

In view of the existence of sporadic, dominant and recessive forms of the lobster claw defect, the propensity for skipped generations, and the fact that there is probably considerable heterogeneity, genetic counselling can be very difficult. The practical implications of this situation have been discussed in detail by Preus and Fraser (1973).

Other ectrodactyly syndromes

(a) A remarkable form of ectrodactyly is present in the Doma people of the Eastern Zambezi valley. Stories of 'ostrich-footed' men in this remote area were considered to be mere travellers tales until Gelfand, Roberts and Roberts (1974) examined an affected male and documented the disorder in detail. The feet were bidactylic, while the hands were normal apart from minimal skin webbing between the third and fourth fingers. A male cousin on the father's side of the family was also known to have the condition. The local population attributed the abnormality to the fact that the individual's mother had imprudently eaten 'two footed' animals during her pregnancy. However, the patient's own family suspected bewitchment. Offering an alternative explanation, the investigators proposed that the condition was an autosomal dominant, with non-penetrance in the father.

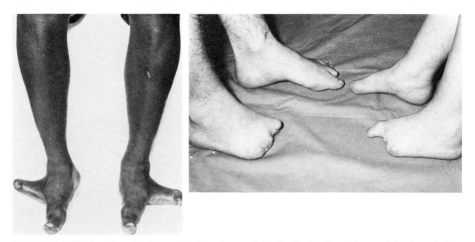

Fig. 15.18. (left) Ectrodactyly; the ostrich-footed man of the Zambezi valley. The condition is probably inherited as an autosomal dominant, with variable penetrance. (From Gelfand, M., Roberts, C. J. & Roberts, R. S. (1974) *Rhodesia History*, 5, 92.)

Fig. 15.19. (right) Ectrodactyly; the feet of a brother and sister, with defects of the toes, fingers and scalp. Members of four generations of the kindred had similar abnormalities.

(b) The association of ectrodactyly, ectodermal dysplasia and cleft lip and palate was reported by Roselli and Guilienetti (1961). Dominant inheritance of 'lobster claw' deformities of the hands and feet together with facial clefts was noted in three kindreds, by Walker and Clodius (1963). Similar features were present in a girl described by Rudiger, Haase and Passarge (1970) under the designation 'EEC syndrome'. The manifestations of this syndrome have been reviewed by Bixler *et al*. (1972). Inheritance is apparently autosomal dominant with variable expression and occasional non-penetrance. Penchaszadeh and de Negrotti (1976) emphasised the variability in phenotypic manifestations in affected members of a kindred. Rosenmann, Shapira and Cohen (1976) described a father and son of Jewish–Polish origin, with ectrodactyly, ectodermal dysplasia and cleft palate. These authors tabulated the features of 18 familial and three sporadic cases of the EEC syndrome and suggested that the condition existed in two forms; one with cleft lip, with or without cleft palate, and the other with cleft palate alone. The ectrodactyly, anodontia and lacrimal duct dysplasia syndrome which was studied in six patients by Pashayan, Pruzansky and Soloman (1974) may be yet another separate disorder. Reed *et al*. (1975) described a mother and daughter with similar abnormalities, terming the condition the 'REEDS' syndrome.

(c) The author has studied a South African kindred, of British stock, in whom ectrodactyly of the feet was associated with minor changes in the fingers and a defect in the scalp. Individuals in four generations were investigated, and it was evident that inheritance was autosomal dominant, with very variable expression. Adams and Oliver (1945) have previously reported a similar kindred, in which dominantly inherited scalp defects occurred in conjunction with variable deficiencies of the extremities. The tibia and fibula were also hypoplastic in one member of this family. It is not known whether these two conditions are the same genetic entity (see Chapter 16).

6. DIGITAL DYSPLASIA SYNDROMES

In view of the semantic problems concerning the term 'ectrodactyly', digital dysplasia syndromes in which the extremities are not usually split are considered in this subsection. In some of these conditions, fused or duplicated digits may be present. In others, extensive bone abnormalities are a feature. It must be emphasised that these disorders have been grouped together for the sake of convenience, and that there is no suggestion that they bear any fundamental genetic or embryological relationship to each other.

(a) Holt–Oram syndrome

In the Holt–Oram syndrome, unilateral or bilateral dysplasia of the thumb is associated with malformations of the heart. Typically, the thumb is a triphalangeal finger-like digit, while the cardiac anomaly is usually an atrial septal defect. Using the designation 'upper limb–cardiovascular syndrome', Lewis *et al.* (1965) pointed out that the thumb may be totally absent and that limb defects, particularly on the radial side, may be extensive. Poznanski, Gall and Stern (1970) drew attention to the presence of extra or abnormal bones in the carpus as an important feature of the Holt–Oram syndrome. The orthopaedic aspects of the syndrome have been reviewed by Letts *et al.* (1976). There have been several reports of generation to generation transmission and inheritance is autosomal dominant, with variable expression of the abnormal gene (Harris and Osborne, 1966; Gall *et al.*, 1966).

Tamari and Goodman (1974) presented an extensive classification of the upper limb–cardiovascular syndromes and discussed the embryologic relationships between the arms and the heart. These authors also described a boy with bifid thumbs, deafness and septal defects, in whom a small extra chromosome was identified. Other cytogenetic abnormalities which have been reported in the Holt–Oram syndrome include a partial deletion of the long arm of a B group chromosome in several members of a large kindred (Rybak *et al.*, 1971) and a balanced translocation (6q−, 7p+) in an affected female (Ferrier *et al.*, 1975). It is not known whether these chromosomal anomalies are genuine components of the Holt–Oram syndrome. In the somewhat unlikely event of a cytogenetic defect turning out to be a consistent feature, antenatal diagnosis by investigation of cultured amniotic fluid cells would be feasible.

(b) Orofacial–digital syndrome

Papillon-Léage and Psaume (1954) described eight females with facial clefts, abnormalities of the mouth and teeth and hand malformations. The disorder was designated the orofacial–digital syndrome (OFD). Digital changes were very variable, but included brachydactyly, syndactyly, clinodactyly, preaxial polydactyly and duplication of the cuneiform bones and hallux. Gorlin and Psaume (1961) subsequently reported 22 cases, all of whom were female, and suggested that the condition was inherited as an X-linked dominant, with lethality in the male. A further large kindred with 15 affected females in four generations was described by Doege *et al.* (1964). Melnick and Shields (1975) reported an additional family, reviewed published pedigrees and confirmed the X-linked dominant mode of transmission. An inconsistent abnormality of chromosome I has been reported by Reuss *et al*. (1962). It is not known if this cytogenetic anomaly has any direct relationship to the syndrome.

Rimoin and Edgerton (1967) suggested that the OFD syndrome might be heterogeneous and proposed that the condition should be subdivided into type I and II, with the eponym 'Mohr' pertaining to the second type. Whelan, Feldman and Dost (1975) disagreed, pointing out that there was considerable phenotypic overlap and contended that the two forms of the disorder should be grouped together as one syndrome. The situation became even more complex when Temtamy *et al*. (1975) described a newborn Philippino girl with phenotypic features which were consistent with a diagnosis of a severe form of the Mohr syndrome or a mild form of the Majewski syndrome, and postulated that these two conditions might represent variable expression of the same genetic disorder.

(c) Oculodental–osseous dysplasia

Case reports concerning this entity have appeared under a variety of titles. The term 'oculodental–digital syndrome' (ODD syndrome) was favoured by Gorlin, Meskin and Geme (1963), but the designation 'oculodental–osseous dysplasia' was preferred in the Paris Nomenclature. In this way, confusion with the oculodentodigital syndrome type II (ODDII), which was described by O'Rourk and Bravos (1969), is avoided. This latter disorder may be distinguished by the presence of preaxial polydactyly and absence of defects of bone modelling.

Patients with oculodental–osseous dysplasia have a characteristic facies, with hypotrichosis, microphthalmia and a narrow nose. Glaucoma is a frequent complication and hypoplasia of the enamel of the teeth is a constant feature. Cutaneous syndactyly and flexion and ulnar deviation of the fourth and fifth fingers are additional components of this syndrome. Undermodelling of the tubular bones, clavicles, ribs and mandible is radiographically evident. Autosomal dominant inheritance is well established. Lightwood and Lewis (1963) described father to son transmission, while Rajic and de Veber (1966) reported seven individuals in three generations.

(d) Aase syndrome

In the Aase syndrome, triphalangeal thumbs and minor abnormalities of the radius are associated with hypoplastic anaemia and variable cardiac defects. Aase and Smith (1969) reported a pair of brothers with the condition and further affected boys were described by Murphy and Lubin (1972) and Jones and Thompson (1973). The mother of the child reported by Murphy and Lubin (1972) had a triphalangeal thumb. It is therefore possible that the condition is X-linked, and that this abnormality was the result of expression of the gene in a heterozygous female.

(e) Aglossia–adactylia syndrome

In the aglossia–adactylia syndrome, defects of the extremities are associated with hypoplasia of the tongue. The skeletal changes are often asymmetrical and range from shortening of a finger or toe to absence of a limb. Mandibular hypoplasia, missing teeth and an aberrant frenulum are common concomitants. In a review of the literature, Hall (1971) pointed out that all 14 reported cases had been sporadic, and that there was no indication of a genetic aetiology. Nevin *et al*. (1975) contended that new dominant mutation could not be ruled out.

(f) Tetramelic monodactyly

Svejcar, Kleinebrecht and Degenhardt (1976) described two German brothers in whom only a single digit was present on each hand and foot. The authors designated the disorder 'tetramelic monodactyly' and speculated that as no other members of the kindred were affected, inheritance could be recessive or the result of germinal mosaicism. A kindred with monodactyly in two generations was mentioned by Temtamy and McKusick (1969) and three families with autosomal dominant inheritance of a similar anomaly were reported by David (1972).

(g) Goltz syndrome (focal dermal hypoplasia)

Digital anomalies are a prominent feature of this disorder. Syndactyly of the third and fourth fingers is frequently present, but other digital abnormalities include polydactyly, brachydactyly and adactyly. The clavicles and ribs may be hypoplastic and vertebral malformations are not uncommon. Radiographically, longitudinal striations are evident in the tubular bones. Areas of dermal atrophy and pigmentation, dystrophy of the nails, colobomata of the eyes and papillomata of the mouth are other components of the syndrome. The condition has been reviewed by Goltz et al. (1970).

The majority of cases have been sporadic, with an overwhelming female preponderance. Transmission through four generations was observed by Goltz et al. (1962). An excessive number of spontaneous abortions had occurred in two kindreds reported by these authors. It is likely that inheritance is X-linked dominant, with a strong tendency to lethality in the male.

REFERENCES

Digital abnormalities
Temtamy, S. & McKusick, V. A. (1969) Synopsis of hand malformation with particular emphasis on genetic factors. *Birth Defects: Original Article Series*, 5/3, 125.

Polydactyly
Ammann, F. (1970) Investigations cliniques et génétiques sur le syndrome de Bardet-Biedl en Suisse. *Journal de Génétique Humaine*, 18, 1.
Atasu, M. (1976) Hereditary index finger polydactyly. *Journal of Medical Genetics*, 13/6, 469.
Cantu, J. M., Del Castillo, V., Cortes, R. & Urrusti, J. (1975) Autosomal recessive postaxial polydactyly: report of a family. *Birth Defects: Original Article Series*, 11/5, 19.
Eaton, G. O. & McKusick, V. A. (1969) A seemingly unique polydactyly–syndactyly syndrome in four persons in three generations. *Birth Defects: Original Article Series*, 5/3, 221.
Hefner, R. A. (1940) Hereditary polydactyly. *Journal of Heredity*, 31, 25.
Holt, S. (1975) Polydactyly and brachymetapody in two English families. *Journal of Medical Genetics*, 12, 355.
James, J. I. P. & Lamb, D. W. (1963) Congenital abnormalities of the limbs. *Practitioner*, 191, 159.
Johnston, O. & Davis, R. W. (1953) On the inheritance of hand and foot anomalies in six families. *American Journal of Human Genetics*, 5, 356.
Laurence, K. M., Prosser, R., Rocker, I., Pearson, J. F. & Richards, C. (1975) Hirschsprung's disease associated with congenital heart malformation, broad big toes and ulnar polydactyly in sibs: a case for foetoscopy. *Journal of Medical Genetics*, 12, 334.
Okeahialam, T. C. (1974) The pattern of congenital malformations observed in Dar-es-Salaam. *East African Medical Journal*, 51/1, 101.
Pashayan, H., Fraser, F. C., McIntyre, J. M. & Dunbar, J. S. (1971) Bilateral aplasia of the tibia, polydactyly and absent thumb in father and daughter. *Journal of Bone and Joint Surgery*, 53B, 495.
Say, B., Feild, E., Coldwell, J. G., Warnberg, L. & Atasu, M. (1976) Polydactyly with triphalangeal thumbs, brachydactyly, camptodactyly, congenital dislocation of the patellas, short stature and borderline intelligence. *Birth Defects: Original Article Series*, 12/5, 279.
Swanson, A. B. & Brown, K. S. (1962) Hereditary triphalangeal thumb. *Journal of Heredity*, 53, 259.

Temtamy, S. A. & Dorst, J. P. (1975) Polydactyly of triphalangeal thumbs with upper limb and pectoral dysplasia. *Birth Defects: Original Article Series*, **11**/5, 340.

Temtamy, S. A. & Rogers, J. G. (1975) A new postaxial polydactyly syndrome? *Birth Defects: Original Article Series*, **11**/5, 344.

Tuncbilek, E. & Say, B. (1976) Polydactyly in Turkey (with special emphasis on associated malformations). In *Fifth International Congress of Human Genetics*, Mexico.

Yasuda, M. (1975) Pathogenesis of preaxial polydactyly of the hand in human embryos. *Journal of Embryology and Experimental Morphology*, **33**/3, 745.

Woolf, C. M. & Woolf, R. M. (1970) A genetic study of polydactyly in Utah. *American Journal of Human Genetics*, **22**, 75.

Syndactyly

Bouvet, J. P., Maroteaux, P. & Briard Guillemot, M. L. (1976) Poland syndrome. Clinical and genetic study. Physiopathological considerations. *Nouveautés médicales*, **5**/4, 185.

David, T. J. (1972) Nature and etiology of the Poland anomaly. *New England Journal of Medicine*, **287**, 487.

Fitch, N., Jequier, S. & Papageorgiou, A. (1976) A familial syndrome of cranial, facial, oral and limb anomalies. *Clinical Genetics*, **10**, 226.

Goldberg, M. J. & Pashayan, H. M. (1976) Hallux syndactyly — ulnar polydactyly — abnormal ear lobes: a new syndrome. *Birth Defects: Original Article Series*, **12**/5, 255.

Temtamy, S. & McKusick, V. A. (1969) Synopsis of hand malformations with particular emphasis on genetic factors. *Birth Defects: Original Article Series*, **5**/3, 125.

Trier, W. C. (1965) Complete breast absence. Case report and review of the literature. *Plastic and Reconstructive Surgery and the Transplantation Bulletin*, **36**, 431.

Symphalangism

Austin, F. H. (1951) Symphalangism and related fusions of tarsal bones. *Radiology*, **56**, 882.

Elkington, S. G. & Huntsman, R. G. (1967) The Talbot fingers. A study in symphalangism. *British Medical Journal*, **1**, 407.

Kassner, E. G., Katz, I. & Qazi, Q. H. (1976) Symphalangism with metacarpophalangeal fusions and elbow abnormalities. *Pediatric Radiology*, **4**/2, 103.

Strasburger, A. K., Hawkins, M. R., Eldridge, R., Hargreave, R. L. & McKusick, V. A. (1965) Symphalangism. Genetic and clinical aspects. *Bulletin of the Johns Hopkins Hospital*, **117**, 108.

Walbaum, R., Hazard, C. & Cordier, B. (1976) Brachydactylia with symphalangism, probably autosomal recessive. *Human Genetics*, **33**/2, 189.

Brachydactyly

Bass, H. N. (1968) Familial absence of middle phalanges with nail dysplasia: a new syndrome. *Pediatrics*, **42**, 318.

Bell, J. (1951) On brachydactyly and symphalangism. *Treasury of Human Inheritance*, **5**, 1.

Bilginturan, N., Zileli, S., Karacadag, S. & Pirnar, T. (1973) Hereditary brachydactyly associated with hypertension. *Journal of Medical Genetics*, **10**, 253.

Christian, J. C., Cho, K. S., Franken, E. A. & Thompson, B. H. (1972) Dominant preaxial brachydactyly with hallux varus and thumb abduction. *American Journal of Human Genetics*, **24**, 694.

Cuevas-Sosa, A. & Garcia-Segur, F. (1971) Brachydactyly with absence of middle phalanges and hypoplastic nails. *Journal of Bone and Joint Surgery*, **53B**, 101.

Davies, A. B. (1975) Stub Thumbs. *Journal of Medical Genetics*, **12**/4, 414.

Gnamey, D., Walbaum, R., Fossati, P. & Prouvost, J. M. (1975) Hereditary brachydactyly type E. Report on a family. *Pédiatrie*, **30**/2, 153.

Goodman, R. M., Adam, A. & Sheba, C. (1965) A genetic study of stub thumbs among various ethnic groups in Israel. *Journal of Medical Genetics*, **2**, 116.

Gorlin, R. J. & Sedano, H. O. (1971) Cryptodontic brachymetacarpalia. In *Clinical Delineation of Birth Defects*, XI, p.200. Baltimore: Williams and Wilkins.

Haws, D. V. (1963) Inherited brachydactyly and hypoplasia of the bones of the extremities. *Annals of Human Genetics*, **26**, 201.

Haws, D. V. & McKusick, V. A. (1963) Farabee's brachydactylous kindred revisited. *Bulletin of the Johns Hopkins Hospital*, **113**, 20.

MacArthur, J. W. & McCullough, E. (1932) Apical dystrophy as inherited defect of hands and feet. *Human Biology*, **4**, 179.

McKusick, V. A. & Milch, R. A. (1964) The clinical behaviour of genetic disease: selected aspects. *Clinical Orthopaedics and Related Research*, **33**, 22.

Mohr, O. L. & Wriedt, C. (1919) A new type of hereditary brachyphalangy, Publ. 295, p.5. Washington: Carnegie Institute.

Newcombe, D. S. & Keats, T. E. (1969) Roentgenographic manifestations of hereditary peripheral dysostosis. *American Journal of Roentgenology*, **106**, 178.

Poznanski, A. K., Garn, S. M. & Holt, J. F. (1971) The thumb in the congenital malformation syndromes. *Radiology*, **100/1**, 115.

Rimoin, D. L., Hollister, D. W. & Lachman, R. S. (1975) Type C brachydactyly with limited flexion of distal interphalangeal joints. *Birth Defects: Original Article Series*, **11/5**, 9.

Robinson, G. C., Wood, B. J., Miller, J. R. & Baillie, J. (1968) Hereditary brachydactyly and hip disease. Unusual radiological and dermatoglyphic findings in a kindred. *Journal of Pediatrics*, **72**, 539.

Sayles, L. P. & Jailer, J. W. (1934) Four generations of short thumbs. *Journal of Heredity*, **25**, 377.

Sorsby, A. (1935) Congenital coloboma of the macula, together with an account of the familial occurrence of bilateral macular coloboma in association with apical dystrophy of hands and feet. *British Journal of Ophthalmology*, **19**, 65.

Temtamy, S. & McKusick, V. A. (1969) Synopsis of hand malformation with particular emphasis on genetic factors. *Birth Defects: Original Article Series*, **5/3**, 125.

Villaverde, M. M. & Silva, J. A. (1975) Distal brachyphalangy of the thumb in mental retardation. *Journal of Medical Genetics*, **12/4**, 401.

Zavala, C., Hernandez Ortiz, J. & Lisker, R. (1975) Brachydactyly type B and symphalangism in different members of a Mexican family. *Annales de Génétique*, **18/2**, 131.

Ectrodactyly

Adams, F. H. & Oliver, C. P. (1945) Hereditary deformities in man due to arrested development. *Journal of Heredity*, **36**, 3.

Bixler, D., Spivack, J., Bennett, J. & Christian, J. C. (1972) The ectrodactyly–ectodermal dysplasia–clefting (EEC) syndrome. *Clinical Genetics*, **3**, 43.

Gelfand, M., Roberts, C. J. & Roberts, R. S. (1974) A two-toed man from the Doma People of the Zambezi valley. *Rhodesian History*, **5**, 93.

Gemme, G., Bonioli, E., Ruffa, G. & Grosso, P. (1976) EEC syndrome: description of two cases in the same family. *Minerva pediatrica*, **28/1**, 36.

Mackenzie, H. J. & Penrose, L. S. (1951) Two pedigrees of ectrodactyly. *Clinical Genetics*, **9**, 347.

Mosavy, S. H. & Vakshuri, P. (1975) Split hands and feet. *South African Medical Journal*, **49**, 1842.

Pashayan, H. M., Pruzansky, S. & Solomon, L. (1974) The EEC syndrome. Report of six patients. *Birth Defects: Original Article Series*, **10/7**, 105.

Penchaszadeh, V. B. & de Negrotti, T. C. (1976) Ectrodactyly–ectodermal dysplasia–clefting (EEC) syndrome: dominant inheritance and variable expression. *Journal of Medical Genetics*, **13/4**, 281.

Preus, M. & Fraser, F. C. (1973) The lobster-claw defect with ectodermal defects, cleft lip–palate, tear duct anomaly and renal anomalies. *Clinical Genetics*, **4**, 369.

Reed, W. B., Brown, A. C., Sugarman, G. I. & Schlesinger, L. (1975) The REEDS syndrome. *Birth Defects: Original Article Series*, **11/5**, 61.

Roselli, D. & Gulienetti, R. (1961) Ectodermal dysplasia. *British Journal of Plastic Surgery*, **14**, 190.

Rosenmann, A., Shapira, T. & Cohen, M. M. (1976) Ectrodactyly, ectodermal dysplasia and cleft palate (EEC syndrome). Report of a family and review of the literature. *Clinical Genetics*, **9**, 347.

Rudiger, R. A., Haase, W. & Passarge, E. (1970) Association of ectrodactyly, ectodermal dysplasia, with cleft lip–palate. *American Journal of Diseases of Children*, **120**, 160.

Stevenson, A. C. & Jennings, L. M. (1960) Ectrodactyly — evidence in favour of a disturbed segregation in the offspring of affected males. *Annals of Human Genetics*, **24**, 89.

Verma, I. C., Joseph, R., Bhargava, S. & Mehta, S. (1976) Split hand and split foot deformity inherited as an autosomal recessive trait. *Clinical Genetics*, **9/1**, 8.

Walker, J. C. & Clodius, L. (1963) The syndromes of cleft lip, cleft palate and lobster claw deformities of hands and feet. *Plastic and Reconstructive Surgery and the Transplantation Bulletin*, **32**, 627.

Digital dysplasia syndromes

(a) Holt–Oram syndrome

Ferrier, P. E., Friedli, B. & Ferrier, S. (1975) The heart and hand syndrome (Holt–Oram). *Semaine des hôpitaux de Paris*, **51/39**, 727.

Gall, J. C. Jr., Stern, A. M., Cohen, M. M., Adams, M. S. & Davidson, R. T. (1966) Holt–Oram syndrome: clinical and genetic study of a large family. *American Journal of Human Genetics*, **18**, 187.

Harris, L. C. & Osborne, W. P. (1966) Congenital absence or hypoplasia of the radius with ventricular septal defect: ventriculo–radial dysplasia. *Journal of Pediatrics*, **68**, 265.

Letts, R. M., Chudley, A. E., Cumming, G. & Shokier, M. H. (1976) The upper limb cardiovascular syndrome (Holt–Oram syndrome). *Clinical Orthopaedics and Related Research*, **116**, 149.

Lewis, K. B., Bruce, R. A., Baum, D. & Motulsky, A. G. (1965) The upper limb–cardiovascular syndrome. An autosomal dominant genetic effect on embryogenesis. *Journal of the American Medical Association*, **193**, 1080.

Poznanski, A. K., Gall, J. C. Jr. & Stern, A. M. (1970) Skeletal manifestations of the Holt–Oram syndrome. *Radiology*, **94**, 45.

Rybak, M., Kozlowski, K., Kleczkowska, A., Lewandowska, J., Sokolowski, J. & Soltysik-Wilk, E. (1971) Holt–Oram syndrome associated with ectromelia and chromosomal aberrations. *American Journal of Diseases of Children*, **121**, 490.

Tamari, I. & Goodman, R. M. (1974) Upper limb–cardiovascular syndromes: a description of two new disorders with a classification. *Chest*, **65**, 632.

(b) OFD syndrome

Doege, T. C., Thuline, H. C., Priest, J. H., Norby, D. E. & Bryant, J. S. (1964) Studies of a family with the oral–facial–digital syndrome. *New England Journal of Medicine*, **271**, 1073.

Gorlin, R. J. & Psaume, J. (1962) Orodigito-facial sysostosis: a new syndrome. *Journal of Pediatrics*, **61**, 520.

Melnick, M. & Shields, E. D. (1975) Orofaciodigital syndrome, type I: A phenotypic and genetic analysis. *Oral Surgery*, **40**/5, 599.

Papillon-Léage & Psaume, J. (1954) Une malformation héréditaire de la muqueuse buccale, brides et freins. *Revue de Stomatologie*, **55**, 209.

Rimoin, D. L. & Edgerton, M. T. (1967) Genetic and clinical heterogeneity in the oral–facial–digital syndromes. *Journal of Pediatrics*, **71**, 94.

Temtamy, S. A., Levin, L. S., Miller, J. D. *et al.* (1975) Severe Mohr syndrome or mild Majewski syndrome? *Birth Defects: Original Article Series*, **11**/5, 342.

Whelan, D. T., Feldman, W. & Dost, I. (1975) The oro–facial–digital syndrome. *Clinical Genetics*, **8**, 205.

(c) ODD syndrome

Gorlin, R. J., Meskin, L. H. & Geme, J. W. (1963) Oculodento–digital dysplasia. *Journal of Pediatrics*, **63**, 69.

Lightwood, J. M. & Lewis, G. M. (1963) The Holmes–Adie syndrome in a boy with acute juvenile rheumatism and bilateral syndactyly. *Archives of Diseases in Childhood*, **38**, 86.

O'Rourk, T. R. Jr. & Bravos, A. (1969) An oculo–dento–digital dysplasia. In *The Clinical Delineation of Birth Defects*, II, p.226. New York: Williams and Wilkins.

Rajic, D. S. & De Veber, L. L. (1966) Hereditary oculodento–osseous dysplasia. *Annals of Radiology*, **9**, 224.

(d) Aase syndrome

Aase, J. M. & Smith, D. W. (1969) Congenital anemia and triphalangeal thumbs: A new syndrome. *Journal of Pediatrics*, **74**, 417.

Jones, B. & Thompson, H. (1973) Triphalangeal thumbs associated with hypoplastic anemia. *Pediatrics*, **52**, 609.

Murphy, S. & Lubin, B. (1972) Triphalangeal thumbs and congenital erythroid hypoplasia: Report of a case with unusual features. *Journal of Pediatrics*, **81**, 987.

(e) Aglossia–adactylia syndrome

Hall, B. D. (1971) Aglossia–adactylia. A case report, review of the literature and classification of closely related entities. *Birth Defects: Original Article Series*, **7**/7, 233.

Nevin, N. C., Burrows, D., Allen, C. & Kernohan, D. C. (1975) Aglossia adactylia syndrome. *Journal of Medical Genetics*, **12**/1, 89.

(f) Tetramelic monodactyly

David, T. J. (1972) Dominant ectrodactyly and possible germinal mosaicism. *Journal of Medical Genetics*, **9**, 316.

Svejcar, J., Kleinebrecht, J. & Degenhardt, K. H. (1976) Identical tetramelic monodactyly in two brothers. *Clinical Genetics*, **9**/2, 143.

Temtamy, S. & McKusick, V. A. (1969) Synopsis of hand malformations with particular emphasis on genetic factors. *Birth Defects: Original Article Series*, **5**/3, 125.

(g) Goltz syndrome (focal dermal hypoplasia)

Goltz, R. W., Peterson, W. C. Jr., Gorlin, R. J. & Ravits, H. G. (1962) Focal dermal hypoplasia. *Archives of Dermatology*, **86**, 708.

Goltz, R. W., Henderson, R. R., Hitch, J. M. & Ott, J. E. (1970) Focal dermal hypoplasia syndrome. A review of the literature and report of two cases. *Archives of Dermatology*, **101**, 1.

16. Limb dysplasias and synostoses

Major epidemiological studies of limb malformations have been undertaken in Denmark by Birch-Jensen (1949), in Brazil by Freire-Maia (1969), in West Germany and England by Henkel and Willert (1969) and in Scotland by Rogala *et al.* (1974). In each of these surveys it was shown that congenital absence of a limb or segment of a limb is usually non-genetic. However, there have been isolated instances of kindreds with aggregation of mixed limb malformations. Limb reduction is also a feature of a few unusual syndromes, which have a proven or possible genetic basis. Finally, synostosis or abnormal bone union is sometimes found in conjunction with limb deficiencies. For this reason, the synostoses are grouped with the limb dysplasia syndromes. The following categories are considered in this section:

1. Non-specific limb reduction
   (a) Phocomelia
   (b) Amelia
   (c) Hemimelia
2. Limb reduction syndromes
   (a) Roberts syndrome
   (b) Acheiropodia
   (c) Grebe syndrome
   (d) Limb reduction — ichthyosis syndrome
   (e) Proximal focal femoral dysplasia
   (f) Tibial hypoplasia — polydactyly syndrome
   (g) Tibial meromelia
   (h) Aplasia cutis congenita — terminal transverse limb defect syndrome
   (i) Fanconi — pancytopenia syndrome
   (j) Radial aplasia — thrombocytopenia syndrome
   (k) Other genetic limb reduction syndromes
3. Synostosis syndromes
   (a) Radioulnar synostosis
   (b) Synostoses of the tarsus
   (c) Synostoses of the carpus
   (d) Hand–foot–genital syndromes
   (e) Multiple synostoses.

In a comprehensive review, Henkel and Willert (1969) discussed the classification of limb defects and defined 'dysmelia' as 'malformations in which there is hypoplasia and partial or total aplasia of the tubular bones of the extremities, ranging from isolated peripheral hypoplasia to complete loss of the extremity.' Similarly, 'ectromelia' implies maldevelopment of the long bones of the limbs and their peripheral rays. The following terms also pertain to limb anomalies:

Phocomelia	hypoplasia of the limbs, so that the hands and feet are attached directly to the limb girdles
Amelia	absence of a limb or limbs
Hemimelia	absence of a longitudinal segment of a limb
Acheira	absence of a hand or hands
Apodia	absence of a foot or feet
Acheiropodia	absence of the hands and feet.

The pattern of defective development may embrace more than one of these anatomical types and multiple mixed abnormalities may be present. In this context, it is relevant that a 'new international terminology for the classification of congenital limb deficiencies' has been proposed, following a conference of delegates from specialised child amputee centres (Kay, 1975). The embryological mechanisms by which limb malformations may be produced have been reviewed by Poswillo (1976) and Wolpert (1976).

A combination of ingenuity and persistance has permitted some individuals to overcome the disability imposed by major limb defects. For instance, Carl Unthan, who was born without arms, achieved international recognition as a concert virtuoso, playing his violin with his toes. Similarly, Eli Bowen, who lacked legs, was a noted acrobat. The non-genetic nature of these abnormalities is exemplified by the fact that both these remarkable men had many normal sibs and offspring.

1. NON-SPECIFIC LIMB REDUCTION

(a) Phocomelia

In phocomelia, the malformed extremities articulate directly with the trunk. This configuration has been likened to the flippers of a seal, hence the designation 'phocomelia'. If all four limbs are involved the abnormality is termed 'tet-

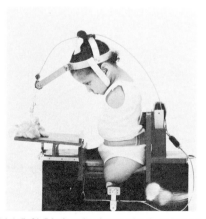 

Fig. 16.1 (left) Limb reduction; technology has much to offer the disabled individual.

Fig. 16.2. (right) Limb reduction; this girl was born without limbs. A vestigial foot is attached to the left hip.

raphocomelia'. The malformation is rare and usually non-genetic. Thaliodomide taken during early pregnancy was responsible for the 'epidemic' of phocomelia which occurred in 1961-2. This tragedy is often quoted as the classic example of the embryopathic action of an environmental agent. Di Battista, Laudizi and Tamborino (1975) reported the birth of a phocomelic infant to a mother who has been treated with an anti-inflammatory drug for a four day period in early pregnancy. The authors emphasised the importance of avoiding potentially toxic agents during the first trimester.

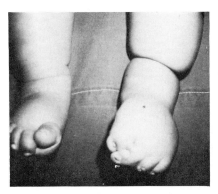 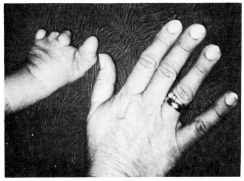

Fig. 16.3. (left) Limb reduction; constriction of the soft tissues of the shin, due to amniotic bands. If bands of this type cause intrauterine amputation, the stump is smooth.

Fig. 16.4. (right) Limb reduction; the presence of rudimentary digits indicates that the abnormality of the left hand is the result of defective development rather than intrauterine amputation.

(b) Amelia (total or partial)

Congenital absence of the forearm or hand is a relatively common abnormality. These defects are usually transverse, and the majority are unilateral. There is no suggestion of any familial tendency. Similarly, absence of part of a leg or foot is usually unilateral and rarely familial. The chromosomes are normal in patients with malformations of this type. However, it is of interest that an embryo with total amelia, spontaneously aborted at the thirteenth week of pregnancy, was shown to lack a C-group chromosome (Pawlowitzki, Cenani and Frischbier, 1973). The relationship between oral contraceptives and congenital limb defects was investigated by Drance (1975). No firm conclusions were reached, but the author drew attention to the VACTERL (vertebral, anal, cardiac, tracheal, oesophogeal, renal and limb) group of anomalies, which are possibly associated with maternal ingestion of progesterone or oestrogen during pregnancy. Limb reduction has been observed in infants of mothers who have used LSD (Assemarry, Neu and Gardner, 1970; Blanc *et al.*, 1971). Nevertheless, there is no firm evidence that this hallucinogen is in fact teratogenic (Long, 1972).

Intrauterine amputation by amniotic bands is thought to be a common cause of amelia. It has been estimated that this abnormality has an incidence of between one in 5,000 and one in 10,000 births (Fisher and Cremin, 1976). The presence of grooves around limbs, or of partial amputation, is evidence for the action of this process. The distal end of a limb which has been amputated in this way is usually smooth. Conversely, if development has been defective, rudimentary digits may be present. In either event, for the purpose of genetic counselling, it is reasonable to offer a very low risk for recurrence.

(c) Hemimelia

Hemimelia, or longitudinal limb deficiency, is the result of defective development of an embryonic ray. The commonest abnormality of this type in the upper limb is

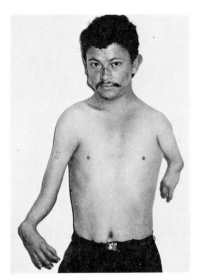
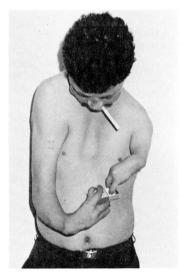

Fig. 16.5. (left) Limb reduction; the bones of both arms are hypoplastic.

Fig. 16.6. (right) Limb reduction; in spite of his handicap, this individual is surprisingly dexterous.

Fig. 16.7. (left) Limb reduction; hypoplasia of the right tibia and fibula. It is unlikely that this unilateral malformation has a genetic basis.

hypoplasia or aplasia of the radius, with corresponding anomalies in the hand. These defects are usually unilateral and non-genetic. However, radial hypoplasia is also a component of a few heritable syndromes (*vide infra*). Defects of the ulna and the medial side of the hand are much less common, the only important disease association being the de Lange syndrome (see Chapter 18).

Tibial hemimelia has been encountered in a pair of sisters (Russell, 1975). Other members of the kindred had hand and foot deformities which could have represented varying phenotypic manifestations of a single dominant gene. Barrie (1976) described two infants with fibular aplasia and limb reduction defects whose mothers had used copper containing intrauterine contraceptive devices. The author speculated that there may have been a causal association between the contraceptives and the fetal maldevelopment.

2. LIMB REDUCTION SYNDROMES

(a) Roberts syndrome

The Roberts syndrome, or pseudophocomelia, is characterised by dysmelia, which ranges from complete phocomelia to minor degrees of deficiency of limb segments, in association with cleft lip and palate. An unusual facies, sparse hair and a variety of visceral abnormalities may also be present. Stillbirth is common and the health of the survivors is precarious. Severe mental deficiency and growth retardation are evident in childhood.

The original case report of Roberts (1919) concerned three affected sibs, born to consanguineous Italian parents. Freeman *et al*. (1974) reviewed the literature and confirmed that the syndrome was inherited as an autosomal recessive. Subsequently, Freeman *et al*. (1975) found chromosomal abnormalities in one patient. Zergollern and Hitrec (1976) studied a Yugoslavian family in which a daughter and two sons had the Roberts syndrome. These authors pointed out that 26 cases have now been reported.

Herrmann *et al*. (1969) described a condition which resembled the Roberts syndrome, with the additional feature of mental deficiency. The designation 'SC syndrome' was taken from the surnames of two affected kindreds. Hall and Greenberg (1972) gave details of a child with the disorder and Lenz, Marquardt and Weicker (1975) mentioned three more cases, using the term 'pseudothalidomide syndrome.' Several investigators, including Gross, Pandel and Wiedemann (1975) have suggested that the Roberts syndrome and the SC syndrome might be the same entity.

(b) Acheiropodia

Acheiropodia is a rare disorder in which all components of the limbs are absent, distal to the middle of the forearms and shins. The condition has been recognised in more than 50 individuals in various parts of Brazil (Freire-Maia, 1970). The affected kindreds, in which consanguinity is frequent, are known as the 'handless and footless families of Brazil'. Toledo *et al*. (1972) summarised information concerning the manifestations of acheiropodia, pointing out that there were no visceral ramifications, and that neither metabolic nor cytogenetic abnormalities could be detected. Pedigree data is entirely consistent with autosomal recessive inheritance and the mutant gene is fully penetrant in the homozygote (Freire-Maia, 1975). It has been estimated that there are a minimum of 25,000 acheiropodia genes in Brazil, all derived from a single mutation (Freire-Maia and Maruyama, 1975).

(c) Grebe syndrome
This condition was delineated by Grebe (1953) when he described two affected German girls, who had consanguineous parents. The term 'achondrogenesis' was used in the title of this paper, and there has been semantic confusion with the lethal short-limb dwarfism syndromes which also bear that name (see Chapter 1). For the sake of clarity, the appellation 'Grebe syndrome' is preferable. The major stigmata are gross shortening of the forearms and shins, with deformity and polydactyly of the extremities. Two Puerto Rican kindreds, each containing two affected children, have been reported by Castro and Comas (1975). These authors commented that 54 similar cases could be recognised in the literature. Inheritance of the Grebe syndrome is undoubtedly autosomal recessive.

(d) Limb reduction — ichthyosis syndrome
Falek *et al*. (1968) reported twin sisters with unilateral reduction defects of the limbs in association with unilateral ichthyosis and major cardiac malformations. Both infants died in the neonatal period. The parents were normal, and there was no admitted consanguinity. The mode of inheritance is unknown.

(e) Proximal focal femoral dysplasia
Proximal focal femoral dysplasia (PFFD) is an unusual form of femoral mal-development. The condition may be bilateral and symmetrical or unilateral. Individuals with the disorder have a remarkable appearance, as their knee joints are in close proximity to the pelvic girdle. Nevertheless, they may have a surprising degree of agility. For instance, Fergusson (1964) described how Hervio Novo had achieved fame as a gymnast, in spite of his femoral deficiencies. There have been no reports of familial aggregation and it seems likely that PFFD is non-genetic. However, a young couple who both have the disorder have recently married, and it is possible that the offspring of this union might provide the answer to the question of heritability (Bailey and Beighton, 1970).

A syndrome in which femoral hypoplasia was associated with other variable limb defects and an unusual facies was delineated by Daentl *et al*. (1975). No genetic basis has been recognised.

(f) Tibial hypoplasia — polydactyly syndrome
Bilateral hypoplasia of the tibia with thickening and bowing of the fibula, in association with polydactyly of the hands and feet, was present in a family studied by Eaton and McKusick (1969). Subsequently, Yujnovsky *et al*. (1974) encountered the same disorder in three generations of a kindred and confirmed that inheritance was autosomal dominant. Pfeiffer and Roeskau (1971) reported a mother and son with similar abnormalities.

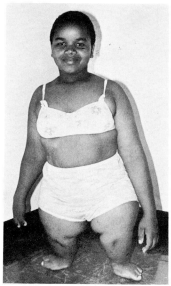

Fig. 16.8. (left) Proximal focal femoral dysplasia; although the right femur is very hypoplastic, the skeleton is otherwise normal.

Fig. 16.9. (right) Proximal focal femoral dysplasia; the femora are short and malformed. The genetic basis of this anomaly is uncertain.

(g) Tibial meromelia

Using the designation 'tibial meromelia', Clark (1975) described a kindred with congenital absence of the tibia. Nine individuals in two generations were affected and the pattern of transmission was consistent with autosomal dominant inheritance.

(h) Aplasia cutis congenita — terminal transverse limb defect syndrome

Adams and Oliver (1945) reported a family in which terminal transverse defects of the limbs were associated with central skull and scalp deficiencies. Eight individuals in three generations were affected, and inheritance was apparently autosomal dominant. Scribanu and Temtamy (1975) described a second kindred with the condition and emphasised that clinical expression was extremely variable.

(i) Fanconi — pancytopenia syndrome

Radial aplasia, absence of the thumb and bone marrow dysfunction constitute the Fanconi pancytopenia syndrome. Small stature and dermal pigmentation are additional features. Haematological problems usually develop in mid-childhood and leukaemia may be a late complication. Bone marrow transplantation from a normal sib produced a dramatic improvement in the blood count of a 15 year old affected boy (Barrett *et al.*, 1977).

More than 150 patients have been reported, and there is abundant evidence for autosomal recessive inheritance. In a review of 48 kindreds, Garriga and Crosby (1959) noted that four potentially heterozygous relatives had developed leukaemia.

Swift and Hirschhorn (1966) suggested that heterozygotes might have a propensity to the development of malignant tumours. Multiple chromosomal breakages have often been observed in homozygotes but so far, there have been no reports of cytogenetic anomalies in obligatory heterozygotes.

(j) Radial aplasia — thrombocytopenia syndrome

Defective development of the radius is the main skeletal abnormality in the radial aplasia–thrombocytopenia syndrome. The ulna may also be hypoplastic and a wide variety of anomalies may be present in the extremities and viscera. A significant proportion of infants die from abnormal bleeding. The haematological status of the survivors improves with the passage of time.

The manifestations have been reviewed by Hall et al. (1969) under the designation 'thrombocytopenia with absent radius, TAR syndrome'. About 20 cases have been described and pedigree data is consistent with autosomal recessive inheritance. An affected fetus has been recognised at the 17th week of an 'at risk' pregnancy by radiological demonstration of absence of the radii (Omenn et al., 1973).

A variety of the TAR syndrome in which tetraphocomelia is associated with neonatal thrombocytopenia has been recognised in two sibs by Pfeiffer and Henke (1975). This disorder seems to be a distinct autosomal recessive entity.

(k) Other genetic limb reduction syndromes

It is likely that other genetic limb reduction syndromes await delineation. For instance, Kucheria, Bhargave and Bamezai (1976) described an Indian brother and sister with tetraphocomelia, cleft lip and palate and multiple minor abnormalities. The authors contended that the condition was a new syndrome, and that inheritance was autosomal recessive. Similarly, Lazjuk et al. (1976) reported amelia, oligodactyly and other malformations in two half-cousins.

3. SYNOSTOSIS SYNDROMES

Synostosis, or bone fusion, is most frequently encountered in the tarsus, carpus and forearms. Coalitions of this type may be isolated anomalies, components of well defined syndromes or discrete genetic entities. Few family studies have been undertaken and there is a paucity of information concerning inheritance of these abnormalities. McCredie (1975) pointed out that congenital fusion is the result of inappropriate organisation of mesenchyme during the fifth week after conception and postulated that the primary defect might lie in the embryonic sensory nerves.

(a) Radioulnar synostosis

Radioulnar synostosis occurs as a sporadic defect, as part of non-genetic multiple malformation syndromes, in the Klinefelter syndrome and as a simple autosomal dominant trait (Hansen and Anderson, 1970). Radioulnar synostosis, in combination with hypoplasia of the digits, has been reported (Davenport, Taylor and Nelson, 1924). Inheritance is autosomal dominant, with considerable variation in expression.

(b) Synostoses of the tarsus

The talus and calcaneum may be united by a fibrous bridge or a bony bar. The only reports of familial occurrence concern a brother and sister (Webster and Roberts, 1951) and a mother and her four offsprings (Diamond, 1974).

Coalition of the calcaneum and the navicular may present clinically as a painful spastic flat foot. This abnormality was recognised in three males in successive generations of an American kindred by Wray and Herndon (1963). Although inheritance was apparently autosomal dominant in this particular family, it is not known whether all cases of calcaneonavicular synostosis have a similar genetic aetiology.

Kindreds with dominant transmission of talonavicular synostosis have been reported by Rothberg, Feldman and Schuster (1935) and Boyd (1944). An English family with coalition of the talus and navicular, in association with anomalies of the fifth fingers, has been studied by Challis (1974).

(c) Synostoses of the carpus

Isolated synostosis of the carpus is uncommon, although bone fusion is found in several specific genetic entities and in association with malformations of the digits. Brug (1975) reported a young man with synostosis of the carpus and postaxial symbrachydactyly and mentioned that the anomaly had been present in the male members of the kindred for nearly 200 years. The situation is consistent with Y-linked inheritance, and if these observations are confirmed, this condition would rank with hairy ear lobes in the exclusive category of Y-linked traits!

(d) Hand–foot–genital syndrome

Stern et al. (1970) noted abnormalities in the extremities and genital tracts of 13 members of four generations of a kindred and suggested that the disorder was a new autosomal dominant entity. Terming the condition the 'hand–foot–uterus' syndrome, Poznanski, Stern and Gall (1970) and Giedion and Prader (1976) have given details of the radiographic changes, including trapezium-scaphoid and cunei-form–navicular synostosis and shortening of the first metacarpals and metatarsals. Poznanski et al. (1975) used pattern profile analysis to evelute their patients, and pointed out that genital anomalies were inconsistent, especially in males. They proposed the alternative designation 'hand–foot–genital' syndrome.

(e) Multiple synostoses

Multiple carpal and tarsal synostosis, with phalangeal involvement and radial head subluxation, was encountered in a mother and daughter by Pearlman, Edkin and Warren (1964). A similar disorder in two generations of a kindred had previously been reported by Bersani and Samilson (1957). A mother and daughter with fusion of several bones in the carpus and tarsus, in association with asymmetry of the metacarpals and metatarsals, and other variable skeletal changes, were described by Christian et al. (1975). The authors suggested that the condition might be inherited as an autosomal dominant. It is possible that these kindreds all had the same genetic disorder, which exhibited variable expression. Equally, they could all have had unique 'private' syndromes. The disorder which was studied in five members of three generations of an Italian family by Ventruto *et al.* (1976) certainly seems to be distinct. These patients had synostoses in the carpus and tarsus, together with symphalangism, brachydactyly, craniostenosis and dysplasia of the hip joints.

REFERENCES

Limb dysplasias and synostoses
Birch-Jensen, A. (1949) *Congenital Deformities of the Upper Extremities*. Copenhagen: Munksgaard.
Freire-Maia, N. (1969) Congenital skeletal limb deficiencies — a general view. *Birth Defects: Original Article Series*, 5/3, 7.
Henkel, L. & Willert, H. G. (1969) Dysmelia. *Journal of Bone and Joint Surgery*, 51B, 399.
Kay, H. W. (1975) Clinical applications of the new international terminology for the classification of congenital limb deficiencies. *Inter-Clinic Information Bulletin*, 14/3, 1.
Poswillo, D. (1976) Mechanisms and pathogenesis of malformation. *British Medical Bulletin*, 32/1.
Rogala, E. J., Wynne-Davies, R., Littlejohn, A. & Gormley, J. (1974) Congenital limb anomalies: frequency and aetiological factors. Data from the Edinburgh Register of the Newborn. *Journal of Medical Genetics*, 11/3, 221.
Wolpert, L. (1976) Mechanisms of limb development and malformation. *British Medical Bulletin*, 32/1.

Non-specific limb reduction
Assemarry, S. R., Neu, R. L. & Gardner, L. I. (1970) Deformities in a child whose mother took LSD. *Lancet*, i, 1290.
Barrie, H. (1976) Congenital malformation associated with intrauterine contraceptive device. *British Medical Journal*, 1/6008, 488.
Blanc, W. A., Mattison, D. R., Kane, R. & Chauhan, P. (1971) LSD intrauterine amputations and amniotic band syndrome. *Lancet*, ii, 158.
Di Battista, C., Laudizi, L. & Tamborino, G. (1975) Phocomelia and agenesis of the penis in a neonate. Possible teratogenic role of a drug taken by mother during pregnancy. *Minerva Pediatrica*, 27/11, 675.
Drance, S. M. (1975) Oral contraceptives and congenital limb defects. *Canadian Medical Association Journal*, 112/5, 551.
Fisher, R. M. & Cremin, B. J. (1976) Limb defects in the amniotic band syndrome. *Pediatric Radiology*, 5, 24.
Long, S. (1972) Does LSD induce chromosomal damage and malformations? A review of the literature. *Teratology*, 6, 75.
Pawlowitzki, I. H., Cenani, A. & Frischbier, H. J. (1973) Autosomal monosomy (45, XX, C−) in a human embryo with total amelia and further malformations. *Clinical Genetics*, 4, 193.
Russell, J. E. (1975) Tibial hemimelia: limb deficiency in siblings. *Inter-Clinic Information Bulletin*, 14/7-8, 15.

Limb reduction syndromes
(a) Roberts syndrome
Freeman, M. V. R., Williams, D. W., Schimke, N. & Temtamy, S. A. (1974) The Roberts syndrome. *Clinical Genetics*, 5, 1.
Freeman, M. V. R., Williams, D. W., Schimke, R. N. *et al.* (1975) The Roberts syndrome. *Birth Defects: Original Article Series*, 11/5, 87.
Grosse, F. R., Pandel, C. & Wiedemann, H. R. (1975) The tetraphocomelia–cleft palate syndrome: description of a new case. *Humangenetick*, 28/4, 353.
Hall, B. D. & Greenberg, M. H. (1972) Hypomelia–hypotrichosis–facial hemangioma syndrome (pseudothalidomide, SC syndrome, SC phocomelia syndrome). *American Journal of Diseases of Children*, 123, 602.
Herrmann, J., Feingold, M., Tuffli, G. A. & Opitz, J. M. (1969) A familial dysmorphogenic syndrome of limb deformities, characteristic facial appearance and associated anomalies: the 'pseudothalidomide' or 'SC-syndrome.' In *The Clinical Delineation of Birth Defects*, III, p.81. New York: National Foundation.
Lenz, W. D., Marquardt, E. & Weicker, H. (1975) Pseudothalidomide syndrome. *Birth Defects: Original Article Series*, 11/5, 97.
Roberts, J. B. (1919) A child with double cleft of lip and palate, protrusion of the intermaxillary portion of the upper jaw and imperfect development of the bones of the four extremities. *Annals of Surgery*, 70, 252.
Zergollern, L. & Hitrec, V. (1976) Three siblings with Robert's syndrome. *Clinical Genetics*, 9, 433.

(b) Acheiropody
Freire-Maia, A. (1970) The handless and footless families of Brazil. *Lancet*, i, 519.
Freire-Maia, A. (1975) Genetics of acheiropodia ('the handless and footless families of Brazil'). VIII. Penetrance and expressivity. *Clinical Genetics*, 7, 98.
Freire-Maia, A., Li, W. H. & Maruyama, T. (1975) Genetics of acheiropodia (the handless and footless families of Brazil). VII. Population dynamics. *American Journal of Human Genetics*, 27/5, 665.
Toledo, S. P. A., Saldanha, P. H., Borelli, A. & Cintra, A. B. U. (1972) Further data on acheiropody. *Journal de Génétique Humaine*, 20/3, 253.

(c) Grebe syndrome
Garcia Castro, J. M. & Perez Comas, A. (1975) Nonlethal achondrogenesis (Grebe–Quelce-Salgado type) in two Puerto Rican sibships. *Journal of Pediatrics*, 87/61, 948.
Grebe, H. (1953) Die Achondrogenesis — ein einfach rezessive Erbmerkmal. *Folia Hereditaria et Pathologica*, 2, 23.
Quelce-Salgado. A. (1964) A new type of dwarfism with various bone aplasias and hypoplasias of the extremities. *Acta Genetica et Statistica Medica*, 14, 63.

(d) Limb reduction — icthyosis syndrome
Falek, A., Heath, C. W., Ebbin, A. J. & McLean, W. R. (1968) Unilateral limb and skin deformities with congenital heart disease in twin siblings. A lethal syndrome. *Journal of Pediatrics*, 73, 910.

(e) PFFD
Bailey, J. A. & Beighton, P. (1970) Bilateral femoral dysgenesis. *Clinical Pediatrics*, 9, 668.
Daentl, D. L., Smith, D. W., Scott, C. I., Hall, B. D. & Gooding, C. A. (1975) Femoral hypoplasia–unusual facies. *Journal of Pediatrics*, 86, 107.
Fergusson, J. (1964) Progress of anatomy and surgery during the present century. *Lancet*, ii, 60.

(f) Tibial hypoplasia — polydactyly syndrome
Eaton, G. O. & McKusick, V. A. (1969) A seemingly unique polydactyly–syndactyly syndrome in four persons in three generations. In *The Clinical Delineation of Birth Defects*, III, p.221. New York: National Foundation.
Pfeiffer, R. A. & Roeskau, M. (1971) Agenesie der Tibia, Fibulaverdopplung und spiegelbildliche Polydaktylie (Diplopodie) bei Mutter und Kind. *Zeitschrift für Kinderheilkunde*, 111, 38.
Yujnovsky, O., Ayala, D., Vinvitorio, A., Viale, H., Sakati, N. & Nyhan, W. L. (1974) A syndrome of polydactyly–syndactyly and triphalangeal thumbs in three generations. *Clinical Genetics*, 6, 51.

(g) Tibial meromelia
Clark, M. W. (1975) Autosomal dominant inheritance of tibial meromelia: report of a kindred. *Journal of Bone and Joint Surgery*, 57/2, 262.

(h) Aplasia cutis congenita — terminal transverse limb defect syndrome
Adams, F. H. & Oliver, C. P. (1945) Hereditary deformities in man due to arrested development. *Journal of Heredity*, 36, 3.
Scribanu, N. & Temtamy, S. A. (1975) The syndrome of aplasia cutis congenita with terminal transverse defects of limbs. *Journal of Pediatrics*, 87/1, 79.

(i) Fanconi — pancytopenia syndrome
Barrett, A. J., Brigden, W. D., Hobbs, J. R., Hugh-Jones, K., Humble, J. G., James, D. C. O., Retsas, S., Rogers, T. R. F., Selwyn, S., Sneath, P. & Watson, J. G. (1977) Successful bone marrow transplant for Fanconi's anaemia. *British Medical Journal*, 1, 420.
Garriga, S. & Crosby, W. H. (1959) The incidence of leukemia in families of patients with hypoplasia of the marrow. *Blood*, 14, 1008.
Swift, M. R. & Hirschhorn, K. (1966) Fanconi's anemia: inherited susceptibility to chromosome breakage in various tissues. *Annals of Internal Medicine*, 65, 496.

(j) Radial aplasia — thrombocytopenia syndrome
Hall, J. G., Levin, J., Kuhn, J. P., Ottenheimer, E. J., Van Berkum, K. A. P. & McKusick, V. A. (1969) Thrombocytopenia with absent radius (TAR). *Medicine*, 48, 411.
Omenn, G. S., Eigley, M. M., Graham, C. B. & Heinrichs, W. Le R. (1973) Prospects for radiographic intrauterine diagnosis: The syndrome of thrombocytopenia with absent radii. *New England Journal of Medicine*, 288, 777.
Pfeiffer, R. A. & Haenke, C. (1975) The phocomelia thrombocytopenia syndrome. A follow-up report. *Humangenetik*, 26/2, 157.

(k) Other genetic limb reduction syndromes

Kucheria, K., Bhargava, S. K. & Bamezai, R. (1976) A familial tetraphocomelia syndrome involving limb deformities, cleft lip, cleft palate and associated anomalies — a new syndrome. In *Fifth International Congress of Human Genetics*, Mexico.

Lazjuk, G. I., Lurie, I. W., Cherstvoy, E. D. & Ussova, Y. I. (1976) A syndrome of multiple congenital malformations including amelia and oligodactyly occurring in half-cousins. *Teratology*, 13/2, 161.

Synostosis syndromes

McCredie, J. (1975) Congenital fusion of bones: radiology, embryology and pathogenesis. *Clinical Radiology*, 26/1, 47.

(a) Radioulnar synostosis

Davenport, C. B., Taylor, H. L. & Nelson, L. A. (1924) Radioulnar synostosis. *Archives of Surgery*, 8, 705.

Hansen, O. H. & Anderson, N. O. (1970) Congenital radioulnar synostosis. Report of 37 cases. *Acta Orthopaedica Scandinavica*, 41, 225.

(b) Synostosis of the tarsus

Boyd, H. B. (1944) Congenital talonavicular synostosis. *Journal of Bone and Joint Surgery*, 26, 682.

Challis, J. (1974) Hereditary transmission of talonavicular coalition in association with anomaly of the little finger. *Journal of Bone and Joint Surgery*, 56/6, 1273.

Diamond, L. S. (1974) Inherited talocalcaneal coalition. *Birth Defects: Original Article Series*, 10/12, 531.

Rothberg, A. S., Feldman, F. W. & Schuster, T. F. (1935) Congenital fusion of astragalus and scaphoid: bilateral: inherited. *New York State Journal of Medicine*, 35, 29.

Webster, F. S. & Roberts, W. M. (1951) Tarsal anomalies and peroneal spastic flat foot. *Journal of the American Medical Association*, 146, 1099.

Wray, J. B. & Herndon, C. N. (1963) Hereditary transmission of congenital coalition of the calcaneus to the navicular. *Journal of Bone and Joint Surgery*, 45A, 365.

(c) Synostosis of the carpus

Brug, E. (1975) A case of familial hereditary malformation of both hands. *Handchirurgie*, 7/3, 125.

(d) Hand–foot–genital syndromes

Giedion, A. & Prader, A. (1976) Hand–foot–uterus (HFU) syndrome with hypospadias: the hand–foot–genital (HFG) syndrome. *Pediatric Radiology*, 4/2, 96.

Poznanski, A. K., Kuhns, L. R., Lapides, J. & Stern, A. M. (1975) A new family with the hand–foot–genital syndrome: a wider spectrum of the hand–foot–uterus syndrome. *Birth Defects: Original Article Series*, 11/4, 127.

Poznanski, A. K., Stern, A. M. & Gall, J. C. Jr. (1970) Radiographic findings in the hand–foot–uterus syndrome (HFUS). *Radiology*, 95, 129.

Stern, A. M., Gall, J. C. Jr., Perry, B. L., Stimson, C. W., Weitkamp, L. R. & Poznanski, A. K. (1970) The hand–foot–uterus syndrome. A new hereditary disorder characterised by hand and foot dysplasia, dermatoglyphic abnormalities, and partial duplication of the female genital tract. *Journal of Pediatrics*, 77, 109.

(e) Multiple synostosis

Bersani, F. A. & Samilson, R. L. (1957) Massive familial tarsal synostosis. *Journal of Bone and Joint Surgery*, 39A, 1187.

Christian, J. C., Franken, E. A., Lindeman, J. P., Lindseth, R. E., Reed, T. & Scott, C. I. Jr. (1975) A dominant syndrome of metacarpal and metatarsal asymmetry with tarsal and carpal fusions, syndactyly, articular dysplasia and platyspondyly. *Clinical Genetics*, 8, 75.

Pearlman, H. S., Edkin, R. E. & Warren, R. F. (1964) Familial tarsal and carpal synostosis with radial-head subluxation (Nievergelt's syndrome). *Journal of Bone and Joint Surgery*, 46A, 585.

Ventruto, V., Girolamo, R., Festa, B., Romano, A., Sebastio, G. & Sebastio, L. (1976) Family study of inherited syndrome with multiple congenital deformities: symphalangism, carpal and tarsal fusion, brachydactyly, craniostenosis, strabismus, hip osteochondritis. *Journal of Medical Genetics*, 13/5, 394.

17. Connective tissue disorders

Successive editions of McKusick's classical mongraph 'Heritable Disorders of Connective Tissue' have brought this group of conditions to the attention of medical science and stimulated investigations into their natural history and pathogenesis. In McKusick's monumental work the interested reader will find a comprehensive account of the manifestations and genetics of these disorders. The skeleton is involved, to some extent, in the majority of them and the following form the subject of this section:

1. Marfan syndrome
2. Ehlers–Danlos syndrome
3. Fibrodysplasia ossificans progressiva
4. Weill–Marchesani syndrome
5. Homocystinuria
6. Alkaptonuria
7. Tight skin syndromes
   (a) Rothmund syndrome
   (b) Stiff skin syndrome
   (c) Syndesmodysplastic dwarfism
   (d) Parana hard skin syndrome
   (e) Moore–Federman syndrome
8. Loose skin syndromes
   (a) Cutis laxa with bone dystrophy
   (b) Leprechaunoid syndrome
   (c) Wrinkly skin syndrome.

McKusick has drawn a noteworthy distinction between those conditions in which connective tissues are primarily at fault, as in the Marfan syndrome and the Ehlers–Danlos syndrome, and those in which involvement is secondary to a metabolic abnormality, such as homocystinuria or alkaptonuria.

Sporadic patients with poorly defined skeletal and connective tissue abnormalities are frequently encountered and there is little doubt that a large number of rare entities await delineation. For example, kindreds in which generalised joint laxity was inherited as an autosomal dominant trait were investigated by Beighton and Horan (1970). In addition to the hypermobility, some of these patients had minor degrees of dermal extensibility and a tendency to stretching of scar tissue, in a manner which was reminiscent of the Ehlers–Danlos syndrome (*vide infra*). Similarly, Greenfield *et al.* (1973) reported an Iraqi–Jewish brother and sister with scoliosis, spondylolisthesis, pelvic distortion, blue sclerae and keratoconus. The parents were consanguineous, and it is likely that the condition was inherited as an autosomal recessive. Again, Frank *et al.* (1973) described a girl from a consanguineous Bedouin kindred who was born with flexion deformities of the digits, clubbing of the feet and marked kyphoscoliosis, in association with megalocornea. Advances in knowledge of the biochemistry of collagen hold great promise for the ultimate definition and categorisation of conditions of this type.

1. MARFAN SYNDROME

The Marfan syndrome is a well known disorder in which excessive height is associated with a variety of connective tissue abnormalities. Since the original reports at the turn of the century, several hundred cases have been described from all parts of the world.

A decade ago, homocystinuria was split off from the Marfan group of disorders, and subsequently other separate entities were recognised, including congenital contractural arachnodactyly and the Marfanoid hypermobility syndrome. McKusick (1975) pointed out that some individuals with the Marfan syndrome are very severely affected and he has proposed that there are 'asthenic' and 'non-asthenic' types of the disorder. The following phenotypically distinct forms of the Marfan syndrome are now recognised:

Type I   asthenic form
Type II   non-asthenic form
Type III   Marfanoid hypermobility syndrome
Type IV   congenital contractural arachnodactyly.

Clinical features

The stigmata of the Marfan syndrome include disproportionately long extremities, a high palate, dislocation of the lens of the eye, thoracic asymmetry, articular hypermobility and a tendency to aneurysm and dissection of the aorta. Robins, Moe and Winter (1975) reviewed the spinal problems in 68 patients and found that more than 50 per cent had significant degrees of scoliosis. Excessive lengthening of the digits, or arachnodactyly, represents one of the cardinal features of the Marfan syndrome. This anomaly is by no means pathognomonic, as arachnodactyly also occurs in isolation and as a component of several other syndromes.

Delineation of the asthenic and non-asthenic forms of the Marfan syndrome is not yet complete. However, children with the asthenic variety have exceptionally severe manifestations and usually do not survive into adulthood. In view of the phenotypic disparity there seems to be little doubt that these conditions are separate entities.

Genetics

In Northern Ireland, a minimum prevalence of 1.5 per 100,000 of the population has been calculated (Lynas, 1958). Autosomal dominant inheritance is well established and many large pedigrees have been published. No consistent biochemical or ultrastructural abnormalities have been identified and the diagnosis is made on a clinical basis. The well known variability of expression is typified by the disparity in the manifestations which was observed in affected monozygous twins by Ambani, Gelherter and Sheahan (1975).

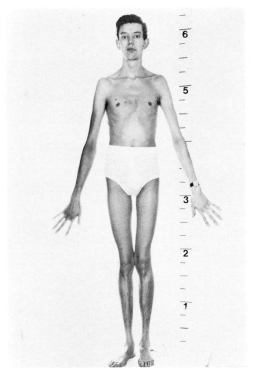

Fig. 17.1. Marfan syndrome; a young man with the Marfan habitus. In addition to excessive limb length, the lenses of the eyes are dislocated.

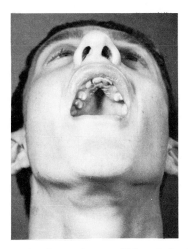

Fig. 17.2. Marfan syndrome; a high arched palate.

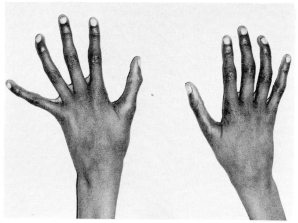

Fig. 17.3. Marfan syndrome; arachnodactyly (spider-like digits).

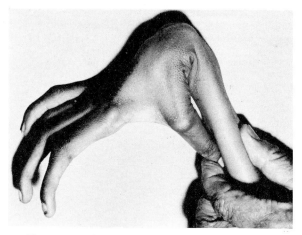

Fig. 17.4. Marfan syndrome; laxity of the wrist joint.

Positive diagnosis is sometimes difficult, but if there are unequivocal cases in the kindred, the presence of a limited number of components of the syndrome may permit recognition. The 'metacarpal index', which is derived from measurements of hand radiographs, or a comparison of upper and lower body segment ratios may be helpful, but these objective criteria are by no means infallible. In the sporadic case, it may be impossible to reach a firm decision for the purposes of genetic counselling. A similar problem arises when it is suspected that the newborn child of an individual with the Marfan syndrome might have inherited the disorder. The length of the middle finger, from the tip to the proximal flexion crease, as a percentage of total hand size, is a useful determinant of arachnodactyly in the neonate and it is sometimes possible to resolve the situation in this way (Feingold and Bossert, 1974). Nevertheless, it may be necessary to defer diagnostic judgement until later childhood.

McKusick (1972) estimated that about 15 per cent of patients with the Marfan syndrome represent new mutations. The paternal age effect has been demonstrated

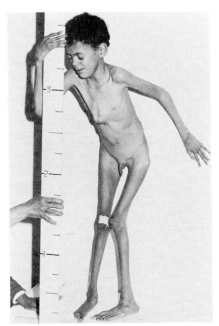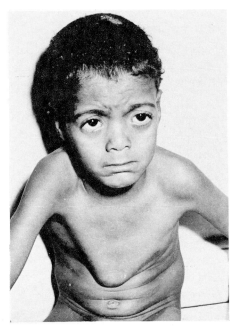

Fig. 17.5. (left) Marfan syndrome; the asthenic form with severe manifestations.

Fig. 17.6. (right) Marfan syndrome; thoracic deformity in the asthenic form.

in a study of sporadic cases (Murdoch, Walker and McKusick, 1972). Lynas and Merrett (1958) undertook linkage studies in kindreds in Northern Ireland using conventional gene markers and further extensive investigations were subsequently carried out in 15 families in the USA by Schleuterman *et al*. (1976). No linkage between the Marfan syndrome locus and the marker loci was demonstrated in either instance.

Tall stature is a notable feature of the Marfan syndrome, and it is likely that individuals with the condition served in the 'regiment of giants', which was raised by King Fredrich I of Prussia. This military unit is said to have made a significant contribution to the gene pool of the town of Potsdam, where they were based for fifty years. Investigations of the prevalence of the Marfan syndrome in that locality might be of interest!

Other forms of the Marfan syndrome
(a) The 'Marfanoid hypermobility syndrome' is characterised by a Marfanoid habitus, extreme joint laxity and extensible skin. A young man with this condition was described by Walker, Beighton and Murdoch (1969). A similar case is recognisable in a previous report by Goodman *et al*. (1965), which was entitled 'the Ehlers–Danlos syndrome occurring with the Marfan syndrome'. The genetic basis of the condition is uncertain.

(b) The clinical manifestations of the congenital contractural arachnodactyly syndrome resemble those of the Marfan syndrome but cardiovascular and ocular problems do not occur. Flexion contractures of the fingers and other joints, severe kyphoscoliosis and small crumpled ears are additional features. Beals and Hecht

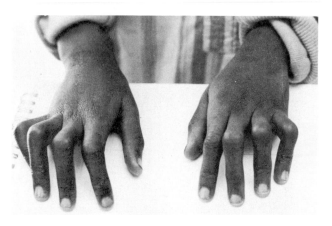

Fig. 17.7. (left) Congenital contractural arachnodactyly; kyphoscoliosis has been corrected at operation. The limbs are long, the digits are flexed and the pinna of the ear is crumpled.

Fig. 17.8. (right) Congenital contractural arachnodactyly; fixed deformities of the digits.

(1971) encountered the syndrome in two generations of two separate families and mentioned that they could recognise 11 other kindreds in the literature. Male to male transmission had taken place, and inheritance was evidently autosomal dominant. The condition is probably common among the Marfan group of patients and other kindreds have been reported by Lowry and Guichon (1972), Macleod and Fraser (1973), Steg (1975), Kontras (1975) and Bjerkeim, Skogland and Trygstad (1976).

(c) Passarge (1975) described two unrelated patients with a 'syndrome resembling contractural arachnodactyly'. These individuals were atypical in that the spine was severely involved, while digital contractures were less marked. The author suggested that this condition might represent a new entity. No genetic background was apparent in either kindred.

(d) The Achard syndrome is characterised by decreased metacarpal width, widespread mild dysostosis and ligamentous laxity. Achard (1902) published his original case report under the designation 'arachnodactyly' and since that time there has been semantic confusion with the Marfan syndrome. In a report of three patients, Duncan (1975) emphasised that the Achard syndrome is clinically distinct from the Marfan syndrome. The mode of genetic transmission is unknown.

2. EHLERS–DANLOS SYNDROME

The Ehlers–Danlos syndrome (EDS) is one of the best known inherited disorders of connective tissue, and more than 600 cases have now been reported. The condition received eponymous recognition at the beginning of the century, but earlier examples can be identified in the literature pertaining to such diverse institutions as the Academy of Leiden and Barnum and Bailey's circus!

The EDS is very heterogeneous and the following seven varities have now been recognised:

Type		Genetics	Basic defect
I	Gravis	AD	?
II	Mitis	AD	?
III	Benign hypermobile	AD	?
IV	Ecchymotic or arterial	AR:AD	Deficient synthesis of type III collagen.
V	X-linked	XL	Lysyl oxidase deficiency?
VI	Ocular	AR	Procollagen lysyl hydroxylase deficiency
VII	Arthrochalasis multiplex congenita	AR	Procollagen protease deficiency

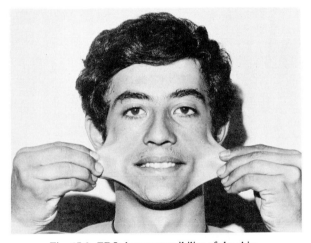

Fig. 17.9. EDS; hyperextensibility of the skin.

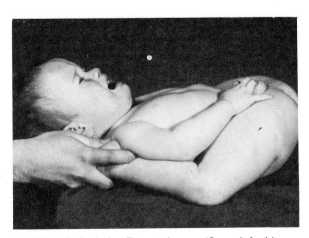

Fig. 17.10. EDS; articular laxity. Presentation as a 'floppy infant' is not unusual.

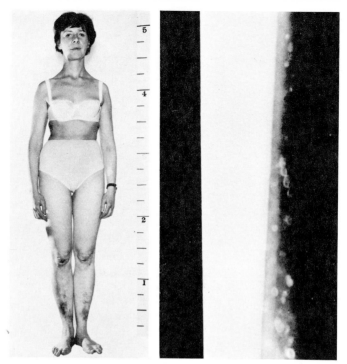

Fig. 17.11. (left) EDS; thin pigmented scars over the knees and shins.

Fig. 17.12. (right) EDS; radiograph of the shin, showing calcified subcutaneous spheroids. (From Beighton, P. & Lea Thomas, M. (1969) *Clinical Radiology*, **20**, 354.)

Clinical features

The major clinical features of the EDS are joint laxity, dermal extensibility and tissue fragility. A bleeding tendency is sometimes present. The skin splits easily and the bony prominences are often covered with distracted papyraceous scars. Hard shotty spheroids can be palpated in the subcutaneous tissues, particularly over the long bones. Hypotonia and articular hypermobility are often the first stigmata to be recognised in early life. Indeed, a professional contortionist with the EDS was heard to lament 'I used to be a floppy infant, but now I'm a loose woman!'

Orthopaedic complications include joint instability, recurrent dislocations, spinal malalignment and pes planus. Serious problems such as dissection of the aorta, rupture of major arteries and spontaneous perforation of the intestine occasionally occur. The phenotypic features of the various forms of the EDS are not clear cut and precise categorisation is not always possible.

Genetics

In a study of 100 patients in Southern England, a minimum prevalence of 1 in 150,000 was determined (Beighton *et al.*, 1969). The autosomal dominant types I, II and III are by far the most common of the various forms of the EDS. They occur in approximately equal frequency, and collectively they constitute over 90 per cent of the total reported cases (Beighton, 1970).

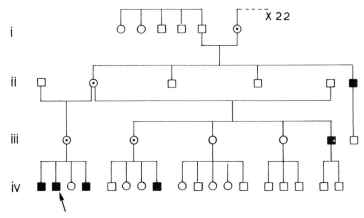

Fig. 17.13. EDS; X-linked inheritance of EDS type V.
Key to pedigree: □ normal male; ○ normal female; ■ affected male; ⊙ carrier female; / deceased.
(From Beighton, P. *British Medical Journal* (1968), **53**, 409.)

Fig. 17.14. (left) Familial joint laxity; a young woman with an autosomal dominant form of familial joint laxity. (From Beighton, P. & Horan, F. T. (1970) *Journal of Bone and Joint Surgery*, 52, 145.)

Fig. 17.15. (right) Familial joint laxity; the range of movements of all joints is excessive. (From Beighton, P. & Horan, F. T. (1970) *Journal of Bone and Joint Surgery*, 52, 145.)

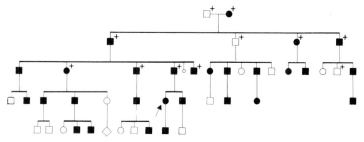

Fig. 17.16. Familial joint laxity; pedigree showing autosomal dominant inheritance.
Key to pedigree: □ normal male; ○ normal female; ■ affected male; ● affected female; / deceased.
(From Beighton, P. & Horan, F. T. (1970) *Journal of Bone and Joint Surgery*, 52, 145.)

The potentially lethal type IV is very rare (Barabas, 1967). Nevertheless, there have been reports which are consistent with both autosomal dominant and recessive inheritance. In view of the serious complications, genetic counselling is of particular importance in this form of the EDS. However, at the present time, these entities cannot be recognised on a clinical basis alone.

The X-linked type was initially delineated in two kindreds in England. No linkage with the genes for the Xg^a blood groups or colour blindness could be demonstrated in these families (Beighton, 1968).

The autosomal recessive background of the uncommon EDS type VI, in which ocular problems predominate, is recognisable in a few early case reports. The autosomal recessive type VII, in which joint laxity is a major feature, is probably underdiagnosed. However, is is noteworthy that there are dominant and recessive forms of familial joint laxity, which are distinct from the EDS (Beighton and Horan, 1973).

Lysyl oxidase is probably deficient in the X-linked type V EDS (diFerrante *et al.*, 1975) but it is not known if this enzyme deficiency is detectable in the female heterozygote. The basic abnormalities in EDS types IV, VI and VII have been elucidated and the biochemical background of the EDS has been reviewed by Martin and Smith (1975). There is little doubt that other forms of the EDS await recognition. With advances in the understanding of the biochemistry of collagen, it is to be anticipated that further delineation will be achieved.

3. FIBRODYSPLASIA OSSIFICANS PROGRESSIVA

Fibrodysplasia ossificans progressiva (FOP), also known as 'myositis ossificans progressiva,' has been the subject of semantic confusion with other conditions in which soft tissue calcification or ossification takes place. FOP has a long history and at the present time, more than 300 cases have been described.

Clinical features
The disorder has its onset in childhood. Episodes of spontaneous localised inflammation occur in the soft tissues, particularly in the fascia and tendons. The shoulder girdle, trunk and proximal regions of the limbs are the sites of predilection. These areas subsequently ossify, movements become progressively limited and by early adulthood there is great disability. Nevertheless, general health remains surprisingly

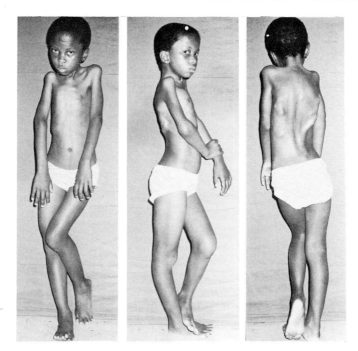

Fig. 17.17. (left) FOP; movements are restricted, due to extensive soft tissue ossification.

Fig. 17.18. (centre) FOP; a release operation in the region of the right shoulder precipitated further ossification, and no long-term benefit was obtained.

Fig. 17.19. (right) FOP; ossified swellings are very obvious over the posterior thorax.

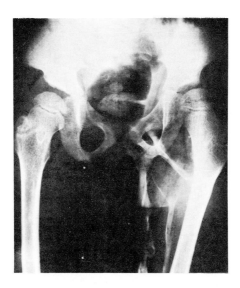

Fig. 17.20. FOP; radiograph showing extensive ossification in the soft tissues of the left thigh.

good and survival into middle-age is not uncommon. Some degree of microdactyly of the thumb and great toe, due to phalangeal hypoplasia and synostosis, may be present at birth.

Ruderman *et al.* (1974) produced evidence to support the concept that the basic defect might lie in a calcium adenosine-five-triphosphate mediated mechanism of calcification. The diphosphate 'EHDP' may have a therapeutic role in FOP (Smith, 1975). However, complications such as demineralisation and disorganised metaphyseal development have occurred following long-term treatment and the value of EHDP in this condition is doubtful.

Genetics
Fibrodysplasia ossificans progressiva seems to be inherited as an autosomal dominant. A few instances of generation to generation transmission can be recognised in the literature, and there have been two reports of FOP in identical twins (Vastine, Vastine and Oriel, 1948; Eaton, Conkling and Daeschner, 1957). Occasionally, phenotypic expression may be limited to microdactyly. However, severely affected individuals rarely reproduce and the majority of patients are probably the result of new dominant mutation. In keeping with this concept, the paternal age effect has been recognised in a study of 23 sporadic cases (Tünte, Becker and Knorr, 1967).

4. WEILL–MARCHESANI SYNDROME

Individuals with Weill–Marchesani or spherophakia–brachymorphia syndrome have short stature, rigid digits and ectopia lentis. The condition casues little disability and general health, intelligence and life span are normal.

Genetics
Affected sibs were mentioned in an early report (Marchesani, 1939). The autosomal recessive inheritance of the condition and the fact that the height of heterozygotes is reduced was recognised by Kloepfer and Rosenthal (1955). An anomalous kindred in which a father and two sons has the Weill–Marchesani syndrome was reported by Gorlin, L'Heureux and Shapiro (1974). As the mother was short, it is possible that she was heterozygous for the condition. If this were so, transmission in this family would be an excellent example of pseudodominant inheritance.

5. HOMOCYSTINURIA

Homocystinuria was separated from the Marfan syndrome following a survey of biochemical abnormalities in the urine of mentally retarded individuals in Northern Ireland (Carson and Neill, 1962). Once the phenotypic features of the condition had been defined, many other patients were recognised and more than 100 cases have now been reported.

Clinical features
The main characteristics of homocystinuria are a Marfanoid habitus and dislocation

of the lens of the eye. Mental deficiency is present in about 70 per cent of homocystinurics and vascular thrombosis is a frequent complication in adulthood. The skeleton is osteoporotic, the long bones are undermodelled, the epiphyses are widened and vertebral collapse may occur (Schedewie, 1973).

Genetics
Family studies have indicated that homocystinuria is inherited as an autosomal recessive (Schimke *et al.*, 1965; Carey *et al.*, 1968). Variations in phenotypic features, notably mental deficiency, and the inconsistent response to pyridoxine therapy, led Carson and Carré (1969) to postulate that homocystinuria was probably heterogeneous. McKusick (1975) discussed the possibility that the genes determining the pyridoxine responsive and unresponsive forms of homocystinuria could be allelic and suggested that cell hybridisation studies might provide the answer to this problem.

Activity of the enzyme cystathionine synthase is defective in cultured skin fibroblasts from homozygotes and asymptomatic heterozygotes (Uhlendorf and Mudd, 1968). As the enzymatic defect is expressed in cultured amniotic fluid cells, Fleisher *et al.* (1974) were able to monitor an 'at risk' pregnancy.

6. ALKAPTONURIA

Alkaptonuria is the prototype upon which Garrod based his concept of inborn errors of metabolism. The term 'ochronosis', which is descriptive of dark pigmentation of the connective tissue, is sometimes applied to the condition. However, this designation is not truly synonymous with alkaptonuria as ochronosis can have an acquired basis, such as poisoning with phenol.

Clinical features
Individuals with alkaptonuria pass urine which is black, or which turns black on standing. The urine can be shown to contain homogentisic acid, which gives a positive reaction to tests for reducing substances. Progressive deposition of black pigment takes place in the connective tissues. By adulthood, this pigmentation is often clinically evident in the pinnae of the ears and in the sclerae. Involvement of the articular cartilages leads to widespread degenerative osteoarthropathy. The spine is often severely affected and patients may be disabled by middle age. Arteriosclerosis, myocardial infarction and involvement of the cardiac valves are common complications.

Radiographically, degenerative changes are evident in the large joints and the spine. Intra-articular calcified loose bodies may be present and synchondrosis of the pubis is not uncommon.

Genetics
Alkaptonuria has a wide geographic distribution and it has been reported in many ethnic groups. In a review of the world literature, O'Brien, La Du and Bunim (1963) were able to identify 520 cases. There is a particularly high prevalence in Dominica (Harrold, 1956) and Czechoslovakia (Červeňanský, Sitaj and Urbánek, 1959).

The autosomal recessive inheritance of alkaptonuria was established by Hogben, Worrall and Zieve (1932) following a formal segregation analysis of pedigree data. A high frequency of parental consanguinity has been recorded (Molony and Kelly, 1970). Pedigrees have been published of large kindreds in which inheritance was apparently dominant (Khachadurian and Feisal, 1958). However, as with other autosomal recessive disorders which have a particularly high frequency in certain populations, this situation has resulted from marriage between affected homozygotes and clinically normal heterozygotes. It is evident that the alkaptonuria gene is not a recent mutation, as the condition has been recognised radiographically in Egyptian mummies (Simon and Zorab, 1961). The basic abnormality in alkaptonuria, defective activity of homogentisic acid oxidase, is not expressed in cultured amniotic fluid cells. Antenatal diagnosis by current techniques is therefore not possible.

7. TIGHT SKIN SYNDROMES

This group of very rare disorders share the common feature of skin thickening, to an extent that joint mobility is impaired. In some, the skeleton and other tissues are also involved.

(a) Rothmund syndrome

The Rothmund or Rothmund–Thomson syndrome is characterised by sclerodermatous skin lesions, cataracts, depression of the nasal bridge and alopecia. Bone reabsorption takes place and the peripheral joints become disorganised, presenting a Charcot-like radiographic appearance. Patients may be severely disabled. Recent

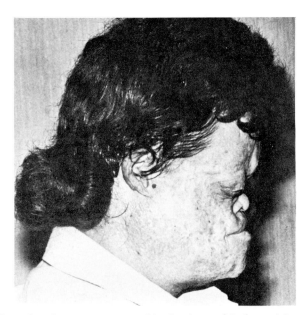

Fig. 17.21. Rothmund syndrome; a young man with scleroderma of the face and depression of the nasal bridge. Alopecia is hidden by a wig. (From Beighton, P. (1976) *Syndrome Identification*, 4, 8.)

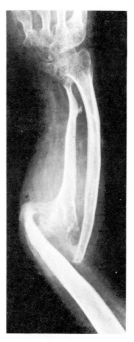

Fig. 17.22. Rothmund syndrome; radiographs of the elbow and wrist of the patient depicted in Fig.17.21. The disorganised joints have a Charcot-like appearance.

case reports have emanated from Kirkham and Werner (1975) and Sivayoham and Ratnaike (1975). Autosomal recessive inheritance is well established.

(b) Stiff skin syndrome

Thickening of the skin with restricted joint movements was recognised in a mother and her two offspring by Esterly and McKusick (1971) and termed the 'stiff skin' syndrome. A similar disorder had been previously encountered in a father and his son and daughter by Pichler (1968) and reported under the designation 'hereditary contractures with scleroderma-like changes'. Inheritance was apparently autosomal dominant in each kindred.

(c) Syndesmodysplastic dwarfism

Laplane *et al*. (1972) described two brothers with short stature and sclerodermatous skin. The authors considered that the condition was a new entity, which they named 'familial syndesmodysplastic dwarfism'. The genetic background is unknown.

(d) Parana hard skin syndrome

Cat *et al*. (1974) described eight individuals in the Parana region of Brazil, in whom widespread dermal thickening appeared in infancy. The condition progressed and the patients gradually become immobilised. Several of them eventually died from respiratory embarrassment. The pedigree data were consistent with autosomal recessive inheritance.

(e) Moore–Federman syndrome

Moore and Federman (1965) reported a kindred with 'familial dwarfism and stiff joints'. The affected members had short stature, ocular problems and tight skin over the forearms. The carpal tunnel syndrome and joint stiffness were troublesome features. Inheritance was apparently autosomal dominant.

8. LOOSE SKIN SYNDROMES

Cutis laxa is a rare disorder which has been the subject of semantic confusion with the Ehlers–Danlos syndrome. In cutis laxa the skin hangs in loose folds and is not fragile, and the joints are not hypermobile. Autosomal dominant and autosomal recessive types of cutis laxa have been recognised (Beighton, 1972). The skeleton is not involved in the classical forms of the disorder but severe bone changes are a feature of at least two cutis laxa variants.

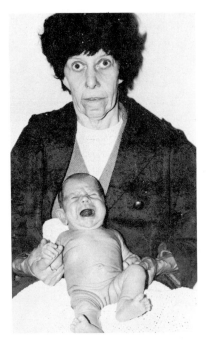

Fig. 17.23. Cutis laxa; a mother and daughter with the autosomal dominant type of cutis laxa. (From Beighton, P. (1972) *Journal of Medical Genetics*, **9**, 216.)

(a) Cutis laxa with bone dystrophy

Debré, Marie and Seringe (1937) reported a prematurely born child who had generalised dermal laxity, genu valgum, abnormality of cranial sutures and loose jointedness in the fingers. No family data was provided in this report. Subsequently, Fittke (1942) described a girl with similar features. Following an extensive study of the girl's kindred, Theopold and Wildhack (1951) published a pedigree showing that the parents were consanguineous and that two of her sisters and a female cousin were also affected. Skeletal changes were severe and included scoliosis, pectus carinatum, congenital dislocation of the hip, wide fontanelles and acrocephaly. Recently, Kaye and Fisher (1975) reported an infant with cutis laxa, skeletal abnormalities and genitourinary anomalies. These disorders might be the same entity. In any event, there is no doubt that they are quite distinct from classical cutis laxa.

(b) Leprechaunoid syndrome — Patterson

Donahue and Uchida (1954) were impressed by the pixie-like facial features of two young sisters, who also had marked dermal laxity. These authors quaintly alluded to the condition as 'leprechaunism'. These girls, who had no bone changes, died in infancy. Subsequently, Patterson and Watkins (1962) reported a male infant with gross laxity of the skin of the extremities, in association with radiographic evidence of defective ossification, marked irregularity of the ends of the tubular bones and a low serum alkaline phosphatase. Patterson (1969) described the development of this child, who became cushingoid and died in his seventh year. Dallaire *et al.* (1976) reported three male infants with the same condition from two related consanguineous Italian kindreds. On this evidence, it is likely that inheritance is autosomal recessive. The 'leprechaunoid syndrome of Patterson' is almost certainly distinct from the 'leprechaunism syndrome' of Donohue.

(c) Wrinkly skin syndrome

Gazit *et al.* (1973) reported two sisters of Iraqi–Jewish stock, who had loose wrinkled skin on the extremities and ventral surfaces, together with kyphosis and winging of the scapula. The parents were consanguineous and it is likely that inheritance was autosomal recessive. This disorder differs from the better known 'Prune belly' syndrome, in which absence of the abdominal muscles is associated with abnormalities of the urinary tract.

REFERENCES

Connective tissue disorders
Beighton, P. & Horan, F. T. (1970) Dominant inheritance of generalised articular hypermobility. *Journal of Bone and Joint Surgery*, **52B/1**, 145.
Frank, Y., Ziprkowski, A., Romano, A., Stein, R., Katznelson, M. B. M., Cohen, B. & Goodman, R. M. (1973) Megalocornea associated with multiple skeletal anomalies: a new genetic syndrome? *Journal de Génétique Humaine*, **21/2**, 67.
Greenfield, G., Romano, A., Stein, R. & Goodman, R. M. (1973) Blue sclerae and keratoconus: key features of a distinct heritable disorder of connective tissue. *Clinical Genetics*, **4**, 8.

Marfan syndrome
Achard, C. (1902) Arachnodactylie. *Bulletins et Mémoires de la Société Médicale des Hôpitaux de Paris*, **19**, 834.
Ambani, L. M., Gelehrter, T. D. & Sheahan, D. G. (1975) Variable expression of Marfan syndrome in monozygotic twins. *Clinical Genetics*, **8/5**, 358.
Beals, R. K. & Hecht, F. (1971) Contractural arachnodactyly, a heritable disorder of connective tissue. *Journal of Bone and Joint Surgery*, **53A**, 987.
Bjerkeim, I., Skogland, L. B. & Trygstad, O. (1976) Congenital contractural arachnodactyly. *Acta orthopaedica Scandinavica*, **47**, 250.
Duncan, P. A. (1975) The Achard syndrome. *Birth Defects: Original Article Series*, **11/6**, 69.
Feingold, M. & Bossert, W. H. (1974) Normal values for selected physical parameters: an aid to syndrome delineation. *Birth Defects: Original Article Series*, **10/13**, 4.

Goodman, R. M., Wooley, C. F., Frazier, R. L. & Covault, I. (1965) Ehlers–Danlos syndrome occurring together with the Marfan syndrome. Report of a case with other family members affected. *New England Journal of Medicine*, **273**, 514.

Kontras, S. B. (1975) Congenital contractural arachnodactyly. *Birth Defects: Original Article Series*, **11/6**, 63.

Lowry, R. B. & Guichon, V. C. (1972) Congenital contractural arachnodactyly: a syndrome simulating Marfan's syndrome. *Canadian Medical Association Journal*, **107**, 531.

Lynas, M. A. (1958) Marfan's syndrome in Northern Ireland; an account of 13 families. *Annals of Human Genetics*, **22**, 289.

Lynas, M. A. & Merrett, J. D. (1958) Data on linkage in man; Marfan's syndrome in Northern Ireland. *Annals of Human Genetics*, **22**, 310.

MacLeod, P. M. & Fraser, F. C. (1973) Congenital contractural arachnodactyly. (A heritable disorder of connective tissue distinct from Marfan syndrome). *American Journal of Diseases of Children*, **126**, 810.

McKusick, V. A. (1975) The classification of heritable disorders of connective tissue. *Birth Defects: Original Article Series*, **11/6**, 7.

Murdoch, J. L., Walker, B. A. & McKusick, V. A. (1972) Parental age effects on the occurrence of new mutations for the Marfan syndrome. *Annals of Human Genetics*, **35**, 331.

Passarge, E. (1975) A syndrome resembling congenital contractural arachnodactyly. *Birth Defects: Original Article Series*, **11/6**, 53.

Robins, P. R., Moe, J. H. & Winter, R. B. (1975) Scoliosis in Marfan's syndrome. Its characteristics and results of treatment in 35 patients. *Journal of Bone and Joint Surgery*, **57/3**, 358.

Schleutermann, A., Murdoch, J. L., Walker, B. A., Bias, W. B., Chase, G. A., Freidhoff, L. B. & McKusick, V. A. (1976) A linkage study of the Marfan syndrome. *Clinical Genetics*, **10**, 51.

Steg, N. L. (1975) Congenital contractural arachnodactyly in a black family. *Birth Defects: Original Article Series*, **11/6**, 57.

Walker, B. A., Beighton, P. H. & Murdoch, J. L. (1969) The marfanoid hypermobility syndrome. *Annals of Internal Medicine*, **71**, 349.

Ehlers–Danlos syndrome

Barabas, A. P. (1967) Heterogeneity of the Ehlers–Danlos syndrome. *British Medical Journal*, **2**, 612.

Beighton, P. (1968) X-linked recessive inheritance in the Ehlers–Danlos syndrome. *British Medical Journal*, **3**, 409.

Beighton, P., Price, A., Lord, J. & Dickson, E. (1969) Variants of the Ehlers–Danlos syndrome. *Annals of Rheumatic Disease*, **28**, 228.

Beighton, P. (1970) *The Ehlers–Danlos Syndrome*, p.115. London: William Heinemann.

Beighton, P. & Horan, F. (1973) Recessive inheritance of generalised articular hypermobility. *Rheumatology and Rehabilitation*, **12**, 47.

Di Ferrante, N., Leachman, R. D., Angelini, P., Donnelly, P. V., Francis, G., Almazan, A., Segni, G., Franzblau, C. & Jordan, R. E. (1975) Ehlers–Danlos type V (X-linked form): A lysyl oxidase deficiency. *Birth Defects: Original Article Series*, **11/6**, 31.

Martin, J. O. F. & Smith, R. (1975) Polymeric collagen of skin in osteogenesis imperfecta, homocystinuria and Ehlers–Danlos and Marfan syndromes. *Birth Defects: Original Article Series*, **11/6**, 15.

Fibrodysplasia ossificans progressiva

Eaton, W. L., Conkling, W. S. & Daeschner, C. W. (1957) Early myositis ossificans progressiva occurring in homozygotic twins; a clinical and pathologic study. *Journal of Pediatrics*, **50**, 591.

Ruderman, R. J., Leonard, F. & Elliot, D. E. (1974) A possible aetiological mechanism for fibrodysplasia ossificans progressiva. *Birth Defects: Original Article Series*, **10/12**, 299.

Smith, R. (1975) Myositis ossificans progressiva: a review of current problems. *Seminars of Arthritis and Rheumatism*, **4/4**, 369.

Tünte, W., Becker, P. E. & Knorr, G. V. (1967) Zur Genetik der Myositis ossificans progressiva. *Humangenetik*, **4**, 320.

Vastine, J. H., Vastine, M. E. & Oriel, A. (1948) Myositis ossificans progressiva in homozygotic twins. *American Journal of Roentgenology*, **59**, 204.

Weill–Marchesani syndrome

Gorlin, R. J., L'Heureux, R. R. & Shapiro, I. (1974) Weill–Marchesani syndrome in two generations: genetic heterogeneity or pseudodominance? *Journal of Pediatric Ophthalmology*, **11**, 139.

Kloepfer, H. W. & Rosenthal, J. W. (1955) Possible genetic carriers in the spherophakia–brachymorphia syndrome. *American Journal of Human Genetics*, **7**, 398.

Marchesani, O. (1939) Brachydaktylie und angeborene Kugellinse als Systemerkrankung. *Klinische Monatsblätter für Augenheilkunde*, **103**, 392.

Homocystinuria
Carey, M. C., Donovan, D. E., Fitzgerald, O. & McAuley, F. D. (1968) Homocystinuria. A clinical and pathological study of nine subjects in six families. *American Journal of Medicine*, **45**, 7.
Carson, N. A. J. & Carré, I. J. (1969) Treatment of homocystinuria with pyridoxine. *Archives of Diseases in Childhood*, **44**, 387.
Carson, N. A. J. & Neill, D. W. (1962) Metabolic abnormalities detected in a survey of mentally backward individuals in Northern Ireland. *Archives of Disease in Childhood*, **37**, 505.
Fleisher, L. D., Longhi, R. C., Tallan, H. H. *et al*. (1974) Homocystinuria: investigators of cystathionine synthase in cultured fetal cells and the prenatal determination of genetic status. *Journal of Pediatrics*, **85/5**, 677.
McKusick, V. A. (1975) Homocystinuria. In *Mendelian Inheritance in Man*, 4th Edition, p.449. Baltimore, London: Johns Hopkins University Press.
Schedewie, H. (1973) Skeletal findings in homocystinuria: a collaborative study. *Pediatric Radiology*, **1**, 12.
Schimke, R. N., McKusick, V.A. Huang, T. & Pollack, A. D. (1965) Homocystinuria: studies of 20 families with 38 affected members. *Journal of the American Medical Association*, **193**, 711.
Uhlendorf, B. W. & Mudd, S. H. (1968) Cystathionine synthase in tissue culture derived from human skin: enzyme defect in homocytinuria. *Science*, **160**, 1007.

Alkaptonuria
Cervenanský, J., Sitaj, S. & Urbánek, T. (1959) Alkaptonuria and ochronosis. *Journal of Bone and Joint Surgery*, **41**, 1169.
Harrold, A. J. (1956) Alkaptonuric arthritis. *Journal of Bone and Joint Surgery*, **38**, 532.
Hogben, L., Worrall, R. L. & Zieve, I. (1932) The genetic basis of alkaptonuria. *Proceedings of the Royal Society of Edinburgh*, **52**, 264.
Khachadurian, A. & Feisal, K. (1958) Alkaptonuria; a report of a family with seven cases appearing in four successive generations with metabolic studies in one patient. *Journal of Chronic Disease*, **7**, 455.
Molony, J. & Kelly, D. J. (1970) Alkaptonuria, ochronosis, and achronotic arthritis. *Journal of the Irish Medical Association*, **63**, 22.
O'Brien, W. M., La Du, B. N. & Bunim, J. J. (1963) Biochemical, pathologic and clinical aspects of alcaptonuria, ochronosis and ochronotic arthropathy; review of world literature. *American Journal of Medicine*, **34**, 813.
Simon, G. & Zorab, P. A. (1961) The radiographic changes in alkaptonuric arthritis. *British Journal of Radiology*, **34**, 384.

Tight skin syndromes

(a) Rothmund syndrome
Kirkham, T. H. & Werner, E. B. (1975) The ophthalmic manifestations of Rothmund's syndrome. *Canadian Journal of Ophthalmology*, **10/1**, 1.
Sivayoham, I. S. S. R. & Ratnaike, V. T. (1975) Rothmund–Thomson syndrome in an oriental patient. *Annals of Ophthalmology*, **7/3**, 417.

(b) Stiff skin syndrome
Esterley, N. B. & McKusick, V. A. (1971) Stiff skin syndrome. *Pediatrics*, **47**, 360.
Pichler, E. (1968) Hereditaere Kontrakturen mit sklerodermieartigen Hautveraenderungen. *Zeitschrift für Kinderheilkunde*, **104**, 349.

(c) Syndemodysplastic dwarfism
Laplane, R., Fontaine, J.-L., Lagardere, B. & Sambucy, F. (1972) Nanisme syndemodysplasique familal. Une entite morbide nouvelle. *Archives françaises de Pédiatrie* (Paris), **29**, 831.

(d) Parana hard skin syndrome
Cat, I., Rodrigues-Magdalena, N. I., Parolin-Marinoni, L., Wong, M. P., Freitas, O. T., Malfi, A., Costa, O., Estieves, L. & Giraldi, D. J. (1974) Parana hard skin syndrome: study of seven families. *Lancet*, **i**, 215.

(e) Moore–Federman syndrome
Moore, W. T. & Federman, D. D. (1965) Familial dwarfism and 'stiff joints'. *Archives of Internal Medicine*, **115**, 398.

Loose skin syndromes
Beighton, P. (1972) The dominant and recessive forms of cutix laxa. *Journal of Medical Genetics*, **9**, 216.

(a) Cutis laxa with bone dystrophy
Debré, R., Marie, J. & Seringe, P. (1937) 'Cutis laxa' avec dystrophies osseuses. *Bulletins et Mémoires de la Société Médicale des Hôpitaux de Paris*, **61**, 1038.
Fittke, H. (1942) Über eine ungewöhnliche Form 'multipler Erbabartung' (Chalodermie und dysostose). *Zeitschrift für Kinderheilkunde*, **63**, 510.
Kaye, C. I. & Fisher, D. E. (1975) Cutis laxa and associated anomalies. *Birth Defects: Original Article Series*, **11/2**, 130.
Theopold, W. & Wildhack, R. (1951) Dermatochalasis im Rahmne multipler Abortungen. *Monatsschrift für Kinderheilkunde*, **99**, 213.

(b) Leprechanoid syndrome
Dallaire, L., Cantin, M., Melancon, S. B., Perreault, G. & Potier, M. (1976) A syndrome of generalised elastic fiber deficiency with leprechaunoid features: a distinct genetic disease with an autosomal recessive mode of inheritance. *Clinical Genetics*, **10**, 1.
Donohue, W. L. & Uchida, I. (1954) Leprechaunism: a euphism for a rare familial disorder. *Journal of Pediatrics*, **45**, 505.
Patterson, J. H. (1969) Presentation of a patient with leprechaunism. In *Clinical Delineation of Birth Defects*, IV, p.117. Ed. Bergsma, D. New York: National Foundation.
Patterson, J. H. & Watkins, W. L. (1962) Leprechaunism in a male infant. *Journal of Pediatrics*, **60**, 730.

(c) Wrinkly skin syndrome
Gazit, E., Goodman, R. M., Bat-Miriam Katznelson, M. & Rotem, Y. (1973) The wrinkly skin syndrome: a new heritable disorder of connective tissue. *Clinical Genetics*, **4**, 186.

18. Primary disturbances of growth

Proportionate or relatively proportionate short stature with low birth weight is a major feature of a number of disorders. Apart from the failure of growth, the skeleton is not usually dysplastic in these conditions. However, as they are frequently confused with other true skeletal dysplasias, they merit consideration in this section.

1. Pituitary dwarfism
   (a) Type I. Isolated growth hormone deficiency
   (b) Type II. Laron type
   (c) Type III. Panhypopituitarism
   (d) Other forms of pituitary dwarfism
2. Cornelia de Lange syndrome
3. Seckel syndrome
4. Russell–Silver syndrome
5. Senility syndromes
   (a) Progeria
   (b) Atypical progeria
   (c) Cockayne syndrome
   (d) Werner syndrome
6. Bloom syndrome
7. Other disorders of growth
   (a) Dubowitz syndrome
   (b) 3M syndrome
   (c) Dwarfism–ichthyosis syndrome
   (d) Nathalie syndrome
   (e) KBG syndrome
   (f) Levi snub-nosed dwarfism
   (g) Fetal alcohol syndrome.

1. PITUITARY DWARFISM

In the early literature, individuals of proportionate short stature were loosely grouped together under the designation 'midget' or as primordial or ateliotic dwarfs. The ateliotic group were conventionally subdivided into 'sexual' and 'asexual' forms. It was subsequently recognised that dysfunction of the pituitary gland was the usual basis of uncomplicated proportionate short stature, and the term 'pituitary dwarfism' gained favour. With the accumulation of genetic and endocrine data, it has become evident that there are several distinct forms of pituitary dwarfism.

Fig. 18.1. Pituitary dwarfism; Bushmen of the Kalahari desert. The average height of adults is under 150 cm. (Courtesy of Dr M. C. Botha, Cape Town.)

Tom Thumb, who was a famous exhibitionist in the circus world, provides an excellent example of pituitary dwarfism. His wife Lavinia Warren was a midget. Her parents and siblings were normal, but as her sister Minnie was also small, it is likely that she had an autosomal recessive form of the disorder.

Current concepts concerning pituitary dwarfism have been reviewed by Rimoin (1976). Delineation is still incomplete, but the following types are recognised:

(a) Pituitary dwarfism type I (Isolated growth hormone deficiency)
Autosomal recessive inheritance of isolated growth hormone deficiency was established by Rimoin, Merimee and McKusick (1966). There is also an autosomal dominant form of the disorder, as a kindred with cases in four generations has been reported (Merimee *et al.*, 1969). Other kindreds with affected members in successive generations have been described by Sheikholislam and Stempfel (1972) and Poskitt and Rayner (1974).

(b) Pituitary dwarfism type II (Laron type)
Laron, Pertzelan and Mannheimer (1966) studied 20 individuals in Israel who had the clinical features of isolated growth hormone deficiency, in the presence of elevated serum growth hormone levels. These patients were all of Sephardic-Jewish stock and the pedigrees were consistent with autosomal recessive inheritance. Pertzelan, Adam and Laron (1968) suggested that the growth hormone in this condition might be functionally incompetent although immunoreactive. Alternative explanations would be end-organ unresponsiveness to growth hormone or defective hepatic generation of somatomedin, the intermediate substance (Laron, 1974). A

Fig. 18.2. Pituitary dwarfism; a Kalahari Bushman with pituitary dwarfism. (From Beighton, P. (1972) *South African Medical Journal*, **46**, 881.)

positive response to treatment with growth hormone has been reported by Clemons, Costin and Kogut (1976).

The Pygmies of the Central African rain forests have been shown to have peripheral unresponsiveness to growth hormone (Rimoin *et al.*, 1969). It is likely that this unique ethnic characteristic has a genetic basis, but the mode of inheritance has not been determined. The Bushmen of the Kalahari desert of Southern Africa are also of slight stature. However, their gracile habitus differs from the stocky build of the Pygmies and ethnically these two groups are not closely related. Joffe *et al.* (1971) demonstrated that the Bushmen have normal growth hormone responses to oral glucose, and speculated that, as with the Pygmies, their small stature is the consequence of end-organ unresponsiveness. This issue is complicated by the fact that a Bushman with the classical clinical features of pituitary dwarfism has been encountered (Beighton, 1972).

(c) Pituitary dwarfism type III (Panhypopituitarism)
Individuals with panhypopituitarism have the stigmata of thyroid, adrenal and gonadal dysfunction in addition to proportionate short stature. This condition is far more common than all other forms of pituitary dwarfism. The vast majority of patients are sporadic, with no genetic basis to their disorder. However, in a minority, inheritance is autosomal recessive or X-linked recessive.

(d) Other forms of pituitary dwarfism
Two sibs with aplasia of the pituitary gland were described by Steiner and Boggs

(1965) and Sadeghi-Hejad and Senior (1974). Two sisters with panhypopituitarism and a minute sella turcica were reported by Ferrier and Stone (1969) and a similar case, with the additional feature of retinitis pigmentosa, was described by Ozer (1974). Although unproven, it is likely that inheritance is autosomal recessive in these conditions.

Rappaport, Ulstrom and Gorlin (1976) encountered three children with proportionate short stature and a single central incisor. The authors obtained information concerning three similar children, demonstrated that growth hormone was deficient in two, and termed the condition 'monosuperocentroincisivodontic dwarfism'. No genetic background was apparent in this report.

Genetic considerations

Genetic counselling is dependant upon accurate recognition of the dominant, recessive and sporadic forms of the disorder. Apart from shortness of stature, individuals with pituitary dwarfism enjoy good health, and it could be argued that genetic distinction is of little practical importance. However, the autobiographical comments of Tom Thumb's midget wife Lavinia Warren are of significance in this context. 'If nature endowed me with any superior personal attraction, it was comparatively small compensation for the inconvenience, trouble and annoyance imposed upon me by my diminutive stature'.

2. CORNELIA DE LANGE SYNDROME

Individuals with the Cornelia de Lange syndrome bear a striking resemblance to one another. The eyebrows are confluent, the anterior and posterior hairlines are low, the nostrils are anteverted and the lips are thin, with down-turned angles. Mental and physical development is retarded. Skeletal abnormalities range from absence of the extremities to minor anomalies such as proximally placed thumbs, fifth finger clinodactyly and limitation of elbow movements.

Genetics

Since Cornelia de Lange's initial description of two affected children in 1933, there have been more than 250 case reports. In their monograph on the syndrome, Berg et al. (1970) quoted the incidence in live births at between 1 in 30,000 and 1 in 50,000. The majority of cases have been sporadic, although sibs with the condition have been reported by Falek, Schmidt and Jervis (1966) and Beratis, Hsu and Hirschhorn (1971). Nevertheless, the accumulated pedigree data does not indicate autosomal recessive inheritance and it is possible that the syndrome is polygenically determined. In this context, in a report of 11 children, Begeman and Duggan (1976) mentioned that minor stigmata were present in the mother of an affected boy. Pashayan et al. (1969) analysed genetic data from 54 families and concluded that the recurrence risk to sibs of patients was between 1.5 per cent and 2.2 per cent. Stevenson and Scott (1976) described male twins who were discordant for the syndrome. Garakushansky and Berthier (1976) reported discordant monozygotic female twins and commented that the situation could be explained by postzygotic mutation of a gene of large effect.

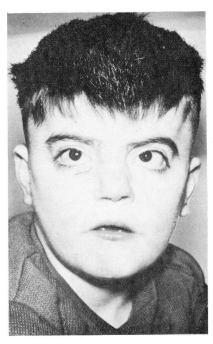
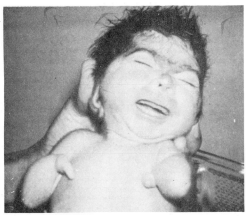

Fig. 18.3. (left) Cornelia de Lange syndrome; confluent eyebrows, a low anterior hair line and thin lips.

Fig. 18.4. (right) Cornelia de Lange syndrome; gross malformations of the extremities. (From Begeman, G. & Duggan, R. (1976) *South African Medical Journal*, 50, 1475.)

There have been many reports of cytogenetic abnormalities in the Cornelia de Lange syndrome. For instance, Craig and Luzzato (1965) found chromosomal anomalies in 11 out of 38 patients. Conversely, the chromosomes were normal in 20 cases investigated by McArthur and Edwards (1967). Although they are inconsistent, the reports of cytogenetic abnormalities are too numerous to be ignored. It is possible that there is an inherant predisposition to the development of non-specific chromosomal changes. Equally, cytogenetic anomalies may be a fundamental facet of the disorder.

For all the abovementioned reasons, the genetic basis of the syndrome remains very uncertain. A statistical analysis of the clinical manifestations of a group of patients was indicative of heterogeneity (Preus and Fraser, 1976). This technique, in which the features of a particular disorder are weighted according to the ratio of their frequency as compared to a control sample of normal individuals, may well prove to be of value in the diagnosis and delineation of other rare genetic conditions.

3. SECKEL SYNDROME

Following a report of a 'microcephalic midget' (Mann and Russell, 1959), Seckel (1960) published a review entitled 'Bird-headed dwarfs'. Subsequently, the eponym 'Seckel' was employed as an alternative to 'bird-headed dwarfism' by Harper, Orti and Baker (1967). The main features of the syndrome are low birth weight, short stature, microcephaly and mental deficiency. The thin pointed face and prominent nose produce an avian appearance. Skeletal abnormalities include scoliosis, absence of ribs, maldevelopment of the radius and fibula, dislocation of the hips, pes planus and talipes equinovarus.

The features of the Seckel syndrome are clearly recognisable in Alfred Chalon's portrait of Carolina Crachani, the Sicilian dwarf, which hangs in the Royal College of Surgeons, London. Carolina died of pulmonary tuberculosis at the age of eight, and her skeleton, which measures less than 55 cm in height, now reposes in the Hunterian Museum of the College. Her condition was attributed to 'maternal impression', as her mother had been bitten on the hand by a monkey during early pregnancy. However, there have been no other similar reports, and it can be confidently assumed that monkey bites are not an important aetiological factor in the Seckel syndrome!

Genetics

About 30 cases have now been described. Multiple affected sibs with normal parents have been reported by Black (1961), Harper *et al*. (1967) and McKusick *et al*. (1967). Parental consanguinity has been noted and there is little doubt that the Seckel syndrome is inherited as an autosomal recessive.

Fitch, Pinsky and Lachance (1970) described a patient with features of the Seckel syndrome, together with premature senility. McKusick (1975) commented that he knew of a similar brother and sister, suggested that inheritance was autosomal recessive and proposed that, in deference to the origins of the first report, this condition should be termed 'Bird-headed dwarfism, Montreal type.'

4. RUSSELL–SILVER SYNDROME

There has been controversy as to whether the initial case reports of Silver (1953) and Russell (1954) pertained to the same disorder or to separate entities. However, the designation 'Russell–Silver sundrome' is now applied to a dwarfing condition in which lateral asymmetry and low birth weight are associated with a triangular face, clinodactyly of the fifth finger and café-au-lait macules on the trunk. Abnormalities of urinary gonadotrophin excretion and sexual development, which were emphasised by Silver (1964), are inconsistent features. Hypoglycaemia may be a problem during infancy, but general health is otherwise good. Treatment with growth hormone has not been successful (Tanner and Ham, 1969).

Genetics

The Russell–Silver syndrome is probably relatively common, although mild cases may remain undiagnosed. Reports of about 100 individuals with the syndrome have now appeared in the literature. In a review of the pattern of growth of 39 affected children, Tanner, Lejarraga and Cameron (1975) commented that contrary to earlier reports, no abnormalities of sexual development had occurred. These authors mentioned that none of the children in this series had affected sibs.

Apart from three sibs in a consanguineous kindred described by Fuleihan *et al*. (1971), and monozygous twins reported by Rimoin (1969), the majority of cases have

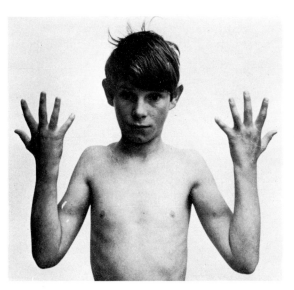 

Fig. 18.5. (left) Russell–Silver syndrome; the face is triangular and the chin is pointed. Clinodactyly is present in the fifth fingers.

Fig. 18.6. (right) Russell–Silver syndrome; in this young woman, short stature and disproportion are unusually severe.

been sporadic. There is no increased frequency of parental consanguinity, which might indicate autosomal recessive inheritance, or of paternal age or birth rank which could point to new mutation of a dominant gene. The sexes are equally affected and chromosomes are normal. On the available evidence, it seems likely that the Russell–Silver syndrome is non-genic. For the purposes of genetic counselling, it would be reasonable to assume that the risks of recurrence are very low.

5. SENILITY SYNDROMES

(a) Progeria

Progeria was delineated at the turn of the century by Gilford (1904). However, the condition had been clearly described more than 150 years previously in the obituary of Hopkin Hopkins, a Welsh fairground exhibitionist. Premature aging is the most notable feature. Patients have a senile appearance in childhood and usually die of arteriosclerotic complications, such as myocardial infarction or cerebrovascular accidents, by the end of the second decade. Skeletal changes include osteoporosis, coxa valga and ovoid vertebral bodies. The manifestations of progeria have been extensively reviewed by De Busk (1972).

About 60 patients have now been described. The majority have been sporadic, and the mode of inheritance is uncertain. Progeric Egyptian sisters with consanguineous parents were reported by Gabr *et al.* (1960) and sets of sibs have been described by Rava (1967) and Franklyn (1976). A pair of male monozygotic twins with the condition have been studied by Viegas, Souza and Salzano (1974). De Busk (1972) found evidence of parental consanguinity in 19 previously reported cases and an increased mean paternal age in 20. These findings are consistent with autosomal recessive inheritance on one hand, and new mutation for a dominant gene on the other. As affected individuals do not reproduce, evidence to support the latter concept will not be forthcoming. It is possible that progeria is heterogeneous.

(b) Atypical progeria

Atypical forms of progeria have been encountered. For instance, Welsh (1975) described a kindred in which two males and two females had an unusual type of premature senility and speculated that this condition might be a new syndrome. Mulvihill and Smith (1975) made a similar suggestion when they reported a patient with progeria, mental deficiency and multiple pigmented naevi.

(c) Cockayne syndrome

The Cockayne syndrome or 'cachectic dwarfism' resembles progeria in that defective growth and progressive loss of adipose tissue produce a senile appearance in late childhood. Mental deficiency, photosensitivity, retinal degeneration and optic atrophy are associated features. About 30 cases have been described, including sibs reported by Cockayne (1946) and Paddison et al. (1963) and a severely affected boy born to consanguineous parents (Pfeiffer and Bachmann, 1973). Inheritance is autosomal recessive.

(d) Werner syndrome

The stigmata of the Werner syndrome include premature senility, shortened stature, cataracts and diabetes mellitus. More than 120 cases have been reported, and the accumulated pedigree data are indicative of autosomal recessive inheritance (Epstein et al., 1966).

6. BLOOM SYNDROME

Bloom (1966) described the association of low birth weight, dwarfism, malar hypoplasia and photosensitive telangectasia of the face. Other abnormalities in the Bloom syndrome include dental defects, digital deformities and talipes equinovarus. The immunoglobulins are deficient and leukaemia and other neoplasms are a common complication.

Genetics

About 50 case reports have now been published and autosomal recessive inheritance is well established. The majority of patients have been of Ashkenazi Jewish stock. As with many genetic disorders which have a high prevalence in the Jewish people, there is a distinctive geographical distribution of the abnormal gene. German (1969) has shown that the majority of kindreds with the Bloom syndrome had their origins in the Ukraine region of North-Western Europe.

Excessive chromosome breakage and rearrangement is a consistent feature (Sawitsky, Bloom and German, 1966). Borgeois *et al*. (1975) undertook trypsin banding studies on chromosomes from an affected Jordanian girl and demonstrated that the cytogenic abnormalities were non-random. Obligatory heterozygotes show similar changes, in lesser degree, in cultured fibroblasts.

Goodman (1975) pointed out that consanguinity was not increased amongst Jewish families with the Bloom syndrome, although several consanguineous non-Jewish kindreds had been reported. The sex ratio is equal in the Jewish cases, but there is a male preponderance in non-Jews. These observations led Goodman (1975) to speculate that the condition might be heterogeneous.

7. OTHER DISORDERS OF GROWTH

There have been a number of reports of individuals with proportionate short stature and a variety of associated abnormalities. As these conditions are probably distinct syndromes, their features are summarised below:

(a) Dubowitz syndrome

Dubowitz (1965) reported two sibs and two unrelated patients with low birth weight and a distinctive facies. Grosse, Gorlin and Opitz (1971) described two more cases and termed the condition 'Dubowitz syndrome'. The stigmata bear some resemblance to the Russell–Silver syndrome but microcephaly, mild mental retardation, hypertelorism and ptosis are additional features. In a review of the syndrome Opitz *et al*. (1973) recognised kindreds with parental consanguinity and multiple affected sibs and concluded that inheritance was autosomal recessive. In a further review Majewski *et al*. (1975) commented that a total of 11 cases had now been described, and drew attention to the similarity of their stigmata to those seen in the fetal alcohol syndrome.

(b) 3M syndrome

Miller, McKusick and Malvaux (1975) reported two sets of sibs with low birth weight dwarfism, a 'hatchet-shaped' cranial configuration, deformity of the thorax and anomalies of the mouth and teeth. The authors termed the condition the '3M syndrome' and suggested that, as the unaffected parents in one of the kindreds were consanguineous, the disorder was probably inherited as an autosomal recessive.

(c) Dwarfism–ichthyosis syndrome

Passwell *et al*. (1973) described a new autosomal recessive syndrome, which was characterised by proportionate dwarfism, ichthyosis, mental retardation and generalised aminoaciduria. Subsequently, Passwell *et al*. (1975) reported a similar disorder in three sibs in a consanguineous Iranian kindred.

(d) Nathalie syndrome

Using the Christian name of the proband in the title of their paper, Cremers, ter Haar and van Rens (1975) described three sisters and a brother with retarded growth, deafness, cataracts, sexual infantilism and electrocardiographic abnormalities. Mild but widespread skeletal dysplasia was present. Two of the girls developed hip joint problems which resembled Perthe's disease, while Scheuermann's disease was diagnosed in the boy. Although the family study did not provide conclusive evidence, it is likely that the Nathalie syndrome was inherited as an autosomal recessive.

(e) KBG syndrome
The KBG syndrome, named from the initials of the probands, consists of short stature, macrodontia and mental retardation. Herrmann *et al*. (1975) described the features of seven patients from three kindreds, and concluded that inheritance was autosomal dominant.

(f) Levi snub-nosed dwarfism
The Levi or 'snub-nosed' type of dwarfism is characterised by proportionate short stature and an unusual facies (Black, 1961). The condition is not well delineated, but it is possible that there are phenotypically similar autosomal dominant and recessive types.

(g) Fetal alcohol syndrome
It is becoming increasingly evident that maternal alcoholism poses a definite threat to the fetus. The main features of the fetal alcohol syndrome are low birth weight, mental retardation, microcephaly, short palpebral fissures and maxillary hypoplasia (Jones and Smith, 1973). As the manifestations are not clear-cut, firm diagnosis is not always easy. Nevertheless, it is likely that the condition is fairly common (Jones and Smith, 1975). The majority of mothers of children with the syndrome have sustained a high alcohol intake throughout pregnancy. However, it is a sobering thought that a single bibulous episode at a crucial stage of fetal development might also be embryopathic!

REFERENCES

Pituitary dwarfism

(a) Pituitary dwarfism type I
Merimee, T. J., Hall, J. G., Rimoin, D. L. & McKusick, V. A. (1969) A metabolic and hormonal basis for classifying ateliotic dwarfs. *Lancet*, **i**, 963.
Poskitt, E. M. & Rayner, P. H. W. (1974) Isolated growth hormone deficiency: two families with autosomal dominant inheritance. *Archives of Diseases in Childhood*, **49**, 55.
Rimoin, S. L., Merimee, T. J. & McKusick, V. A. (1966) Growth hormone deficiency in man: an isolated, recessively inherited defect. *Science*, **152**, 1635.
Sheikholislam, B. M. & Stempfel, R. S., Jr. (1972) Hereditary isolated somatotropin deficiency: effects of human growth hormone administration. *Pediatrics*, **49**, 362.

(b) Pituitary dwarfism type II
Beighton, P. (1972) Pituitary dwarfism in a Kalahari Bushman. *South African Medical Journal*, **46**, 881.
Clemons, R. D., Costin, G. & Kogut, M. D. (1976) Laron dwarfism: growth and immunoreactive insulin following treatment with human growth hormone. *Journal of Pediatrics*, **88/3**, 427.
Joffe, B. I., Jackson, W. P. U., Thomas, M. E., Toyer, M. G., Keller, P., Pimstone, B. L. & Zamit, R. (1971) Metabolic responses to oral glucose in the Kalahari Bushmen. *British Medical Journal*, **4**, 206.
Laron, Z., Pertzelan, A. & Mannheimer, S. (1966) Genetic pituitary dwarfism with high serum concentration of growth hormone. A new inborn error of metabolism? *Israel Journal of Medical Science*, **2**, 152.

Laron, Z. (1974) The syndrome of familial dwarfism and high plasma immunoreactive human growth hormone. In *Clinical Delineation of Birth Defects*, Volume XVI, p.231. Ed. Bergsma, D. Baltimore: Williams and Wilkins.

Pertzelan, A., Adam, A. & Laron, Z. (1968) Genetic aspects of pituitary dwarfism due to absence or biological inactivity of growth hormone. *Israel Journal of Medical Science*, **4**, 895.

Rimoin, D. L., Merimee, T. J., Rabinowitz, D., Cavalli Sforza, L. L. & McKusick, V. A. (1969) Peripheral subresponsiveness to human growth hormone in the African pygmies. *New England Journal of Medicine*, **281**, 1383.

(d) Other forms of pituitary dwarfism

Ferrier, P. E. & Stone, E. F., Jr. (1969) Familial pituitary dwarfism associated with an abnormal sella turcica. *Pediatrics*, **43**, 858.

Ozer, F. L. (1974) Pituitary dwarfism with retinitis pigmentosa and small sella turcica. In *Clinical Delineation of Birth Defects*, Volume XVI, p.354. Ed. Bergsma, D. Balitmore: Williams and Wilkins.

Rappaport, E. B., Ulstrom, R. & Gorlin, R. J. (1976) Monosupercentroincisivodontic dwarfism. *Birth Defects: Original Article Series*, **12**/5, 243.

Sadeghi-Hejad, A. & Senior, B. (1974) A familial syndrome of isolated aplasia of the anterior pituitary. Diagnostic studies and treatment in the neonatal period. *Journal of Pediatrics*, **84**, 79.

Steiner, M. M. & Boggs, J. D. (1965) Absence of pituitary gland, hypothyroidism, hypoadrenalism and hypogonadism in a 17-year-old dwarf. *Journal of Clinical Endocrinology*, **25**, 1591.

Cornelia de Lange syndrome
Begeman, G. & Duggan, R. (1976) The Cornelia de Lange syndrome: A study of nine affected individuals. *South African Medical Journal*, **50**, 1475.

Beratis, N. G., Hsu, L. Y. & Hirschhorn, K. (1971) Familial de Lange syndrome. Report of three cases in a sibship. *Clinical Genetics*, **2**, 170.

Berg, J. M., McCreary, B. D., Ridler, M. A. C. & Smith, G. F. (1970) The de Lange syndrome (*Institute for Research into Mental Retardation, Monograph No.2*) Oxford: Pergamon Press.

Craig, A. P. & Luzzatto, L. (1965) Translocation in de Lange's syndrome. *Lancet*, **2**, 445.

Falek, A., Schmidt, R. & Jervis, G. A. (1966) Familial de Lange syndrome with chromosome abnormalities. *Pediatrics*, **37**, 92.

Garakushansky, G. & Berthier, C. (1976) The de Lange syndrome in one of twins. *Journal of Medical Genetics*, **13**/5, 404.

McArthur, R. G. & Edwards, J. H. (1967) de Lange syndrome: report of 20 cases. *Canadian Medical Association Journal*, **96**, 1185.

Pashayan, H., Whelan, D., Gittman, S. & Fraser, F. C. (1969) Variability of the de Lange syndrome: report of 3 cases and genetic analysis of 54 families. *Journal of Pediatrics*, **75**, 853.

Preus, M. & Fraser, F. C. (1976) A methodology for establishing a diagnostic index for syndromes of unknown etiology. *Clinical Genetics*, **10**/5, 249.

Stevenson, R. E. & Scott, C. I. (1976) Discordance for Cornelia de Lange syndrome in twins. *Journal of Medical Genetics*, **13**/5, 402.

Seckel syndrome
Black, J. (1961) Low birth weight dwarfism. *Archives of Diseases in Childhood*, **36**, 633.

Fitch, N., Pinsky, L. & Lachance, R. C. (1970) A form of bird-headed dwarfism with features of premature senility. *American Journal of Diseases of Children*, **120**, 260.

Harper, R. G., Orti, E. & Baker, R. K. (1967) Bird-headed dwarfs (Seckel's syndrome). A familial pattern of developmental, dental, skeletal, genital, and central nervous system anomalies. *Journal of Pediatrics*, **70**, 799.

Mann, T. P. & Russell, A. (1959) Study of a microcephalic midget of extreme type. *Proceedings of the Royal Society of Medicine*, **52**, 1024.

McKusick, V. A., Mahloudji, M., Abbott, M. H., Lindenburg, R. & Kepan, D. (1967) Seckel's bird-headed dwarfism. *New England Journal of Medicine*, **277**, 279.

McKusick, V. A. (1975) Bird-headed dwarfism, Montreal type. In *Mendelian Inheritance in Man*, 4th Edition, p.366. Baltimore, London: Johns Hopkins University Press.

Seckel, H. P. G. (1960) *Bird-headed Dwarfs*, p. 241. Springfield, Ill.: Charles C. Thomas.

Russell–Silver syndrome
Fuleihan, D. S., Vazken, B. A., Kaloustian, M. & Najjar, S. S. (1971) The Russell–Silver syndrome: report of three siblings. *Journal of Pediatrics*, **78**, 654.

Rimoin, D. L. (1969) The Silver syndrome in twins. In *Clinical Delineation of Birth Defects*, Volume II, p.183. Ed. Bergsma, D. New York: National Foundation.

Russell, A. (1954) A syndrome of 'intra-uterine' dwarfism recognisable at birth with craniofacial dysostosis, disproportionately short arms and other anomalies. *Proceedings of the Royal Society of Medicine,* **47**, 1040.

Silver, H. K. (1953) Syndrome of congenital hemihypertrophy, shortness of stature and elevated urinary gonadotrophins. *Pediatrics,* **12**, 368.

Silver, H. K. (1964) Asymmetry, short stature, and variations in sexual development. A syndrome of congenital malformations. *American Journal of Diseases of Children,* **107**, 495.

Tanner, J. M. & Ham, T. J. (1969) Low birth weight dwarfism with asymmetry (Silver's syndrome): treatment with human growth hormone. *Archives of Disease in Childhood,* **44**, 231.

Tanner, J. M., Lejarraga, H. & Cameron, N. (1975) The natural history of the Silver–Russell syndrome: a longitudinal study of 39 cases. *Pediatric Research* (Baltimore), **9/8**, 611.

Senility syndromes

(a) Progeria

De Busk, F. L. (1972) The Hutchinson–Gilford progeria syndrome. *Journal of Pediatrics,* **80**, 697.

Franklyn, P. P. (1976) Progeria in siblings. *Clinical Radiology,* **27/3**, 327.

Gabr, M., Hashem, N., Hashem, M., Fahmi, A. & Safouh, M. (1960) Progeria, a pathologic study. *Journal of Pediatrics,* **57**, 70.

Gilford, H. (1904) Progeria: a form of senilism. *Practitioner,* **73**, 188.

Rava, G. (1967) Su un nucleo familiare di progeria. *Minerva medica,* **58**, 1502.

Viegas, J., Souza, P. L. R. & Salzano, F. M. (1974) Progeria in twins. *Journal of Medical Genetics,* **11/4**, 384.

(b) Atypical progeria

Mulvihill, J. J. & Smith, D. W. (1975) Another disorder with prenatal shortness of stature and premature aging. *Birth Defects: Original Article Series,* **11/2**, 368.

Welsh, O. (1975) Study of a family with a new progeroid syndrome. *Birth Defects: Original Article Series,* **11/5**, 25.

(c) Cockayne syndrome

Cockayne, E. A. (1946) Dwarfism with retinal atrophy and deafness. *Archives of Disease in Childhood,* **21**, 52.

Paddison, R. M., Moossy, J., Derbes, V. J. & Kloepfer, H. W. (1963) Cockayne's syndrome. A report of five new cases with biochemical, chromosomal, dermatologic, genetic and neuropathologic observations. *Dermatologica Tropica et Ecologica Geographica,* **2**, 195.

Pfeiffer, R. A. & Bachmann, K. D. (1973) An atypical case of Cockayne's syndrome. *Clinical Genetics,* **4**, 28.

(d) Werner syndrome

Epstein, C. J., Martin, G. M., Schultz, A. L. & Motulsky, A. G. (1966) Werner's syndrome. *Medicine,* **45**, 177.

Bloom syndrome

Bloom, D. (1966) The syndrome of congenital telangiectatic erythema and stunted growth. *Journal of Pediatrics,* **68**, 103.

Borgeois, C. A., Claverley, M. H., Forman, L. & Polani, P. E. (1975) Bloom's syndrome: a probable new case with cytogenetic findings. *Journal of Medical Genetics,* **12/4**, 423.

German, J. (1969) Bloom's syndrome. Genetical and clinical observations in the first 27 patients. *American Journal of Human Genetics,* **21**, 196.

Goodman, R. M. (1975) Genetic disorders among the Jewish people. Bloom's syndrome. In *Modern Trends in Human Genetics,* 2nd Edition, p.285. Ed. Emery, A. London, Boston: Butterworths.

Sawitsky, A., Bloom, D. & German, J. (1966) Chromosomal breakage and acute leukaemia in congenital telangiectatic erythema and stunted growth. *Annals of Internal Medicine,* **65**, 487.

Disorders of growth

(a) Dubowitz syndrome

Dubowitz, V. (1965) Familial low birth weight dwarfism with an unusual facies and a skin eruption. *Journal of Medical Genetics,* **2**, 12.

Grosse, R., Gorlin, J. & Opitz, J. M. (1971) The Dubowitz syndrome. *Zeitschrift für Kinderheilkunde,* **110**, 175.

Majewski, F., Michaelis, R., Moosmann, K. & Bierich, J. R. (1975) A rare type of low birth weight dwarfism: the Dubowitz syndrome. *Zeitschrift für Kinderheilkunde*, **120/4**, 283.
Optiz, J. M., Pfeiffer, R. A., Hermann, J. P. R. & Kushnick, T. (1973) Studies of the malformation syndromes of man. XXIV B: the Dubowitz syndrome. Further observations. *Zeitschrift für Kinderheilkunde*, **116**, 1.

(b) 3M syndrome
Miller, J. D., McKusick, V. A. & Malvaux, P. *et al.* (1975) The 3M syndrome: a heritable low birth weight dwarfism. *Birth Defects: Original Article Series*, **11/5**, 39.

(c) Dwarfism–ichthyosis syndrome
Passwell, J., Ziprkowski, M., Katznelson, D., Szeinberg, A., Crispin, M., Pollack, S., Goodman, R., Bat-Miriam, M. & Cohen, B. E. (1973) A syndrome characterised by congenital ichthyosis with atrophy, mental retardation, dwarfism and generalised aminoaciduria. *Journal of Pediatrics*, **82**, 466.
Passwell, J. H., Goodman, R. M., Ziprkowski, M. & Cohen, B. E. (1975) Congenital ichthyosis, mental retardation, dwarfism and renal impairment: A new syndrome. *Clinical Genetics*, **8**, 59.

(d) Nathalie syndrome
Cremers, C. W. R. J., ter Haar, B. G. A. & van Rens, T. J. G. (1975) The Nathalie syndrome. A new hereditary syndrome. *Clinical Genetics*, **8**, 330.

(e) KBG syndrome
Herrmann, J., Pallister, P. D., Tiddy, W. & Opitz, J. M. (1975) The KBG syndrome: A syndrome of short stature, characteristic facies, mental retardation, macrodontia and skeletal anomalies. *Birth Defects: Original Article Series*, **11/5**, 7.

(f) Levi snub-nosed dwarfism
Black, J. (1961) Low birth weight dwarfism. *Archives of Disease in Childhood*, **36**, 633.

(g) Fetal alcohol syndrome
Jones, K. L. & Smith, D. W. (1973) Recognition of the fetal alcohol syndrome in early infancy. *Lancet*, **ii**, 999.
Jones, K. L. & Smith, D. W. (1975) The fetal alcohol syndrome. *Teratology*, **12**, 1.

Glossary

This short glossary is concerned with genetic, anatomical and orthopaedic terms which are sometimes a source of confusion. Emphasis has been placed upon simple explanation rather than precise definition.

ANATOMICAL AND ORTHOPAEDIC TERMS

Acromelia	Shortening in the distal segment of a limb, i.e. hand or foot
Acromesomelia	Shortening of the middle and distal segment of a limb, i.e. forearm or shin and hand or foot
Amelia	Absence of a limb
Anomalad	A malformation with its subsequently derived structural changes, e.g. Klippel–Feil anomalad
Calvarium	The vault of the skull
Cortex	The compact outer part of the shaft of a bone
Deformity	Alteration in shape of a previously normal part, e.g. scoliosis
Diaphysis	The shaft of a long bone
Dysostosis	A malformation of an individual bone, either singly or in combination, e.g. mandibulofacial dysostosis
Dysplasia	A generalised abnormality, e.g. achondroplasia
Ectrodactyly	Maldevelopment of digits. Used loosely for a split-hand or split-foot malformation
Epiphysis	Portion of a long bone derived from a centre of ossification distinct from that of the shaft of the bone
Genu valgum	Knock-knees
Genu varus	Bowlegs
Gibbus	Localised sharp backwards angulation of the spine
Hemimelia	Absence of a segment of a limb
Kyphoscoliosis	Combined backwards and sideways curvature of the spine
Kyphus	Backward curvature of the spine
Lordosis	Forewards curvature of the spine
Madelung deformity	'Dinner-fork' deformity of the forearm
Malformation	Primary structural defect due to failure of normal development, e.g. phocomelia
Medulla	The inner part or marrow cavity of a bone
Mesomelia	Shortening in the forearm or shin

Metaphysis	The region between the epiphysis and the diaphysis of a long bone
Micromelia	Shortening of all segments of a limb
Phocomelia	Absence of segments of a limb or limbs, so that 'flipper-like' hands and feet articulate directly with the trunk
Rhizomelia	Shortening of the proximal segment of a limb
Scoliosis	Sideways curvature of the spine
Symphalangism	Union of the bones of the phalanges
Syndactyly	Union of the digits, either by bone or soft tissues
Synostosis	Abnormal union of bones
Talipes equinovarus	Club foot

GENETIC TERMS

Allele	Alternative forms of a gene at the same site or locus on a particular chromosome
Ascertainment	The tracing of an individual or family with a genetic condition
Autosome	Any chromosome other than the X or Y sex chromosomes. In every human somatic cell nucleus there are 44 autosomes and two sex chromosomes
Chromosome	A body made up from DNA which is contained in a cell nucleus, and which is visible by light microscopy in suitable preparations. Each bears tens of thousands of genes
Clone	Cells which are derived from a single parent cell and which therefore contain identical genetic material
Concordance	The presence of a particular abnormality or genetic disorder in both members of a pair of twins
Congenital	An abnormality which is present at birth. This is not necessarily genetic
Consanguinity	Relatedness. A consanguineous marriage is one in which the partners have a common ancestor
Dermatoglyphics	The pattern of skin ridges on the extremities, i.e. finger and palm prints
Dizygous twins	Twins produced by the union of two sperms with two ova
Embryo	The product of conception before the end of the eighth week of pregnancy
Expression	The degree of severity of clinical manifestations of an abnormal gene
Fetus	The product of conception from the end of the eighth week to the moment of birth
Gene	A portion of the DNA molecule which determines the structure of a polypeptide chain

Heterozygote	An individual who possesses a pair of dissimilar genes
Homozygote	An individual who possesses a pair of similar genes
Karyotype	A photograph of an individual's chromosomes, arranged according to size and shape
Locus	The site of a particular gene on a chromosome
Multifactorial inheritance	Inheritance determined by the interaction of environmental factors and multiple genes
Mutation (gene)	An alteration in a gene, which is perpetuated in subsequent generations
Penetrance	The presence or absence of clinical manifestations of an abnormal gene
Phenocopy	An abnormality produced by an environmental factor which mimics a particular genetic disorder
Phenotype	Clinical features which reflect the basic genetic constitution
Polygenic	Determined by multiple genes
Private syndromes	A syndrome which is apparently confined to a single kindred
Proband	The first affected individual in a kindred to come to the notice of the investigator
Sex chromosomes	The two chromosomes which are designated XX in the famale and XY in the male
Sib	Brother or sister
X-linked	A gene which is situated on an X chromosome
Zygote	A fertilised ovum.

Appendix 1

Relevant reviews or monographs
BIRTH DEFECTS ORIGINAL ARTICLE SERIES, The National Foundation, March of Dimes.
Editor D. Bergsma.
Limb Malformations, Vol. X, No.5, 1974.
Malformation Syndromes, Vol. X, No.7, 1974.
Skeletal Dysplasias, Vol. X, No.9, 1974.
Skeletal Dysplasias, Vol. X, No.12, 1974.
Disorders of Connective Tissue, Vol. XI, No.6, 1975.
CONGENITAL MALFORMATIONS, Josef Warkany. Year Book Medical Publishers, Inc., Chicago,
1971.
HERITABLE DISORDERS OF CONNECTIVE TISSUE, 4th Edition, Victor A. McKusick. The C.
V. Mosby Company, St. Louis, 1972.
DISPROPORTIONATE SHORT STATURE, Joseph A. Bailey II. W. B. Saunders Company,
Philadelphia–London–Toronto, 1973.
PROGRESS IN PEDIATRIC RADIOLOGY, INTRINSIC DISEASES OF BONES, Volume 4. Editor
H. J. Kaufmann. S. Karger, Basel–München–Paris–New York–Sydney, 1973.
BONE DYSPLASIAS, Jürgen W. Spranger, Leonard O. Langer, H.-R. Wiedemann. Gustav Fischer
Verlag, Stuttgart, 1974.
MALADIES OSSEUSES DE L'ENFANT, Pierre Maroteux. Flammarion Médecine-Sciences, Paris,
1974.
THE GENETICS OF LOCOMOTOR DISORDERS, C. O. Carter, T. J. Fairbank. Oxford University
Press, London–New York–Toronto, 1974.
MENDELIAN INHERITANCE IN MAN, 4th Edition, Victor A. McKusick. The Johns Hopkins
Press, Baltimore–London, 1975.
RECOGNISABLE PATTERNS OF HUMAN MALFORMATION, 2nd Edition, David W. Smith. W.
B. Saunders Company, Philadelphia–London–Toronto, 1976.
FAIRBANK'S ATLAS OF GENERAL AFFECTIONS OF THE SKELETON, 2nd Edition, Ruth
Wynne-Davies, T. J. Fairbank. Churchill Livingstone, Edinburgh–London–New York, 1976.
SYNDROMES OF THE HEAD AND NECK, 2nd Edition, Robert J. Gorlin, Jens J. Pindborg, M.
Michael Cohen, Jr. McGraw-Hill Book Company, London–New York–Toronto, 1976.

Appendix 2
Paris nomenclature for constitutional disorders of bone—1970

CONSTITUTIONAL (INTRINSIC) DISEASES OF BONES

Constitutional diseases of bones with unknown pathogenesis
Osteochondrodysplasias (abnormalities of cartilage or bone growth, or both, and development)

1. Defects of growth of tubular bones or spine, or both
   A. Manifested at birth
      1. Achondrogenesis
      2. Thanatophoric dwarfism
      3. Achondroplasia
      4. Chondrodysplasia punctata (formerly stippled epiphysis or chondrodystrophia calcificans congenita), several forms
      5. Metatropic dwarfism
      6. Diastrophic dwarfism
      7. Chondroectodermal dysplasia (Ellis–van Creveld)
      8. Asphyxiating thoracic dysplasia (Jeune)
      9. Spondyloepiphyseal dysplasia congenital
      10. Mesomelic dwarfism
         (a) Nievergelt type
         (b) Langer type
      11. Cleidocranial dysplasia (formerly cleidocranial dysostosis)

   B. Manifested in later life
      1. Hypochondroplasia
      2. Dyschondrosteosis
      3. Metaphyseal chondrodysplasia, Jansen type
      4. Metaphyseal chondrodysplasia, Schmid type
      5. Metaphyseal chondrodysplasia, McKusick type (formerly cartilage–hair hypoplasia)
      6. Metaphyseal chondrodysplasia with malabsorption and neutropenia
      7. Metaphyseal chondrodysplasia with thymolymphopenia
      8. Spondylometaphyseal dysplasia (Kozlowski)
      9. Multiple epiphyseal dysplasia (several forms)
      10. Hereditary arthro-ophthalmopathy
      11. Pseudoachondroplastic dysplasia (formerly pseudoachondroplastic type of spondyloepiphyseal dysplasia)
      12. Spondyloepiphyseal dysplasia tarda
      13. Acrodysplasia
         (a) Trichorhinophalangeal syndrome (Giedion)

  (b) Epiphyseal (Thiemann)
  (c) Epiphysometaphyseal (Brailsford)
2. Disorganised development of cartilage and fibrous components of the skeleton
  1. Dysplasia epiphysealis hemimelica
  2. Multiple cartilagenous exostoses
  3. Enchondromatosis (Ollier)
  4. Enchondromatosis with hemangioma (Maffucci)
  5. Fibrous dysplasia (Jaffe–Lichtenstein)
  6. Fibrous dysplasia with skin pigmentation and precocious puberty (McCune–Albright)
  7. Cherubism
  8. Multiple fibromatosis
3. Abnormalities of density, of cortical diaphyseal structure or of metaphyseal modelling, or both
  1. Osteogenesis imperfecta congenita (Vrolik, Porak–Durante)
  2. Osteogenesis imperfecta tarda (Lobstein)
  3. Juvenile idiopathic osteoporosis
  4. Osteopetrosis with precocious manifestations
  5. Osteopetrosis with delayed manifestations
  6. Pycnodysostosis
  7. Osteopoikilosis
  8. Melorheostosis
  9. Diaphyseal dysplasia (Camurati–Englemann)
  10. Craniodiaphyseal dysplasia
  11. Endosteal hyperostosis (Van Buchem and other forms)
  12. Tubular stenosis (Kenny–Caffey)
  13. Osteodysplasty (Melnick–Needles)
  14. Pachydermoperiostosis
  15. Osteoectasia with hyperphosphatasia
  16. Metaphyseal dysplasia (Pyle)
  17. Craniometaphyseal dysplasia (several forms)
  18. Frontometaphyseal dysplasia
  19. Oculodental-osseous dysplasia (formerly oculodentodigital syndrome)

Dysostoses (malformation of individual bone, singly or in combination)
1. Dysostoses with cranial and facial involvement
  1. Craniosynostosis, several forms
  2. Craniofacial dysostosis (Crouzon)
  3. Acrocephalosyndactyly (Apert)
  4. Acrocephalopolysyndactyly (Carpenter)
  5. Mandibulofacial dysostosis (Treacher-Collins, Franceschetti and others)
  6. Mandibular hypoplasia (includes Pierre Robin syndrome)
  7. Oculomandibulofacial syndrome (Hallermann–Streiff–François)
  8. Nevoid basal cell carcinoma syndrome
2. Dysostoses with predominant axial involvement
  1. Vertebral segmentation defects (including Klippel–Feil)
  2. Cervico-oculo-acoustico syndrome (Wildervanck)

   3. Sprengel deformity
   4. Spondylocostal dysostosis (several forms)
   5. Oculovertebral syndrome (Weyers)
   6. Osteo-onychodysostosis (formerly nail–patella–syndrome)
3. Dysostoses with predominant involvement of extremities
   1. Amelia
   2. Hemimelia (several types)
   3. Acheiria
   4. Apodia
   5. Adactyly and oligodactyly
   6. Phocomelia
   7. Aglossia–adactylia syndrome
   8. Congenital bowing of long bones (several types)
   9. Familial radioulnar synostosis
  10. Brachydactyly (several types)
  11. Symphalangism
  12. Polydactyly (several types)
  13. Syndactyly (several types)
  14. Polysyndactyly (several types)
  15. Campodactyly
  16. Clinodactyly
  17. Biedl–Bardet syndrome
  18. Popliteal pterygium syndrome
  19. Pectoral aplasia–dysdactyly syndrome (Poland)
  20. Rubinstein–Taybi syndrome
  21. Pancytopenia–dysmelia syndrome (Fanconi)
  22. Thrombocytopenia–radial–aplasia syndrome
  23. Orofaciodigital (OFD) syndrome (Papillon–Léage)
  24. Cardiomelic syndrome (Holt–Oram and others)

Idiopathic osteolyses
   1. Acro-osteolysis
     (a) Phalangeal type
     (b) Tarso–carpal form with or without nephropathy
   2. Multicentric osteolysis

Primary disturbances of growth
   1. Primordial dwarfism (without associated malformation)
   2. Cornelia de Lange syndrome
   3. Bird-headed dwarfism (Virchow, Seckel)
   4. Leprechaunism
   5. Russell–Silver syndrome
   6. Progeria
   7. Cockayne syndrome
   8. Bloom syndrome
   9. Geroderma osteodysplastica
  10. Spherophakia–brachymorphia syndrome (Weill–Marchesani)
  11. Marfan syndrome

Constitutional diseases of bones with known pathogenesis

I. Chromosomal aberrations

II. Primary metabolic abnormalities

    1. Calcium phosphorous metabolism

        1. Hypophosphatemic familial rickets

        2. Pseudodeficiency rickets (Royer, Prader)

        3. Late rickets (McCance)

        4. Idiopathic hypercalciuria

        5. Hypophosphatasia (several forms)

        6. Idiopathic hypercalcemia

        7. Pseudohypoparathyroidism (normo and hypocalcemic forms)

    2. Mucopolysaccharidosis

        1. Mucopolysaccharidosis I (Hurler)

        2. Mucopolysaccharidosis II (Hunter)

        3. Mucopolysaccharidosis III (Sanfilippo)

        4. Mucopolysaccharidosis IV (Morquio)

        5. Mucopolysaccharidosis V (Ullrich–Scheie)

        6. Mucopolysaccharidosis VI (Maroteaux–Lamy)

    3. Mucolipidosis and lipidosis

        1. Mucolipidosis I (Spranger–Wiedemann)

        2. Mucolipidosis II (Leroy–Opitz)

        3. Mucolipidosis III (pseudo-Hurler polydystrophy)

        4. Fucosidosis

        5. Mannosidosis

        6. Generalised Gml gangliosidosis (several forms)

        7. Sulfatidosis with mucopolysacchariduria (Austin, Thieffry)

        8. Cerebrosidosis including Gaucher's disease

    4. Other metabolic extraosseous disorders

Bone abnormalities secondary to disturbances of extraskeletal systems

    1. Endocrine

    2. Hematologic

    3. Neurologic

    4. Renal

    5. Gastrointestinal

    6. Cardiopulmonary.

REFERENCE

McKusick, V. A. & Scott, C. I. (1971) *Journal of Bone and Joint Surgery*, **53**, 978.

Index